PLASTICS EXTRUSION TECHNOLOGY

Second Edition

ALLAN L. GRIFF

Consultant
Edison Technical Services, Inc.
New York, N.Y.

ROBERT E. KRIEGER PUBLISHING COMPANY
MALABAR, FLORIDA

Original Edition 1968
Reprint 1976, 1979, 1981

Printed and Published by
ROBERT E. KRIEGER PUBLISHING CO., INC.
KRIEGER DRIVE
MALABAR, FLORIDA 32950

4 5 6 7 8 9

Library of Congress Cataloging in Publication Data

Griff, Allan L
Plastics extrusion technology.

Reprint of the ed. published by Van Nostrand Reinhold, in series: Plastics applications series.
Includes bibliographical references and index.
1. Extrusion (Plastics) I. Title.
[TP1175.E9G7 1976] 668'.413 75-45329
ISBN 0-88275-386-X

Printed in the United States of America

Plastics Applications Series

When Reinhold published its first book on extrusion in 1957, the venture was thought to be a mistake by many in the plastics industry. Extrusion, they said, was an art and could not be treated as a technology in a book. However, the title was so well received that a second book, "Plastics Extrusion Technology," by Allan Griff, was introduced in 1962 as one of the first process books in the Reinhold Plastics Applications Series. Now that extrusion has developed and expanded much further, this new 1968 edition is necessary to bring information on the subject up to date.

The series, however, will continue to feature descriptions of all types of plastics presented in a style that will make the books valuable to a variety of readers. The series should be most useful to design engineers, equipment manufacturers, producers of packages and packaging machinery, students at technical schools and, of course, all people in the plastics industry—material manufacturers, molders, extruders, and fabricators. In addition, it is hoped that the books will be of value to readers in specialized categories. For example, plastics from which fibers are made should be of interest to tire and fabric manufacturers, and materials used in the production of plastic sheet will interest handbag and luggage designers.

With many new features of extrusion constantly becoming available, it is hoped that this new book on the subject will be fully as valuable as its predecessors.

HERBERT R. SIMONDS
Editor

Preface to Second Edition

In only five years since the publication of the first edition, the extrusion industry has made great technical and commercial advances. Much of the first edition is already out of date, and therefore this new and fully rewritten second edition has become necessary. It includes such recent developments as the vacuum sizing trough for pipe, five-roll high-output sheet extrusion lines, expandable mandrels for tubular film, equipment for cross-linking wire covering and making oriented narrow ribbons for twine and sack weaving, and the automatic continuous screen changer—as well as current market statistics and the latest standards for extruded products.

Like the first edition, the new book will be limited to actual manufacturing of extruded products. Consequently, the extrusion involved in compounding and blow molding is excluded, as is the extrusion that is now commonly used to heat and melt plastics in injection molding. These are certainly important applications of the extruder, but they serve different markets and thus fall outside the scope of the book.

The annotated bibliography, a special feature of this book, has been completely revised and rearranged. In the first edition, all the references were collected in one place at the end of the book, and divided into sections, with each section further subdivided into topics. In this new edition, the references that pertain to each chapter are listed at the end of that chapter, still subdivided into topics such as die design, take-off equipment, testing, market reviews, etc. All general or additional references are given in a "Supplementary Bibliography" following Chapter 11.

The entire bibliography, which comprised over 15% of the pages of the first edition, has been updated and enlarged even further. Many old references have been dropped, while over 60% are new articles written between 1962 and 1967, including many from foreign journals. As before, almost all references include brief comments as to their content and usually show the company affiliations of the authors. These comments on content will help the reader decide which references he needs, and the affiliations will enable direct contact with many of the authors. Also, to facilitate contact with the journals themselves, a list of their names and addresses is provided in the Supplementary Bibliography.

Although the book is necessarily based on technology as practiced in the United States, considerable space is devoted to the methods and machinery in other countries. All dimensions are given in both English and metric units, because of the increasing use of the metric system in English-speaking countries (which the author strongly supports) as well as the expected use of this book in countries where the metric system is the basis of measurement. The first English edition found its way to almost every corner of the world and was also translated into Italian; several such translations are already planned for the new edition. Communications are improving among the peoples of the world, tariffs are being lowered or eliminated, and the interchange of technical information will be of increasing value in international progress and cooperation.

In the preparation of the revised edition, valuable assistance was received from Rick Honstrater of Gemini Plastic Enterprises (Chapter 5), R. Flowers of Black-Clawson (Chapter 5), H. A. Arbit (Chapter 6), Frank DeMelio of Union Carbide (Chapter 7), and A. Knight of Amacoil (Chapter 7). The machinery suppliers were very helpful in providing photographs, for which credits are given underneath the pictures. Finally, the author wishes to express appreciation and thanks to his wife, Mariana, for her encouragement as well as her assistance in preparing the manuscript.

Fords, New Jersey ALLAN L. GRIFF
December 1, 1967

Preface to First Edition

This book, in keeping with the aims of the Reinhold Plastics Applications Series, is intended to provide useful information to a rather large audience with varying technical skills. Such readers might include the student who is learning about the plastics industry; the industrial worker who wants to learn more about the processes with which he works; the executives and investors who must learn something about extrusion in a short time; newcomers to the extrusion business; purchasers of extruded products; plastics engineers who wish to learn something outside their special field and thus prepare themselves for advancement; and even the everyday consumer of plastics, who is curious enough to want to know how his telephone wire or his refrigerator door is made.

Perhaps this large potential audience is encouraging, but it is also a challenge, as each group of readers will have different interests and different reasons for reading the book. It has been the author's aim to get as much extrusion technology into as small a space as possible, while still keeping it reasonably understandable. Analysis and criticism of the extrusion practices have been avoided, even where deserved, and theory has been kept to a minimum. Instead, emphasis has been placed on telling the reader what actually is being done, and why it is done that way. Quite often, the reasons for things are economic, not technical. The author realizes that the extrusion industry is a business as well as an application of science, and has tried to present both the economic and technical aspects of current practice.

Technology has not been restricted to that practiced in the United States alone. Foreign literature has been reviewed and foreign books read, and several weeks were spent in Europe for the purpose of learning the practices of that area. Such trips soon make it obvious that America and Europe have much to learn from each other, to our mutual benefit, in the long run.

The subject matter stresses the technology of producing actual shapes such as rod, film, sheet, pipe, etc. The use of extruders in compounding of thermoplastics is felt to be basically different, usually being performed at a different market level, and is not included here. Also omitted is the use of extruders to feed blowmolding machines, as this field is sufficiently important to warrant a separate book recently published in this Series. Users of extruders in both these related fields should still find value in the chapters on Principles and Parts, Operation, and Extrusion Fundamentals, and in the Bibliography.

The heart of the book is in the seven chapters on Applications. Ten or twenty pages on an application will not make an expert out of a novice. However, they should provide a useful start, giving him a good perspective on what is important and why it is important. A minimum of space is devoted to specialty plastics and unusual, but low-volume, techniques. Emphasis has been placed instead on the volume thermoplastics—vinyls, polystyrenes, and polyethylenes—which go into the vast majority of extruders today and which are the ones most likely to concern the readers of this book.

Books of this size are often introductory volumes, to be read once, and then shelved in favor of more detailed writings. It is hoped that for some this book may find service as a continuous reference, to be read in small portions as needed. With this in mind, the Bibliography has been prepared in a rather unique way. Very few references are given in the text; rather, they are arranged by application in the Bibliography along with the author's comments on each article. In this way, the reader is not only given a list of available information but also some criteria for choosing which articles or books to consult. This should be a welcome aid for busy people.

Sincere acknowledgment is due to the personnel of the Union Carbide Plastics Company, particularly Messrs. R. Bostwick and W. T. Higgins, for their support and assistance throughout the past year. Special appreciation is due to Mr. B. H. Maddock, who not only helped with valuable information and suggestions, but also provided most of the author's foundation in extrusion principles over the past six years. Also much appreciated were the criticism and comments of the following Union Carbide specialists in various fields of extrusions: Messrs. H. A. Arbit, W. Egan, A. J. Fischer, W. A. Haine, W. H. Joyce, F. D. Marrow, H. J. Pazinski, G. L. Richon, H. F. Tomfohrde III, J. E. Wankow, and J. C. Wright. The author gratefully acknowledges the help in preparation of the manuscript of Miss Mary

Gallo and her staff, Mrs. Barbara Cochilla and Mrs. Barbara K. Gallant. The contribution of the personnel of the Union Carbide International Company and their various affiliates, who assisted the study of European technology, is sincerely appreciated. Finally, too numerous to name, many other companies both here and abroad—either with actual extrusion operations or engaged in manufacture of extrusion machinery—deserve the author's thanks for their kind cooperation.

ALLAN L. GRIFF

Metuchen, N. J.
May 15, 1962

Contents

ONE

the extruder— components and principles

Extrusion of thermoplastics consists of the melting and compressing of plastic particles while they are being forced through a long passage by a screw conveyor turning in that passage. The opening at the end of the passage takes various shapes depending on the desired cross section; for example, a circular opening makes a continuous rod, an annular opening makes a tube, a narrow slit makes a flat belt or sheet.

Figures 1.1 and 1.2 are photographs of modern commercial extruders. The components of an extruder are shown in Figure 1.3. These components and their functions are described in the following pages, and are further referred to in Chapter 10, which covers operation of the machinery.

POWER SOURCE

This is an electric motor, usually AC, varying from 1 to 200 or more horsepower. Each horsepower can produce from 5 to 15 lbs ($2\frac{1}{2}$–$6\frac{1}{2}$ kg) of extruded product per hour.

SPEED VARIATION

Most machines have continuous variation of screw speed; in other words, the speed at which the extruder screw rotates can be con-

Figure 1.1. Medium-sized "compact" extruder, with built-in motor and speed variator. Note pressure gauge at head. (*Courtesy of Prodex division, Koehring Co.*)

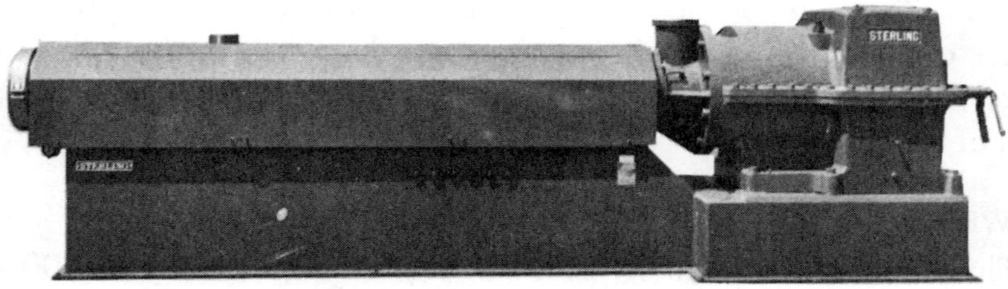

Figure 1.2. Large (6-inch barrel) vented extruder. (*Courtesy of Sterling Extruder Corp.*)

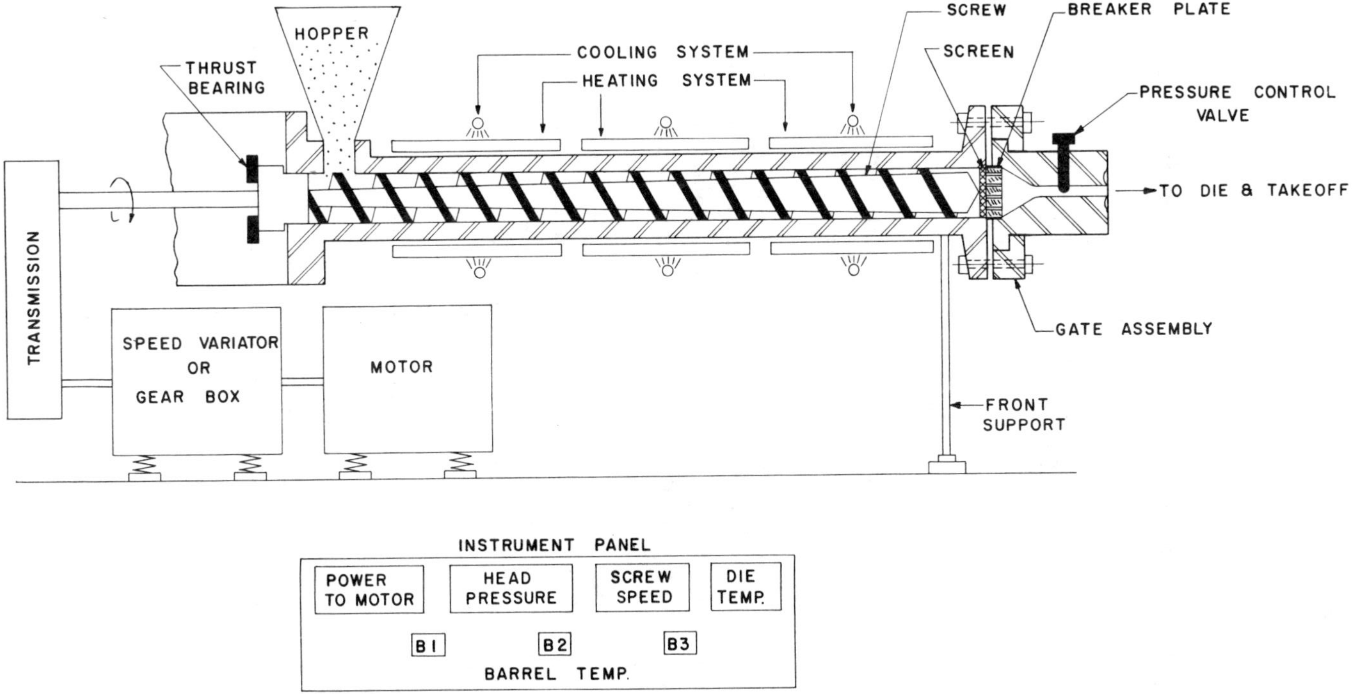

Figure 1.3. Diagram of a single-screw extruder showing its components. (*Illustration by author*)

tinuously increased or decreased, within certain limits, by merely turning a knob or pushing a button. Magnetic, electrical, and mechanical devices are available to do this, and these three systems are used about equally. Hydraulic speed variation is also used—particularly in blow molding and injection systems. These methods of speed variation are further discussed and compared in Chapter 10. A few manufacturers, mostly European, make their extruders without continuous speed variation, using different sizes of gears in order to obtain a number of fixed speeds. This simplified construction reduces initial and maintenance costs, and is often adequate, especially if speeds can be changed without stopping extrusion.

SPEED REDUCTION AND POWER TRANSMISSION

The power source and speed variators usually produce rotational speeds well above those needed for screw extrusion. Common extrusion speeds (screw speeds) are from 10 to 200 rpm on a single-screw extruder, although no one machine normally operates over this entire range except perhaps in an experimental laboratory. The reduction of the output speed of the variator to these lower values is done by a set of gears ending in a large driving gear coaxial with, and surrounding, the shank of the screw. Sometimes belts are used between the motor and these gears, giving further reduction. Twin-screw machines normally run at even slower speeds (5 to 40 rpm), and they too need this reduction. The gears are oil-lubricated, either by immersion or by forced circulation.

Many machines are now offered with interchangeable gears or pulleys which permit changing of the speed *range*. This is important, because in many drive systems not all the horsepower is available at all speeds; consequently, production is sometimes slowed down and in extreme cases must be stopped completely, if the material or the operating conditions are changed very much. This problem is common enough to warrant further explanation in Chapter 10, p. 293.

EXTRUDER SUPPORT

Extruders are mounted in a welded-steel or cast-iron frame and base, designed to be bolted to a level concrete floor. The motor and speed variator are often supported independently. A few specialized smaller machines are not fixed but can roll or pivot into various positions as needed, and some roll on rails.

The delivery end of the extruder is also supported. Such support should allow linear expansion (elongation) of the barrel when heated, and also should be designed to avoid any heat loss from the barrel at this point. Radial barrel expansion is less important, but should be considered, as well as the effect of heavy dies mounted on the end of the barrel.

SCREW SUPPORT

The screw fits into a thrust bearing just behind the hopper, as shown in Figure 1.4. The rated life of this bearing at extrusion conditions has an important effect on the condition and output capacity

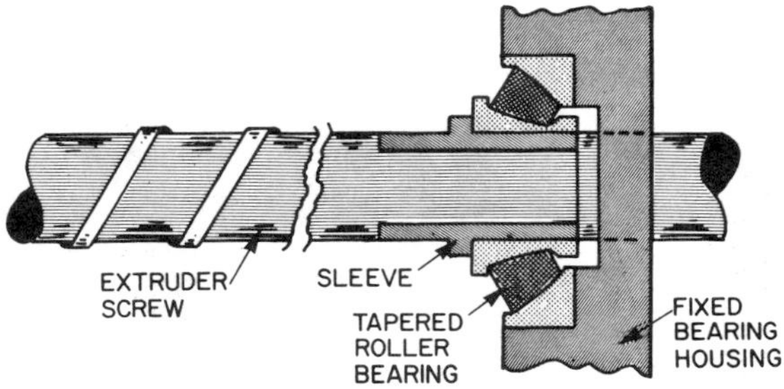

Figure 1.4. Typical thrust bearing as used in single-screw extruders. (*Illustration by author*)

of the extruder. These bearings are usually purchased from established bearing manufacturers and have specified ratings applicable to extruder operation. Ball bearings are not considered adequate, and roller bearings—especially spherical (barrel-shaped) rollers—are commonly used. Several bearings are sometimes used in line, one behind the other, for twin-screw machines where bearing diameter is very limited; they are arranged such that each unit receives a portion of the load. This need for small-diameter bearings was formerly a serious disadvantage of some twin-screw machinery, but now most of the manufacturers of twin-screw machines have found bearing systems that are satisfactory if the machines are operated under the conditions for which they are designed. Twin-screw extruders run with less melt pressure and at a lower speed than single-screw extruders of equivalent output, and this alleviates the problem.

FEED SECTION

In almost all extruders there is an opening in the barrel wall at the driven end, through which the plastic enters. The hopper, a simple sheet-metal enclosure, is mounted above that opening and holds about an hour's capacity of material. If the plastic needs to be preheated, the hopper can be fitted with a device to blow hot air through the pellets. Preheating in trays kept in separate ovens is an older method, and is obsolete except for a few polymers that are very hard to dry—but even for these there are now automatic hopper-dryers that circu-

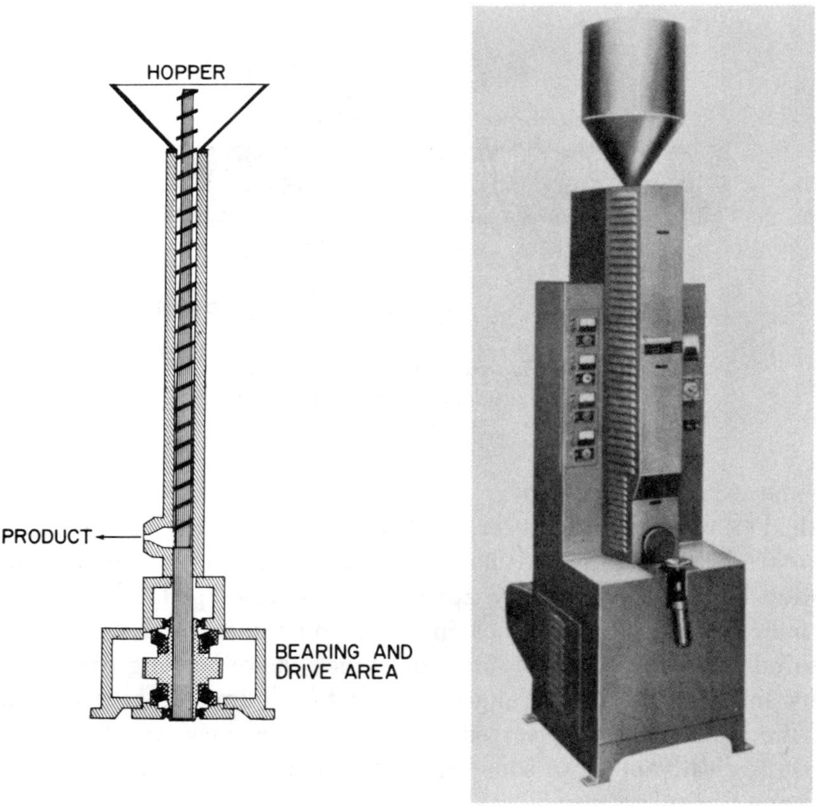

Figure 1.5. A vertical extruder in which the drive is mounted at the *delivery* end of the screw; the feed end extends up through the hopper. (*Photograph courtesy of Aragon Products, Inc. Drawing by Aragon Products, Inc., adapted by author*)

late dehumidified air. In some hoppers agitators are fitted, especially for powder feed, and still other attachments are sold to ensure proper feeding of chopped film, foam scrap, etc., as well as powder. A special air-tight hopper is sometimes used, with vacuum drawn on it to eliminate air entrapment in powder or fluffy feed. This method of air extraction is not as popular nor efficient as venting, discussed later in this chapter.

Many designs of feed openings have been tried, but now all openings are directly over the screw(s). They are normally one screw-diameter wide, and more than one diameter long. Cold water is circulated around the barrel in the feed section to prevent sintering of the plastic particles, which would block further feeding. This is not necessary with twin-screw extruders, which provide a more positive feed.

At least two manufacturers offer vertical extruders in which the hopper surrounds the feed end of the screw (see Figure 1.5). This makes feeding of difficult materials easier, and the hopper can be sealed to allow application of vacuum or addition of gaseous or volatile liquid additives.

THE EXTRUDER SCREW

This is the heart of the extruder. Figure 1.6 illustrates some typical screws, and Figure 1.7 shows a screw being made. There is considerable variation in design, the most important variable being the depth of the channels. Despite much desire for a universal screw, it is advisable to use a different design for each family of materials to achieve the best results. In fact, some design features are changed even further to fit the type of product being extruded. It should be clearly understood that almost any screw that fits the extruder can be used to produce a product; however, the extrusion industry is so competitive that it is important to use a good design to get the best quality at the highest output.

Today most screws for single-screw machines are full-flighted—i.e., they have turns over their entire length. These turns are usually made with a constant pitch, equal to the screw diameter—partly for easier and cheaper machining, partly to reduce the variables in screw design, and partly because of successful field experience. Some years ago, a number of different types of "torpedo" ends were popular. Most of these have become obsolete now, except perhaps for saran, but a

Figure 1.6. Different screw designs. From top to bottom: (A) Shallow-channel metering, as used for polyethylene and high-impact polystyrene, and—with deeper channels—for high-density polyethylene and ABS. (B) Sudden compression, as used for nylon and high-density polyethylene. (C) Deep channel, little or no metering, as used for rigid PVC. (D) Two-stage vented screw, used for all materials, with different channel depths required for each. Note nose assembly of screw (A), which comprises a valve; the resistance to flow is changed by the axial movement of the screw. This type of nose assembly is applicable to all screw designs shown. (*Illustration by author*)

Figure 1.7. Completing the manufacture of an extruder screw on a special automatic lathe designed for this purpose. (*Courtesy of Waldron-Hartig Division of Midland-Ross Corp.*)

few new "mixing-head" screw designs have been developed and are claimed to give excellent mixing at high rates. One of these is illustrated in Figure 1.8.

At the drive end of the screw is the assembly that engages the drive. This is often a single keyway, but multiple keys and splined shanks are also used. There is, unfortunately, no standardization among manufacturers here, although this would be useful. Screw lengths and diameters are supposedly made to standard sizes, but in practice they vary slightly—enough that a screw for one machine will seldom fit in the same nominal size barrel of another make.

Much is known about channel depth in a single-screw extruder, especially at the delivery end. The job of designing a screw centers around deciding how deep these channels should be, and in what way and at what points the depth should change. For example, it might be decided that for a screw of 120 mm ($4\frac{1}{2}$ in.) diameter, for high-impact polystyrene sheet, the first quarter of the screw length should have a constant depth of 15 mm (0.6 in.), the depth should decrease regularly through the second quarter to 3.75 mm (0.15 in.), and it

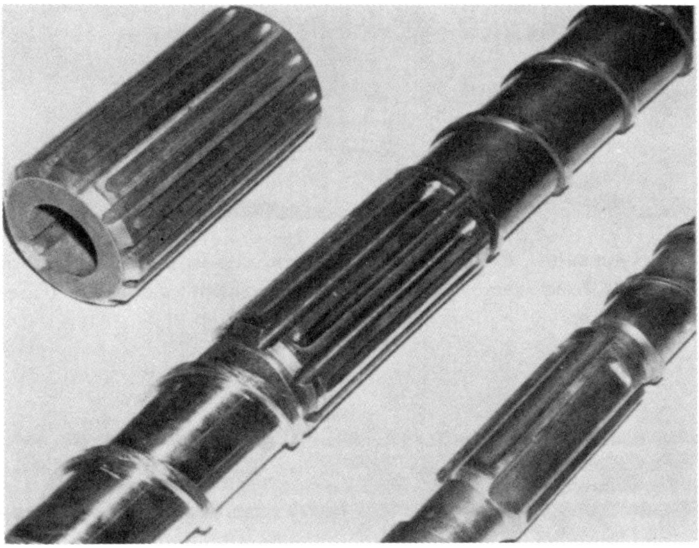

Figure 1.8. Variations of a special screw design in which this fluted section provides intensive mixing. (*Courtesy of SPE Journal and Union Carbide*)

should remain constant at this latter value for the entire second half of the screw. This delivery zone has been thoroughly studied, but there is still much to learn about the first stages of extrusion. These are more difficult to analyze because the plastic is not completely melted and mathematical treatment is extremely complex and difficult to relate to practice.

With twin-screw extruders (Figures 1.9 and 1.10) the screws are more individual in design. Most types have intermeshing screws, but some do not. The pitch, flight depth, and even the thread width may be varied in the same extruder; for example, some machines have three distinct sections, each with progressively decreasing pitch. The pitch is usually *less* than the diameter, and the channel depths are usually deeper than those in a single-screw extruder. The two screws

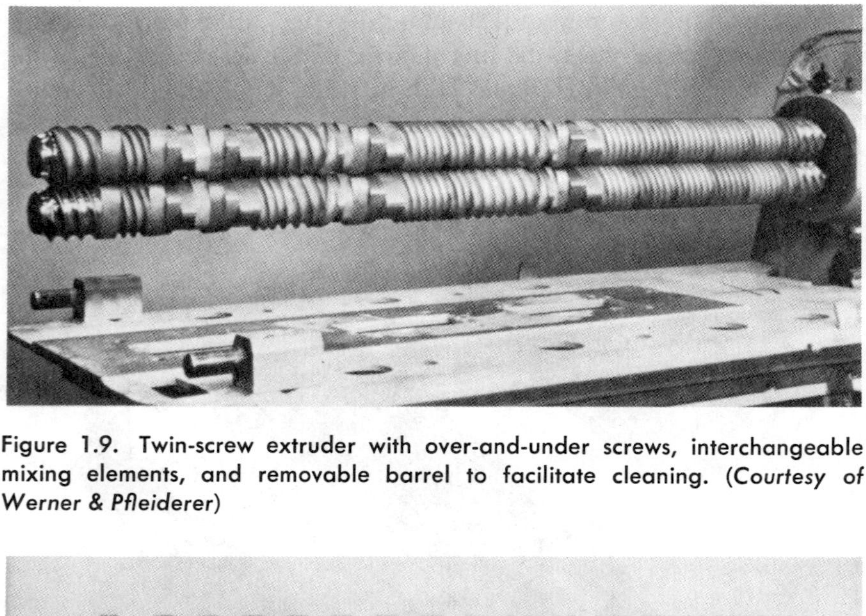

Figure 1.9. Twin-screw extruder with over-and-under screws, interchangeable mixing elements, and removable barrel to facilitate cleaning. (*Courtesy of Werner & Pfleiderer*)

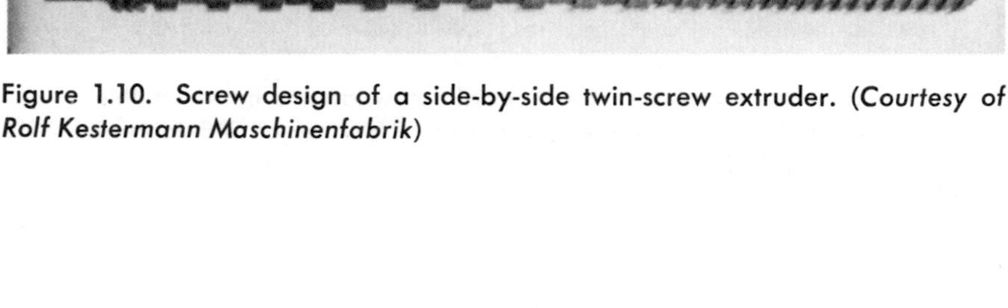

Figure 1.10. Screw design of a side-by-side twin-screw extruder. (*Courtesy of Rolf Kestermann Maschinenfabrik*)

turn in opposing directions, forming a bite between them—except for one type in which the screws turn in the *same* direction, producing high shear at the bite. There is only one manufacturer of twin-screw machinery in the United States, compared to over fifteen in Europe. Most all the technical work has accordingly been done in Europe, where twin-screw extruders are commonly used for polyvinyl chloride (PVC)—both for compounding powder mixes into pellets for re-extrusion, and for one-step extrusion of these mixes into finished products such as pipe. The positive-displacement nature of a twin-screw machine allows easy feeding of the powder mix, while the screws are designed for good mixing at low total shear. This keeps temperature down, which is important to prevent decomposition of the PVC. For other materials, single-screw extruders are preferred in Europe as in the U.S., although they too can be run on a twin-screw extruder. The latter is more expensive per unit output, however, and is thus bought only where its advantages of low-temperature mixing and positive feeding are most wanted.

Most extruder screws are made of low-carbon alloy steel such as SAE 4140. The tips of their flights are hardened to prevent wear, either by welding on a covering of an appropriate metal alloy, by flame-hardening the flights, or by nitriding the entire screw. The first two methods are most common in the United States, and represent "premium" and "regular" treatment, respectively. Nitriding is still common in Europe (see page 285). Screws for vinyl polymers are sometimes chromium-plated for corrosion resistance, and screws for saran are made of special nickel alloys. Occasionally, stainless steel screws are made to prevent corrosion, but they are not suitable for saran. Screws for other materials are sometimes chromium-plated too, as a plated surface makes cleaning of the screw easier. Some manufacturers regard any plating as ineffective, since it will wear off at the flights, and the exposed interface may then be subject to chipping and peeling, and eventual corrosion.

Screw surfaces have another importance—their effect on actual extrusion performance. It is known that plastic pellets are conveyed fastest where there is minimum friction between them and the screw, and maximum friction between them and the barrel wall. But, as the plastic melts, its behavior becomes harder to predict. The effect of screw surface on hot, molten plastic is still not fully known, and its study might well lead to improvements and changes in materials of screw construction.

It is common practice in single-screw machines to make the screw hollow through most of its length to allow control of its temperature with water or steam. This control has in the past been a *cooling* of the screw to improve mixing and to reduce extrudate temperature. This practice of screw cooling is not now popular, because of significant losses in output rate, as well as discovery of better means of achieving the above improvements (higher melt pressure and better barrel cooling, respectively). However, *control* at higher temperatures is sometimes done, and an extruder buyer usually wants his screws bored because he cannot do this later if the need arises. This could distort the dimensions of the finished screw; this problem is avoided by boring before the flights are ground down to final dimensions.

The possible need for a bored screw seems to outweigh the advantages of a solid screw (lower cost, stronger root and higher torque limit). For certain products, screw control is the accepted technique; for example, with unplasticized PVC, hot oil is used to control screw temperature and thus prevent decomposition on overheated internal metal surfaces, especially at the screw tip. The boring must therefore extend through to the very tip of the screw. Steam can be used instead of oil, but requires more equipment, unless steam is already available throughout the plant. Also, it is corrosive to the steel bore. Screws for twin-screw machines are seldom made hollow, as adequate temperature control can be obtained externally.

Screw noses generally come to a point (e.g., 120°). Many early machines had flat ends, but a blunt point soon became more desirable, and a sharper point (e.g., 60 to 90°) is sometimes seen. The point displaces a possible stagnation area in the extruder head, and it can also serve as a valve when moved axially into or out of its seat (see Figures 1.6 and 1.20). A chisel-shaped nose has also been used successfully.

Screw construction is so important that it is mentioned in several places in this book. In Chapter 11 (pages 318–325) a brief description is given of what actually happens to the plastic in the screw, and aids are given to the choosing of a screw design. Further construction and design information appears in Chapter 10 on Extruder Operation, and also wherever relevant in the seven chapters on applications.

SCREW REMOVAL

A mechanical device is offered with most extruders which attaches to the drive end of the screw and pushes it out of the barrel at the other end. This simplifies shutdown and protects the screw from damage inflicted by eager workmen when the screw appears stuck in the barrel. On a few machines, mostly twin-screw types, the screw need not be pushed out at all; instead, the barrel rides forward on rails, and is pulled off the stationary screw(s) (see Figure 1.9).

BARREL

The extruder barrel must take internal pressures of 50 to 400 atmospheres (atm) and this pressure may even reach 1000 atm in extrusion-preplasticizers for injection molding. Therefore, heavy steel cylinders are used, usually with a smooth interior and—for single-screw machines—made in one piece. Rifling or other grooves have been tried in a few cases, but with little commercial success. A few extruders are now sold with non-cylindrical barrels (and screws), which usually get smaller in diameter at the delivery end. This reduction can be either gradual or sudden.

The interior surface of the barrel is either nitrided for hardness, or lined with a very hard and abrasion-resistant metal alloy, centrifugally cast into the barrel cylinder. Such alloys are used in almost every extruder made in the United States, and are also used by some manufacturers in Europe and Japan. The most popular is a ferrous alloy (Rockwell Hardness C60–78) which is principally iron and carbon, plus small amounts of nickel and boron. Another type is a complex alloy of nickel, cobalt, chromium and molybdenum, with small amounts of silicon, carbon and boron, and is used for maximum corrosion resistance. It is suitable for saran, where special corrosion-resistant metals are needed for all parts that contact the hot plastic. Nitrided barrels are still common outside the United States, but are generally not as permanent as the alloy linings.

The long extruder barrel is now generally accepted. Early, shorter designs were based on rubber machinery, but it soon became evident that longer barrels give more uniform heating and cooling, more latitude for using different screws, and higher outputs. Barrel lengths are now usually 20 or 24 times their diameter (this number is commonly called the L/D ratio). In very large machines (200 mm =

8 in. diameter or more) shorter barrels are sometimes adequate, since the material takes a long enough path just going around the big screw a few times. Barrels with L/D ratios of 24 to 28 are common for vented extruders. Extra long barrels of L/D = 30 or more present unique engineering problems and have not been widely accepted. Some are properly built now, with increased cost justified by increased output. The longest barrels in commercial use have L/D ratios of 40 or more, and are used for extruding foamed polystyrene sheeting. The foaming agent is injected at a point near the middle of the barrel and the melt is mixed and cooled as it passes through the remaining portion.

Modern machines have one-piece barrels, but a few older segmented barrels are still working despite leakage problems. Twin-screw barrels are often built in more than one section, but since they operate at lower pressures, leakage is less of a problem. Special segmented-barrel extruders are found in many laboratories, and enable study of the effects of barrel length.

The end of the barrel lining at the delivery end is exposed, and usually forms a seal against leakage by its contact with the breaker plate. The barrel end must therefore be ground exactly perpendicular to the barrel axis.

Twin-screw machines are usually shorter than L/D = 20, as there is more plastic material contained per unit length, and the pitch is smaller. The L/D ratios vary from 12 to 16. The barrel, of course, is more complex and costly than the single-screw cylinder. It is, in fact, two intersecting or adjacent cylinders with axes that are usually parallel; the distance between the axes depends on actual screw design.

VENT

Numerous systems have been devised for venting the barrel—that is, extracting air or other gases from the plastic after melting. The most common system is a two-stage extrusion, which is really two extruders on a single shaft (see Figure 1.6). The plastic softens, melts, and mixes in the first stage, and then suddenly passes into a deep-channeled zone open to the atmosphere or to a vacuum. Gases flash out of the hot plastic, which then moves into the second stage and is carried to the delivery end of the extruder to form the final product.

This system requires careful screw design to allow versatility in operation while avoiding extrusion of material through the vent. A valve in the extruder head, and sometimes another valve at the end of the first stage, are used to control pressures and to permit uniform extrusion. Operating conditions are more critical and can be varied less, compared to a modern unvented machine.

Other venting systems include: extraction of volatiles through a hole drilled in the root of the hollow screw at a critical point; bypass systems which pass the material over or around a vented flight; and the vacuum-hopper—a hopper held under vacuum so that the air cannot get into the extruder at all. This last system allows use of any unvented extruder and ordinary screw, but its venting ability is limited to air and low-boiling liquids. It may not be sufficient to extract all absorbed moisture from some very hygroscopic plastics. Finally, the screw shaft must be perfectly sealed, and powder leaks may impair this seal.

Many vented machines can be operated with their vent blocked, as single-stage long-barrel extruders. A special insert is bolted into the vent opening, which is usually, but not always, on the top side of the barrel.

Some twin-screw machines have one or more places vented to the atmosphere along the screws. They do not necessarily operate as two-stage extruders, but the screws are designed to convey the plastic through the venting zone at near zero pressure, thus avoiding the escape of plastic through the vent.

HEATING SYSTEM

Most extruders are electrically heated, and can reach at least 300°C. Band-type resistance heaters around the barrel, with either mica or cement insulation, are the least expensive method (see Figure 1.11). Cast heaters (Figure 1.12) are very popular; they are insulated resistance wires cast in blocks of metal shaped to fit around the barrel. Tubular resistance heaters wrapped around the barrel (Figure 1.13) are also used, and one manufacturer supplies resistance heaters set in segmented ceramic blocks. Induction heating (Figure 1.14) is offered by several manufacturers; with this method the heat is actually produced in the barrel walls. This is rather expensive, but offers several unique advantages, particularly quick and uniform heat, and ease of cooling. Oil and steam have also been used

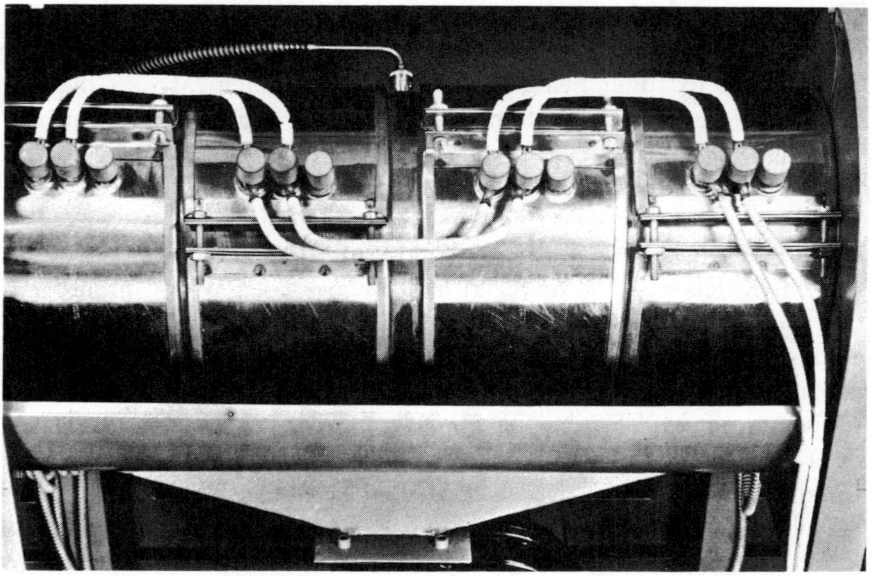

Figure 1.11. The cover of this extruder has been removed, showing the band heaters around the barrel.

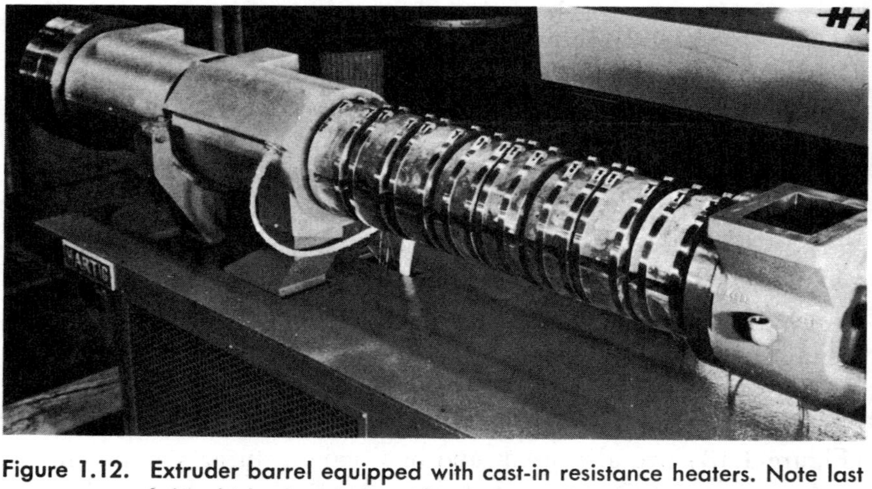

Figure 1.12. Extruder barrel equipped with cast-in resistance heaters. Note last two zones at left, which already have housings on them through which cooling air can be circulated. (*Courtesy of Waldron-Hartig Division of Midland-Ross Corp.*)

Figure 1.13. Tubular resistance heaters, connected in groups, completely surround the barrel of this extruder. (*Courtesy of John Royle & Sons*)

Figure 1.14. Induction heating of an extruder is illustrated here. The induction coils surround the barrel, and electricity passed through them induces current in the barrel walls themselves, making these walls hot. (*Courtesy of Union Carbide*)

to heat extruders. They provide very uniform heat within their limited ranges, but they are not very common today.

Heating systems remain necessary despite a trend to extrude with as much heat as possible added by internal friction in the plastic. They are needed for start-up, to ensure against over-cooling (and overloading the drive) and to achieve temperatures above those naturally produced by frictional working of the plastic. In extrusion-coating, for example, a low-viscosity polyethylene, which builds up relatively little frictional heat, must often be heated to as high as 300°C to promote adhesion to the substrate.

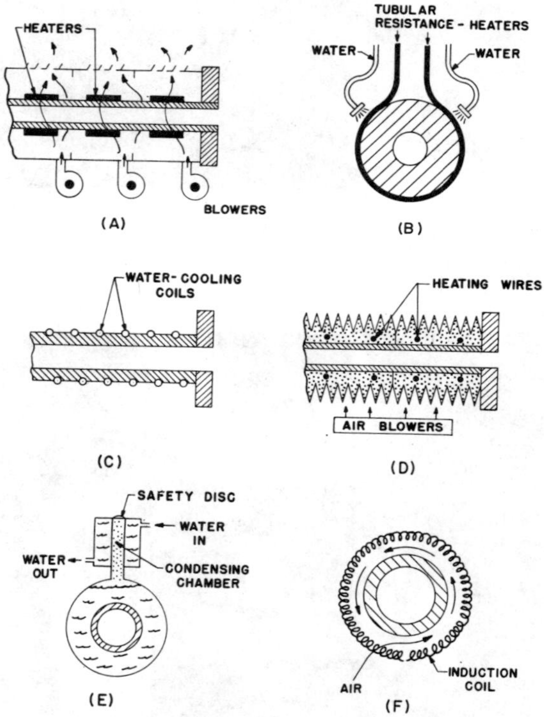

Figure 1.15. Different cooling systems: (A) Fans blowing over heaters. (B) Water-fog sprayed over heaters. (C) Copper tubing for cooling water set in grooves cut in barrel. (D) Finned metal heater sections, with air blown over the fins. (E) Liquid vaporization with water cooling of vapor—air can also be used as a vapor coolant. (F) Air blown between induction heaters and barrel walls. (*Illustration by author*)

COOLING SYSTEM

Quite often the cooling system is even more important than the heating system because of the trend toward faster speeds, longer machines, and more internal mixing—all of which increase melt temperature. If the temperature gets too high, this may cause polymer degradation, difficulty in take-off, or both. Several cooling systems are illustrated in Figure 1.15.

A basic cooling system includes fans that blow room air over or around the barrel and its heaters. Where cooling is likely to be a problem it is advisable to use a more efficient system, such as: (a) fins on the barrel or on the heaters to speed up heat transfer to the air; (b) a water-fog spray over the barrel or heaters; (c) continuous, controlled vaporization of a liquid (usually water); and (d) blowing of air directly over the barrel between it and its induction heaters. Copper tubing carrying circulating cold water around the barrel (Figure 1.16) is sometimes used, and is rather inexpensive although not always uniform.

The choice among these many heating and cooling systems is dictated by several factors, such as the specialty of the extruder supplier, the needs of the products and the plastics involved, the personal preference of the buyer, and, of course, the funds available.

Figure 1.16. Photograph of an extruder barrel, showing cooling coils set in grooves cut into the steel barrel. The coils are copper tubing. (*Courtesy of the Waldron-Hartig Division, Midland-Ross Corp.*)

Twin-screw extruders are seldom fitted with cooling systems. Since there is much less shear of the plastic, less heat is produced in the material, and adequate control can be achieved with the barrel heaters.

INSTRUMENTATION

Temperature. A heating and cooling system is only as good as the devices which control it. These are usually the "proportioning" type. Some merely operate the heaters, allowing only natural cooling through radiation, while others are three-position controllers that actuate both heating and positive cooling. They all normally proportion the heating *time*, not the heater temperature, current, etc. Both the barrel and the die are controlled with the same types of controllers, but the die control is usually more critical. To measure temperature and to provide signals for the controllers, bimetallic thermocouples are set in the barrel and die walls near the plastic stream. Another system which claims great precision employs thermistors as the sensing elements. Another improved system uses two thermocouples for each controller, with the pair compensating for some of the error caused by the need to transfer heat through the mass of the barrel. A thermocouple or other sensing element is often immersed in the flowing plastic to measure its actual temperature, but in only a few cases can this reading be used to control the process automatically.

Sometimes these time-proportioning controllers are replaced by the costlier stepless controllers, which do not keep switching on and off but rather adjust the heater voltage to that needed to give the desired temperature all the time. These stepless controllers are becoming more common now, but are still used only in the best and most precise systems, and then usually on the die rather than the barrel.

On the simplest extruders, the only control is a bank of rheostats which vary the voltage on the heaters; their dials are set by hand. Also, the gate and head of many more sophisticated machines are heated by similar devices. It is common to have several available in an extrusion plant, too, for emergency heaters on dies, especially in the trial stages of a die.

Pressure. Automatic pressure *control* is not commonly used, but its construction has been discussed and some manufacturers will supply it on special order. Pressure is commonly and routinely *measured,* however, with a grease-filled Bourdon gauge mounted near the head just behind the screens. It must tap the underside of the barrel so that the relatively fluid grease will not leak into the plastic stream. If it did leak, it would be displaced by molten plastic which would soon plug the gauge. A better system is a transducer, connected to a metal diaphragm set flush with the barrel inside wall (see Figure 1.17).

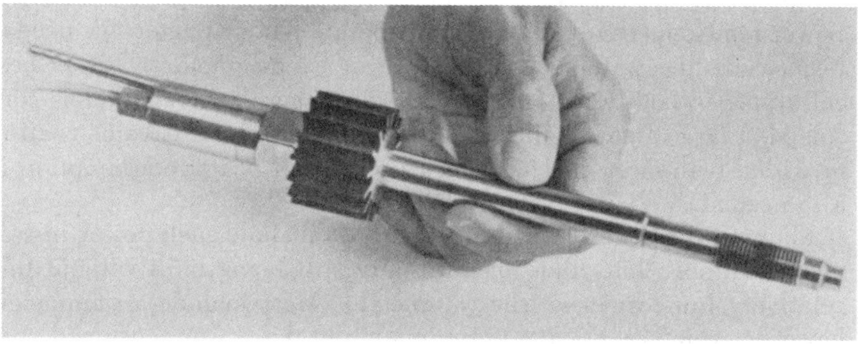

Figure 1.17. Transducer-type gauge for measuring melt pressure inside the barrel. This gauge requires no cooling. (Courtesy of *American-Standard*)

The diaphragm can be connected to the transducer mechanically, or (lately) through a mercury tube. This gauge can be mounted in any position, can feed a recorder, actuate a control device, and is very accurate. Unfortunately, it is more complex to use and is several times the cost of the Bourdon gauge. It has therefore found use only in critical work and in experimental laboratories. There are still other gauges offered at intermediate cost: grease-type or pneumatic gauges with a diaphragm between the grease or air column and the plastic at the barrel wall to prevent leakage and mixing. Any system that produces an electrical or pneumatic signal can be connected to an alarm which rings when the pressure goes above a preset value.

Why measure material temperature and pressure? A good answer is an analogy to the temperature and pulse of a hospital patient. These do not in themselves tell us what is wrong with the patient, but they are essential information for the diagnosis and cure of most ills. Likewise, the temperature and pressure in an extruder are invaluable aids to the analysis and solution of most production problems. Another reason for pressure measurement is the increased use of high pressures for blending and coloring of plastics, where the exact pressure becomes a most important variable.

Power. Quite often, the available power is the limit to production rate. It should be the limit, in fact, in most well-balanced extruder systems. This means that every horsepower that is present, and that was designed and paid for, is being used, within reasonable safety limits. This, unfortunately, is not always true. To understand the power requirements of an extrusion operation a wattmeter * is needed to measure the wattage (power) drawn by the motor. This is best calibrated in both watts and horsepower. Where heater power is consumed in large amounts during operation, such as in extrusion-coating or with a twin-screw machine, measurement of heater consumption is also needed.

Many of the better extrusion systems do include such power instrumentation now, but other machines are still being built without this relatively low-cost item (the wattmeter). Most include an ammeter, however, which allows an approximation of the power consumed, and also acts as a safety device, telling when the machine is electrically overloaded. A better understanding of power requirements and their significance among extrusion engineers and plant foremen should result in more frequent requests for complete power instrumentation.

Screw Speed. Most machines have a tachometer which indicates screw speed in revolutions per minute. Not all of them remain accurate, but if they are occasionally checked and calibrated, they can be trusted.

Location of Instruments. Instruments are usually mounted on a separate panel, except for the Bourdon pressure gauges, which are set in the extruder head. However, if the panel is integral with the extruder body, the instruments should be of a type that is unaffected by the vibrations and heat produced by the extruder, or they should be mechanically insulated against such vibrations and heat.

SCREENS AND BREAKER PLATE

It is customary to have several layers of wire screen at the output end of the screw across the flow path. Their main function is to strain out contaminants from the plastic melt—wood chips, bits of metal,

* An ammeter/voltmeter combination will do, but will require more calculation.

pieces of cloth, and the like. (See Fig. 1.18) Such contamination is especially likely if much reground scrap is used. The screens collect this foreign material until pressure in the extruder becomes too high; then the head is opened and the screens are changed. Screens are usually stainless steel and, although expensive, are seldom used over again because of the difficulty in cleaning them. Also, they often get damaged when the extruder is opened. The screen pack contains several mesh sizes arranged with the heaviest ones farthest downstream, thus supporting the lighter ones and preventing them from being broken and extruded out of the die. A typical screen pack might consist of a 20-, 40-, and 80-mesh screen (numbers refer to openings per inch).

A breaker plate is used to support the screen pack. This steel disc also serves as the mechanical seal between the barrel lining and the die mounted in the head. Its thickness is about 20 per cent of the barrel diameter, and it is drilled with many holes about 3 to 6 mm (1/8 to 1/4 in.) in diameter. The holes are frequently counter-sunk to provide more streamlined flow and eliminate stagnation between them. Several designs are shown in Figure 1.18, together with some screens. The special breaker plate for flat sheet and film is made with narrow slots instead of holes. It must always be installed with the slots parallel to the plane of the flat sheet. In addition to supporting the screens and sealing the barrel to the die, the breaker plate also breaks up the spiral flow pattern coming off the end of the screw, thus making a more uniform melt.

The breaker plate and screens add to the head pressure (e.g., about 15 atm), and they add much more than this if a heavy and fine screen pack is used, or if the screens have not been changed for some time. However, this method is *not* recommended to achieve high pressure (often needed for mixing and coloring) because the fine screens clog easily, require frequent changing, do not hold a constant melt pressure, and cannot easily be adjusted during operation. A valve in the head is a much better method as it has none of these disadvantages.

Several devices are available to change screen packs without disassembling the extruder. Some of these devices work on the principle of a slide projector (Figure 1.19); the extruder is momentarily stopped to relieve head pressure and a fresh screen pack and breaker plate are forced into the head to replace the old ones. Although idle time is reduced by this method, the chance of leakage may be increased and there is a possibility of creating stagnant spots and conse-

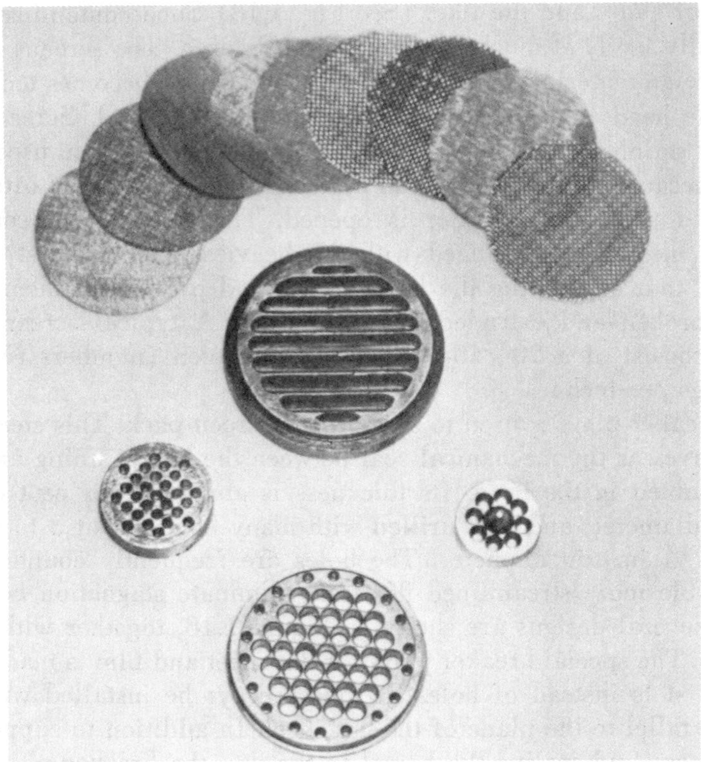

Figure 1.18. Breaker plates of different designs and sizes, shown with typical screens that they support. Note the slotted breaker plate, for use in flat film and sheet extrusion. (*Courtesy of Union Carbide*)

quent decomposition, unless the unit is built very well. The cost of such a unit is rather high—from $2000–$4000. Some designs are even more streamlined but are also more expensive. The above problems have kept the screen changer from being used with many extruders. However, for reclaiming and compounding, they are highly desirable because of the large amount of contamination in the feed; also, large extruders are used, and normal screen changing is correspondingly more difficult. Automatic changers are now frequently used for such operations as quality film and monofilament extrusion, where it is important to change the screens often in order to keep contamination to a minimum.

Figure 1.19. Two types of screen changers. At top, a hydraulically operated unit; when the screens become clogged, the clean assembly visible at the right is quickly pulled into position. (*Courtesy of Prodex Division, Koehring* Co.) Below, the new continuous changer; a strip of screen is continuously fed from the reel at the right and moves across the flow path. As it comes out of the other side it carries all foreign matter with it. The screen is moved entirely by the internal melt pressure, and its speed can be varied by controlling the temperatures of the unit. (*Courtesy of Process Developments Ltd.*)

The continuous screen changer (Figure 1.19) is an important new development; instead of changing individual screens, it passes a continuous strip of screen across the flow path. The "change" can take only a few minutes, but the usual procedure is to move it very slowly (a few inches per hour or per day), all the time, so that there is no progressive clogging, and melt pressure is not affected.

Other screen-changing devices in use include: (a) two identical gates, which require a short shutdown to change from one to the other; and (b) a "vacuum-cleaner" device which scans the upstream surface of the screen pack with a moving tube leading to the atmosphere, thus cleaning the screens without removing them. In a few cases, such as rigid PVC, neither screens nor breaker plate are used—viscosity is already so high that every possible restriction to flow is eliminated. Also, PVC may decompose behind the breaker plate unless it is very streamlined. When no screens are used, extra care must be taken to keep the plastic feed clean.

GATE ASSEMBLY

Extruders are usually fitted with a thick metal gate bolted to the head of the barrel. The gate holds the extrusion die in place by means of an adapter, a connecting passage attached to both gate and die. This adapter is also the usual location for a pressure-control valve. The gate is hinged and pinned to the extruder head to make it easier to open and clean. On large machines, hydraulic opening and closing of the gate is available. The gate and the head are securely fitted together to prevent leakage. Many gates are heated, using resistance heaters and manual rheostat controls.

Some machines, particularly the smaller ones, use a segmented ring in place of a hinged gate. It is generally less expensive and opens quickly, since only one bolt must be loosened. However, it may be a nuisance to clean and fasten, and it is likely to leak unless carefully made and maintained. Another system, less common, but more effective, is a breech-block head, where the gate is rotated with respect to the head, and internal wedges bring the two into tight contact.

Most modern single-screw extruders can be operated at 300 atm internal pressure, and some are rated as high as 650 atm. Not many extruders run at such high pressures today, but there are enough (e.g., wire coating, rigid PVC) to make most extruder manufacturers design their machines to withstand such a load.

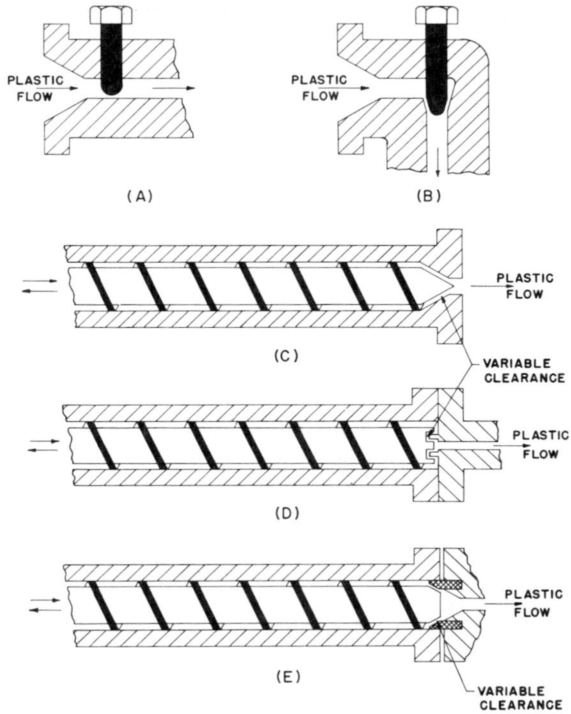

Figure 1.20. Different designs of valves used to control internal pressure in an extruder. Most common is (A), and many variations are found of this simple plug. Design (B) is most suited for dies with elbows. Designs (C) and (D), adjustment of the screw nose into a seat, are good for all plastics because of their streamlining. Design (E) offers a unique way of increasing pressure without a major increase in axial load on the screw. (*Illustration by author*)

Twin-screw extruders do not run at such high pressures because: (1) the plastic is positively conveyed along the screws, with much less pressure drop, (2) thrust bearings of twin-screw extruders are more limited in size and hence limited in pressure-bearing capacity, and (3) screws are shorter and channels usually deeper, producing less pressure than a conventional single screw with a given plastic.

The pressure limit on any machine should be accurately pre-set by including in the gate assembly a shear pin, shear bolt, or similar device designed to fail at a known value.

VALVE

To produce artificially high pressures, a variable mechanical restriction is placed in the head in the path of the plastic flow. Various designs are shown in Figures 1.20, 1.21 and 1.22. High pressures (200 to 400 atm) are often desired to improve coloring or blending in the extruder, or sometimes just to improve temperature uniformity of the extrudate (hence the dimensional uniformity of the product). The restriction is usually a valve which allows pressure change during extrusion. A particularly unique design (Figure 1.20e) is the restriction in the nose of the screw such that the load on the thrust bearing is hardly increased despite great increase in pressure on the melt. The most streamlined valves are necessary for plastics such as PVC which tend to decompose in stagnating areas. On the other hand, a simple plug-type valve is entirely adequate for most polyethylene and polystyrene extrusions.

A special type of valve is sometimes used halfway down the barrel between the stages of a two-stage (vented) extruder. This usually

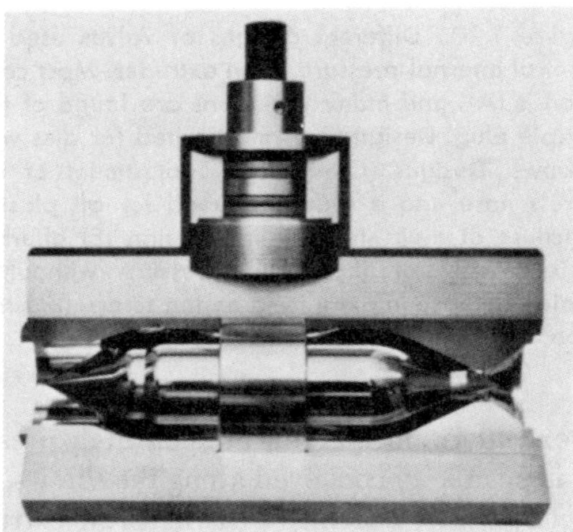

Figure 1.21. Cutaway illustration of a streamlined valve for temperature-sensitive plastics. Adjustment of nut at top produces axial movement of nose into its seat. (*Courtesy of Sterling Extruder* Corp.)

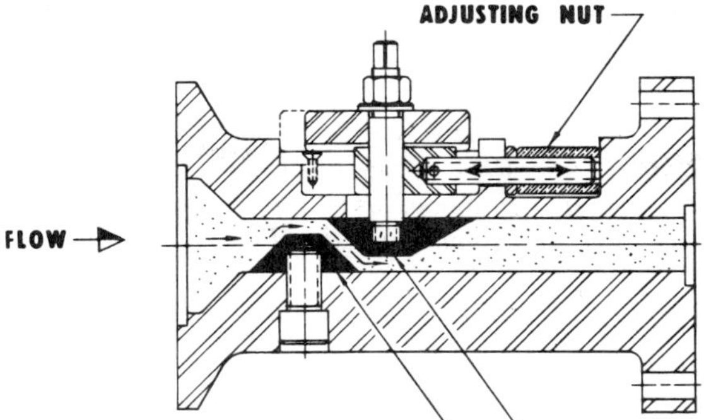

Figure 1.22. Drawing of another type of streamlined valve. The adjusting nut moves the upper gate horizontally, opening or closing the gap between the gates. The channel has a circular cross section. (*Courtesy of Waldron-Hartig Division of Midland-Ross* Corp.)

takes the form of a bulge or ring on the screw mating with a corresponding groove around the circumference of the barrel. Axial movement of the screw changes the gap, hence the resistance at the end of the first stage. This gives more versatility in extrusion, allowing a greater variation in conditions, materials and die resistances. For polyethylene and other less heat-sensitive plastics, this intermediate valving can be done by passing the melt out of the barrel, through a conventional valve, and back into the barrel just before the vent.

DIES AND TAKE-OFFS

These specialized pieces of equipment will be covered in greater detail in the various chapters on applications (see Chapters 3 through 9).

REFERENCES

NOTE: General references such as books and comprehensive articles are listed at the end of this book. Also, for information on screw and die design, see the bibliography at the end of Chapter 11.

General

"The compact extruder," P. N. Richardson (du Pont), *SPE Journal*, February 1965. Results of questionnaire defining this term, sent to extrusion engineers.

"Dual-screw extruders," F. C. Schutz (Welding Engineers, currently a consultant), *SPE Journal*, September 1962. Description of types and commercially available products.

"Extrusion," H. E. Sponaugle (Hale & Kullgren, now Black Clawson), *Plastics Design & Processing*, June 1961. Brief discussion of the various parts of an extruder, with special mention of use of screw nose in seat as a valve. French translation appears in *Industrie des Plastiques Modernes*, October 1961.

"Trends in extruder design," J. Badonsky (Waldron-Hartig), *Modern Plastics*, February 1961. Discussion of extruder components: speed reducers, barrel material, drives, thrust bearings, heating and cooling systems, etc.

"Praktische Erkenntnisse bei Zweischneckenpressen und ihre konstruktive Auswertung" (Practical knowledge about twin-screw extruders, with explanation of construction), K. Tanner, *Kunststoffe*, September 1956. Principles of construction for twin-screw extruders. Many pictures, long article, no mathematics.

Drive Systems

"Application of coordinated drive systems for plastics extrusion lines," R. H. Soderberg (Louis Allis), *SPE Journal* and *SPE Regional Conference Papers*, Boston, both March 1967. Advantages and disadvantages of different drive types, including coordinated drive and takeoff.

"SCR drives for the plastics industry," W. H. Welsh (General Electric), *SPE Regional Conference Papers*, Boston, March 1967. Advantages and disadvantages of solid-state DC drives.

"Elektrische Antriebe für Schneckenpressen in der Gummi- und Kunststoff-Industrie" (Electric drives for rubber and plastic extruders), S. Zaborosch, *Gummi, Asbest, Kunststoffe*, **18**, 2, 1965. Discussion of motors and speed variators.

"Electric drives for rubber and plastics extruders," *Kunststoff und Gummi*, September 1963.

"Extruder drives," P. Moroz (du Pont), *SPE Journal*, March 1963. Tells how they work, principles, advantages. Written from point of view of power engineer, not plastics man.

"Your guide to roller and thrust bearings," N. P. Chironis, *Product Engineering*, January 1961. Helps understanding of bearings used in extruders; article deals with such bearings in general.

"Drives for extruders," G. D. Campbell, Jr. (Reliance), *SPE Journal*, November 1958. Principles and advantages of the three main types of continuous speed variators. Still useful, although some drives have changed—particularly DC systems.

"How to measure torque on an extruder," W. J. Allen and D. A. Hillman (ICI), *Plastics,* October 1957. Describes apparatus and gives some data. For designers and testing specialists.

Feeding Systems

"Feeding the single screw extruder," R. L. Miller (NRM), *SPE Journal,* November 1964. Covers comparison of pellets and powder, hopper design, feed opening; discusses forces in feed channels.

"Advances in dry-blend extrusion," N. T. Flathers *et al.* (Escambia), *Modern Plastics,* May 1961. Discusses use of vacuum hopper to improve appearance and reduce porosity in dry-blended vinyl products. Also appears in *Wire and Wire Products,* June 1961 and June 1960, and *ANTEC* 1960. Shorter version in *SPE Journal,* October 1959.

Screw and Barrel Construction

"Bimetallic vs. nitrided cylinders," W. W. McCandless (Xaloy), *Plastics World,* November 1966. Shows advantages of bimetallic cylinders for extrusion and injection molding.

"Mechanical aspects of extruder cylinders," L. J. Mattek (Industrial Research Labs. = Xaloy), *SPE Journal,* January 1965. Methods of manufacture and properties of bimetallic cylinders, analysis of stresses encountered. Also presented at *ANTEC* 1964.

"Hard surfacing of extruder screws," L. I. Naturman (*SPE Journal* staff, now employed by Predicasts, Inc.), *SPE Journal,* September 1963. Impartial review of advantages and disadvantages of various methods, considering cost, metallurgical differences, properties and service life.

"Extruder screw production," *International Plastics Engineering,* January 1963. Description of screw manufacture at W. Platt & Sons, Ltd.; materials, sequence of steps, problems.

"Materials of construction for extruders and extruder dies," H. Seiwert (Hüls), *Kunststoffe,* September 1962. Short article, mainly deals with nitrided steel, notes Xaloy (bimetallic) as "special."

Venting

"Extruding rigid PVC powder? Then vent your extruder," J. M. Georgiana (Diamond Alkali), *Plastics Technology,* March 1967. Costs and technique for venting normal extruder.

"Improved method and apparatus for venting in extrusion," N. Okazaki (Consultant), *ANTEC* 1966. Discusses various systems with pictures. He proposes system in which melt is extruded into enclosed chamber as strands, vented there, then picked up by second screw.

"Review of vented extruders," L. G. P. Smits (du Pont Nederland), *SPE Journal,* September 1964. Incomplete but impartial review of both European and American methods. Useful for its impartial criticism.

"Vented single screw extruders," G. Schenkel (Troester, now with Stuttgart Technical College), *International Plastics Engineering,* September-October-November 1962. German version in *Der Plastverarbeiter,* November and December 1962, and January 1963. Major article, much mathematics, 49 references. Deals with venting at open discharge, then with valving (resistance) at one or two points, then compares various practical experiences.

"Some comments on vented extruder design," L. B. Ryder (Celanese), *SPE Journal,* August 1961. How the two-stage extruder works.

"Theory and operation of the two-stage, single-screw extruder," B. H. Maddock and P. P. Matzuk (Union Carbide), *SPE Journal,* April 1962 and *ANTEC* 1962. Discussion of vented machines, and proposal to add a valve at the end of the first as well as the second stage.

"Developments in extrusion machinery: valved and vented extrusion," D. Grant (Bone Brothers), *Plastics Institute Transactions,* August 1961. Further discussion of venting theory and practice.

"The vacuum-extruder screw," E. C. Bernhardt (du Pont), *SPE Journal,* March 1956, and *ANTEC* 1956. Principle and operation of a screw that draws volatiles out through its core. Developed for acrylics, but not in general use today.

Heating and Cooling Systems

"Heizsysteme für Kunststoffmaschinen," W. Berger, *Kunststoffe—Plastics,* No. 2, 1965. Advantages and details of induction heating and fluid heating (steam, oil). Similar article in *Der Plastverarbeiter,* January 1964.

"Heating and temperature control for injection and extrusion equipment," E. Miebach (Eckert & Ziegler), *International Plastics Engineering,* January and February 1963. Two-part article, covering simple principles of cylinder heating, types of resistance heaters, and types of controllers and sensors.

"Extruder heating and cooling systems," P. H. Squires (du Pont), *SPE Journal,* August 1962. Survey of the state of technology at this time.

"Induction heating and high screw speeds in extruder design," A. Lord, *Rubber and Plastics Age,* December 1959. Advantages of induction heating; e.g., better cooling, which in turn may allow higher speeds.

"Induction heating for plastics extruders," W. J. Goodwin (Hale & Kullgren, now Black Clawson), *SPE Journal,* February 1958. What it is, how it works, why it is good.

"Heating systems for extrusion cylinders," W. H. Willert (F. W. Egan), *SPE Journal,* June 1957. Brief discussion of principles and operation of resistance and evaporation (Willert) systems.

"Cast-in heaters," W. H. Norton (Thermel), *ANTEC* 1956. Complete description of these heaters, which are used on many extruder barrels.

Instrumentation

"Instrumentation for temperature control," A. M. Willer (West division of Gulton Industries), *Plastics Design and Processing,* February 1967. Discusses each type of controller, with advantages and disadvantages.

"Apparatus for measurement of pressure in melts," D. D. Ryabinin (U.S.S.R.), *Plasticheskii Massy,* October 1966. For extrusion applications.

"Druck und Temperatur-messungen an einen Einschnecken-Extruder beim Verarbeiten von Polyäthylen weich" (Pressure and temperature measurement in a single-screw extruder, using low-density polyethylene), G. Menges and J. Müller, *Der Plastverarbeiter,* October 1966.

"Temperature control of plastics processing machinery," *British Plastics,* October 1966. Describes practical aspects of control circuits, how control signals are generated, how power is proportioned to heaters, various heating methods in use.

"Some principles of temperature control for plastics machinery," R. F. Taylor and J. I. Hopley (Honeywell Controls Ltd.), *British Plastics,* September 1966. Describes basic function of control systems, primary sensors and temperature measurement.

"Gegenüberstellung der Methoden zur Messung von Druck und Temperatur bei Spritzguss und Extrusion" (Review of methods of measuring temperature and pressure in molding and extrusion), G. Scholz and H. Patzschke (East Germany), *Plaste und Kautschuk,* August 1966. Medium-sized artcle but with 79 references.

"The measurement of melt pressure," R. A. Lolley and H. Wykes (ICI), *Transactions of the Plastics Institute,* August 1966. Use of pneumatic diaphragm gauges.

"Temperatur-Messungen in Extrusionsverfahren," G. Schenkel (Stuttgart Technical College), *Kautschuk und Gummi, Kunststoffe,* No. 4, 1966. Temperature measurement systems described.

"New pressure gauge for plastic melts," J. Schijf (TNO, Netherlands), *Plastics,* February 1966. Description of diaphragm and plunger grease gauges.

"A new concept in melt pressure measurement," M. R. Harris (Dynisco), *SPE Journal,* February 1966. New models of strain-gauge transducers for pressure measurement inside extruders described.

"New pressure gauge for extruders," *International Plastics Engineering,* September 1965. Describes pneumatic gauge developed by ICI and made in England by Bowyer, Smyth & Partners, Ltd.

"Schaltende Temperaturregelung an Kunststoffverarbeitungsmaschinen" (Intermittent temperature control of plastics processing machinery), R. Berger

(Hartmann & Braun AG), *Kunststoffe,* June 1963. Tells how a controller works, with good explanation of problems in placing thermocouples.

"Neue Wege der Temperaturregelung an Spritzgussmaschinen und Extrudern" (New methods of temperature control on injection molding machines and extruders), F. Leider, *Kunststoffe,* April 1963. Discusses various kinds of controls, including electronic.

"Use of radiation gauges in plastics," G. I. Doering (Industrial Nucleonics), *Plastics Technology,* April 1958, and *ANTEC* 1958. What a beta-ray gauge is, how it can be used to measure and record thickness, and why better measurement often means better quality, lower costs, or both. For further information in this area, see references after Chapters 5 and 6.

"Application guide to extruder pressure measurements," R. L. Eckman and G. A. Pettit (Barber-Colman), *International Plastics Engineering,* May 1961, and *ANTEC* 1961. Study of the transducer pressure gauge. More information on these gauges is found in *ANTEC* 1959 (also by Pettit), and *ANTEC* 1960, *SPE Journal,* August 1959 and March 1960 (these last three articles by du Pont personnel).

"Advantages of dual thermocouples," R. K. West (West), *SPE Journal,* April 1957. Study of thermocouple location in extruder barrels. This principle (dual thermocouples) is still not common today.

"Stepless and three-position temperature controllers," R. K. West (West), *SPE Journal,* June 1956, and *ANTEC* 1956. Differences between the two types, and advantages and principles of each.

"Stock thermocouples, pressure gauges and rupture disks for use on plastics extruders," E. C. Bernhardt (du Pont), *ANTEC* 1955. Description of this equipment, and instruction on how to build and use it. This is the primary introductory article in this field.

Screens and Breaker Plates

"Screen pack changers," B. R. Voight (Davco Industries), *Modern Plastics,* August 1966. Review of various types, emphasizing slide-projector type.

"A stationary plate for mixing," O. Ingles (Commonwealth Science and Industry Research Organization), *International Plastics Engineering,* August 1963. Angled holes in breaker-plate type discs.

"Extruder screening devices," F. C. Schutz (Welding Engineers, now private consultant), *SPE Journal,* June 1963. Survey of products of seven manufacturers, including Welding Engineers.

Valves

"Dynamic valving vs. static valving," H. E. Sponaugle (Black Clawson), *ANTEC* 1962. Description of unique valve in nose of screw. Axial motion of screw changes gap, which is streamlined and non-seizing.

"Fixed-orifice pressure control for extruders," B. H. Maddock (Union Carbide), *Plastics Technology,* September 1960. How to design and build a low-cost, pressure-producing restriction for extruder heads. It takes the place of a valve; however, a valve is far more versatile and not much more expensive.

"Important advances in extrusion technology," R. D. Sackett and E. H. Hankey (Monsanto), *SPE Journal,* November 1957. Comments on vents and valves, including a special pressure head in which the screw fits into a seat and moves axially to change resistance. Pressure is produced with minimum increase of thrust.

"Valved extrusion," E. C. Bernhardt (du Pont), *SPE Journal,* February 1957, and *ANTEC* 1957. Basic studies and calculations, concerning continuous and intermittent valving.

Miscellaneous

"Variable-depth cooling for screws," E. Lowey (Gene Lowey Inc.), *Plastics Technology,* January 1966. Discusses advantages and methods of partial cooling of core of screw.

"Revised nomenclature for the screw extruder," *SPE Journal,* October 1963 and March 1962. Original list in March 1962 issue; some new additions in October 1963.

"Flexible conduits in extrusion," E. C. Bernhardt and W. M. Edwards (du Pont), *SPE Journal,* September 1959. How to convey plastic melts from an extruder through steel or fluoroplastic flexible hoses.

TWO

markets for extruded products

Extruded products are divided into seven areas of application, each with its own rather distinct technology, as follows:

(1) Pipe
(2) Sheet
(3) Film
(4) Extrusion-coating
(5) Wire and cable covering
(6) Monofilaments
(7) Miscellaneous contours

The following paragraphs describe the most important applications for extruded plastics in these categories. Further discussions of product types and dimensions and special techniques and uses are given for each category in its specific chapter.

Pipe (Figures 2.1 and 2.2). Plastic pipe is used in industry, construction, and agriculture. Its major advantages over other materials are resistance to corrosion and chemical attack, light weight and ease of installation, and toughness. In industry, PVC pipe is widely used in chemical plants to transport products, and it is used in many factories to carry process water and to remove waste liquids. In con-

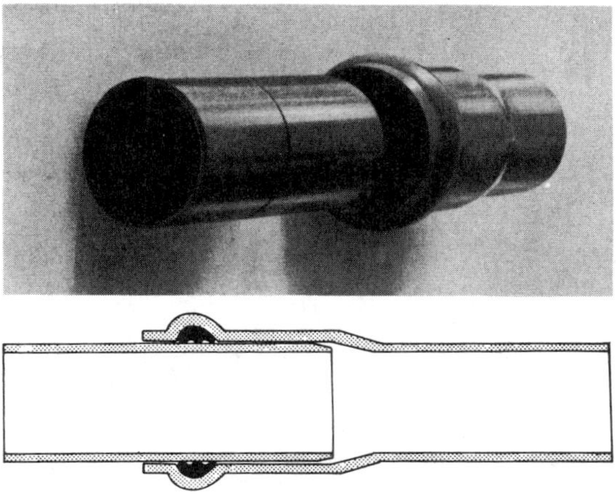

Figure 2.1. Rigid PVC pipe such as this has helped expand the use of plastics in construction. This type has a flexible ring seal that will take a working pressure of 160 psi (11 atmospheres). (*Courtesy of U.S. Anger, Inc. and Johns-Manville*)

struction, PVC and ABS (acrylonitrile-butadiene-styrene) are rapidly growing in DWV (drain-waste-vent) applications. PVC and filled polystyrene are used for electrical conduit, and thin-wall PVC is used on houses to carry off rainfall. In agriculture, polyethylene is used in wells and to water livestock. Both polyethylene and PVC are used in irrigation. Other important uses for plastic pipe are in oil fields, and for the distribution of gas to homes and factories and water to rural communities.

Sheet (Figures 2.3 and 2.4). Most extruded sheet is eventually thermoformed. Packaging provides the largest outlet, and is growing fast. High-impact polystyrene is formed into disposable cups, containers for cottage cheese and ice cream, cigarette packages, and even milk containers; use of foamed polystyrene is well established in fruit, produce, meat, and egg trays, and this material is already well known in candy and cake packaging. PVC thin sheet, made by both extrusion and calendering, is expected to have a good future. Another large outlet is the appliance industry; e.g., thermoformed refrigerator door liners, TV receiver backs, and various housings. Most of these products are made of high-impact polystyrene or of ABS. Other im-

Figure 2.2. Polyethylene pipe being installed for water supply in rural location. (*Courtesy of Union Carbide*)

Figure 2.3. Thermoformed items such as these cottage cheese containers have become well known to the consumer. (*Courtesy of Lily-Tulip Cup Corp.*)

Figure 2.4. The door liners of most refrigerators are thermoformed from extruded sheet, usually high-impact polystyrene or ABS. (*Courtesy of Union Carbide*)

portant uses for extruded sheet include: ABS and polyethylene interior panels for automobiles; ABS luggage; PVC house siding and exterior paneling; lighting fixtures of PVC, acrylic and polystyrene; acrylic and cellulosic signs; and PVC, polystyrene, and cellulosic displays, toys, and novelties.

Film (Figures 2.5 and 2.6). The packaging industry is the largest consumer of extruded film. Most of this is polyethylene, but polypropylene, oriented polystyrene and polyester, and PVC are also used. Vegetables and fruits, baked goods, meats, snacks, and textiles (soft goods) are among the leading items packaged in plastic. Covers for laundered shirts and dry-cleaned garments use thin polyethylene film, and boil-in-bag pouches are a growing use for laminates with polyester. Many multi-layer films involve both film extrusion and extrusion-coating in their manufacture. An important application is the heavy-duty sack used for fertilizers, sugar, flour, animal feed,

Figure 2.5. Heavy-duty polyethylene sacks like these are being used for packaging fertilizer and other agricultural products, chemicals, and even plastics themselves. (*Courtesy of U.S.I. Film Products*)

seeds, chemicals, and even plastic pellets themselves. The construction industry uses polyethylene film for moisture barriers and temporary protective enclosures and covers. Extruded black polyethylene is used in agriculture as a mulch and as protection for produce, silage, and machinery. Transparent polyethylene is also used for greenhouse glazing and as "tunnels"—supported covers over rows of growing plants in the fields. Other uses for extruded film include household wrap (polyethylene and saran), separation film in the rubber industry (polyethylene), document protection (PVC), Christmas and other decorations and novelties (PVC), tapes of all kinds (PVC, cellulosics), and twine and rug backing cord (oriented polypropylene).

Extrusion-coating (Figure 2.7). Polyethylene is used as coating on paper, paperboard, cellophane, oriented polyester, and aluminum foil—mainly for the packaging industry. Food packaging, in particular, often requires a combination of the properties of two or more materials for the best result. Polyethylene allows heat-sealing and

Figure 2.6. Film applications. *Above:* Tray-packs of cans over-wrapped with heat-shrinkable PVC film are beginning to replace traditional corrugated shipping cases. (*Courtesy of Reynolds Metals* Co.) *Below:* Polyethylene film finds much use in construction, such as this protective layer over which a concrete road will be poured. (*Courtesy of Kalle AG*)

Figure 2.7. Milk cartons such as this one are the biggest application of extrusion coating. A thin polyethylene coating on both sides provides economical, glossy, non-flaking protection for the contents. (*Courtesy of Modern Packaging Magazine*)

provides certain barrier properties. Milk cartons are by far the most important of the extrusion-coating applications; others include meat and fish packages, cartons for frozen foods, portion-packages of sugar and other foods, multiwall sacks, and wrapping paper for moist or greasy items.

Wire and Cable Covering (Figure 2.8). Plastic coatings range in size from the flexible PVC covering on tiny telephone wires to the heavy PVC jacketing used to protect bundles of wire several centimeters in diameter. Other important applications are high-voltage power transmission, undersea long-distance telephone cable, TV lead-in wire, appliance wire, instrument panel wiring, and all

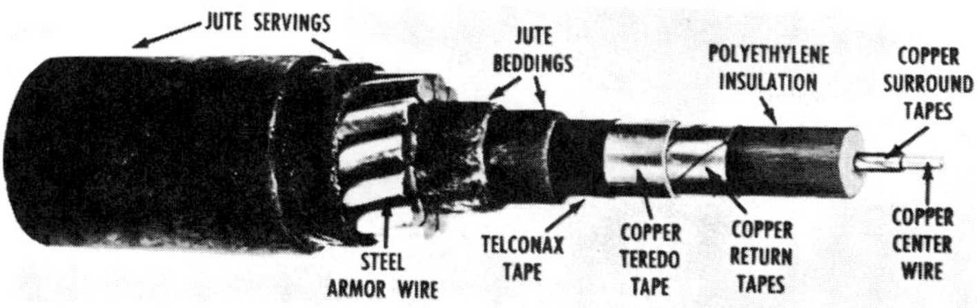

Figure 2.8. This undersea cable for telephone transmission is a complex combination of metals and extruded plastics. (*Courtesy of Western Electric*)

kinds of military applications. Some wires are coated with two different plastics, each offering its special advantages. Polyethylene and plasticized PVC are the most important materials and are consumed in almost equal amounts. PVC offers high flexibility and flame resistance, and polyethylene offers exceptional insulating ability, light weight, chemical resistance, and ease of extrusion, all at reasonable cost. Nylon is occasionally used where exceptional abrasion resistance is needed. All plastics offer moisture resistance.

Monofilaments (Figures 2.9 and 2.10). Much natural bristle and fiber have been replaced by extruded plastic, which is controllable in quality and size, unaffected by importing problems, and resistant to

Figure 2.9. Polystyrene (top) and natural fiber brushes, after undergoing identical wear tests. (*Courtesy of Union Carbide*)

Figure 2.10. Plastic monofilaments, especially polypropylene, are widely used in ropes of all sizes. They can be color coded, they float, they are stronger than natural-fiber ropes of equal diameter, and they do not become waterlogged or rotted in use. (*Courtesy of American Manufacturing Company*)

moisture, rot, and most chemical environments. A large amount of extruded monofilament is used to make rope, most of which is polypropylene. It is extremely strong, suitably resistant to the elements, and reasonably cheap. Its light weight (it floats) makes it particularly suitable for marine applications. High-density polyethylene was used first, but has been largely replaced by polypropylene in the United States; a similar change is now beginning to take place in Europe. Nylon is used when abrasion resistance and very high strength are needed in a non-floating filament, such as in fishing line. Textiles are a large outlet for monofilaments; for example, outdoor furniture and automobile seat covers are made of polypropylene. Saran and PVC bristles are used for nonflammable Christmas trees, while polystyrene, nylon, and polypropylene bristles are used for all kinds of brushes, from tooth brushes to house brooms to huge sweepers for city streets. The orientation process makes the otherwise brittle polystyrene tough enough to compete in this field. Polyolefin ribbon yarns, made by slitting and orienting extruded films, find uses in woven sacks, textiles, rug backing, and twine. Finally, non-woven netting is extruded in a wide variety of sizes, and competes for markets in multiple packaging (jars, fruits, vegetables), bottle protectors, separators in corrugated cases, garment bases, fencing, filters, and many more. Polyolefins and foamed polystyrene are the leading materials.

Miscellaneous Contours (Figures 2.11 and 2.12). This category includes extrusions of innumerable cross sections, usually made on a custom basis. Markets include automotive and shoe welting, weatherstripping, window and door frames, gaskets for refrigerator doors, carrier troughs for electrical wires, and laboratory tubing. Among the largest cross sections are lighting fixtures and waterstops, and seals between poured concrete sections of constructions such as dams. PVC, both flexible and rigid, is the most common material, but all thermoplastics find some use.

Figure 2.11. Extruded vinyl waterstop (arrow) placed in position before the pouring of concrete. The plastic forms a continuous seal between sections of concrete. It is flexible enough to withstand expansion and contraction of the concrete, yet it stays water-tight. (*Courtesy of Union Carbide*)

MARKET FIGURES

In 1966 the extrusion market totaled about 1.4 million long tons in the United States, about 1.2 million tons in Europe, and about 400,000 tons in Japan. The following figures (all in long tons) are given as a general guide to the usage in the United States:

PIPE, excluding fittings, total		100,000
Polyethylene, all densities	33,000	
PVC	36,000	
ABS	12,000	
Polystyrene-based	15,000	
Others	4,000	
SHEET, total		160,000
Polystyrene, including foam	115,000	
Others, including ABS	45,000	
FILM, total		570,000
Polyethylene		
Food Packaging	165,000	
Non-Food Packaging	175,000	
Construction	25,000	
Agriculture	25,000	
Other Uses	95,000	
Other Materials	85,000	
EXTRUSION-COATING, total		148,000
Polyethylene on paperboard	85,000	
Polyethylene on paper	35,000	
Polyethylene on other substrates	25,000	
All other materials	3,000	
WIRE and CABLE COVERING, total		263,000
Polyethylene	145,000	
Vinyls	115,000	
Other materials	3,000	
MONOFILAMENTS, including twine and ribbon, total		47,000
Polypropylene	32,000	
Other materials	15,000	
MISCELLANEOUS CONTOURS, all		100,000
GRAND TOTAL		1,388,000 tons

EXTRUSION INDUSTRY MARKET PATTERNS

Raw Material Suppliers. These are large companies capable of investing in large plants to produce the polymers and of supporting

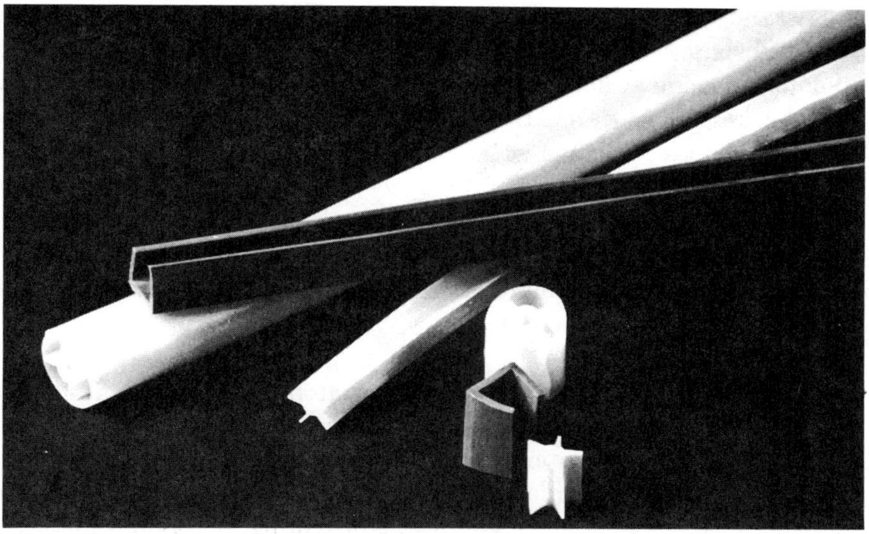

Figures 2.12. Samples of the great variety of profiles that can be made by extrusion. (*Courtesy of duPont*)

the technical staffs needed to develop and improve these polymers and their end-uses. Lately, markets and prices for the major materials have become more stable. Prices have reached a competitive minimum and have even slightly risen in a few cases. Volume of sales of these plastics is high and profit margins are low; accordingly, the suppliers' development and customer service efforts have become more restricted. To increase their "added value"—hence their profit—the suppliers have begun more and more to manufacture end-products such as oriented and unoriented film and sheet, house siding, rain gutters, formed milk containers, and even drinking straws.

Captive Processors. These are large manufacturers who find it more economical to extrude their own products rather than to buy them from independent suppliers. This is common in the appliance and communications industries, but less frequent in the construction and automotive fields. Sometimes a user will maintain a certain amount of control over his supply by buying a portion of an extrusion firm, or by setting up a jointly owned company. Wire and cable coaters are all captive processors, as they cannot sell their coverings separately (except for tape wrappings).

Custom Processors. These firms buy raw materials and produce standard shapes such as pipe, sheet, film, and miscellaneous contours, often to a specific order. Certain standardized sizes of these extrusions may be kept in stock. Many of these firms, including most of the largest ones, are divisions of material suppliers.

Distributors. These firms buy and stock the standard shapes, and sell mainly in small lots. Buyers of larger lots often deal directly with the processor. The distributor, in effect, adds to the processors' sales forces.

Fabricators. These are companies that process extruded shapes into finished products. Important among these are thermoformers of sheet, converters of film and coated paper, and makers of rope, textiles and brushes from filaments. These operations are often captive branches of processors, or of end-users, providing the savings of vertical integration. This is very common with sheet, but less so with film and coated paper.

THREE
pipe

PRINCIPLE

Pipe is extruded by forcing molten plastic horizontally through an annular opening into one of several devices for cooling it and fixing its dimensions. Typical pipe extrusion lines are shown in Figures 3.1 and 3.2. Elements required after the extruder include a die, a sizing device, a cooling bath, a puller, and a cutter or winder.

MATERIALS

Polyethylene and PVC are the leading pipe materials. Among the polyethylenes, high-density and medium-density grades are most important, and some low-density polyethylene is used in blends with higher densities. In PVC, both unmodified (Type I) and high-impact (Type II) are extruded from unplasticized formulations. Recently, dual-purpose PVC compounds have been introduced which meet the requirements for both Type I and Type II. Polystyrene mixtures and ABS polymers are also important pipe materials, while polypropylene and polyacetal find specialized uses, and polyvinyl dichloride may join this group in the future. Even thermosets such as phenolics, epoxies, and polyesters are "extruded" into pipe on special equipment, often with glass-fiber reinforcement.

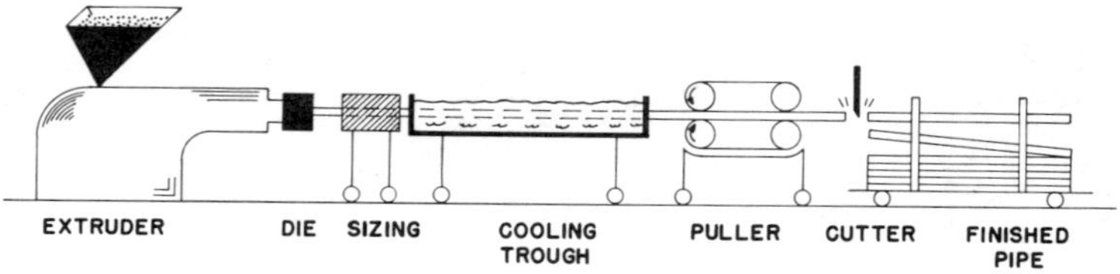

Figure 3.1. Diagram of a pipe extrusion plant. (*Illustration by author*)

Figure 3.2. Pipe extrusion line making 160-mm (6½-inch) PVC pipe. Extruder at extreme left feeds a vacuum sizer, and the sized pipe is pulled through a water trough by the caterpillar-type puller in right foreground. A saw to the right of the puller is not shown. (*Courtesy of Ets. A. Olier*)

PRODUCT AND EQUIPMENT SIZES AND LIMITS

Plastic pipe is any round, self-supporting, continuous, hollow cylinder at least 12 mm (½ in.) in inside diameter. This is not an official classification, but it is in general use. Anything smaller is called tubing (see Chapter 9). Thermoplastic pipes up to 150 mm (6 in.) in diameter are quite common, while sizes up to 1 meter (40 in.) are commercially available. Most of the smaller pipe (50 mm = 2 in. and under) is polyethylene, to take advantage of its flexibility, but all sizes of the major materials find some use.

The extruder output is usually set by the cooling capacity of the system, especially in the case of polyethylene. Beyond this limit, the pipe would be extruded faster than the line could cool it. Outputs of 100 to 160 kg (220 to 350 lb) per hour are common for both polyethylene and PVC, and a few extruders run even faster. For these rates, 90- and 120-mm ($3\frac{1}{2}$- and $4\frac{1}{2}$-in.) extruders are commonly used. The linear speed varies with pipe diameter, of course: 15 meters (50 ft) per minute is considered very good for polyethylene 25 mm (1 in.) in inside diameter, but many processors accept 6 to 10 meters (20 to 33 ft) per minute at this size. Greater outputs from larger machines are possible without increasing the linear speed by having two or more pipes coming out of the same extruder at once, but this is not often done. The largest pipes, of course, require large machines. Other limits to output might be:

(1) Available extruder power—most likely with very viscous materials such as PVC and ABS.

(2) Insufficient mixing in the extruder—remedied by use of a different screw or higher melt pressure in the extruder.

(3) Nonuniform output (surging) at high speed—same remedies as item (2).

(4) For PVC, excessive shear at high speed may raise melt temperature high enough to initiate decomposition.

EQUIPMENT

Dies. Dies can be either "straight-through" or offset (see Figure 3.3). A straight-through die consists of a single cylinder plus a concentric core or pin set in a "spider"—the name given to the supporting arms which help to hold the core in place while allowing the molten plastic to pass around them. The spider must be set some distance back from the exit end of the die in order to develop reasonable pressure (50 to 100 atm) at the spider arms. This pressure is necessary to enable the split streams of plastic to weld together again before leaving the die. Poor strength along such weld lines is in fact a common cause of pipe failure. A good rule is to have the land length (length where walls of die and core are parallel or nearly parallel) at least 20 times the width of the annular opening (wall thickness at exit).

Offset dies were formerly very common, but are now becoming obsolete. In such a die, the plastic is diverted through a heated neck or elbow, and then enters the die assembly from the side. The core is supported at the rear of the die, away from the plastic flow, and can therefore pass through the entire die without the spider arms to

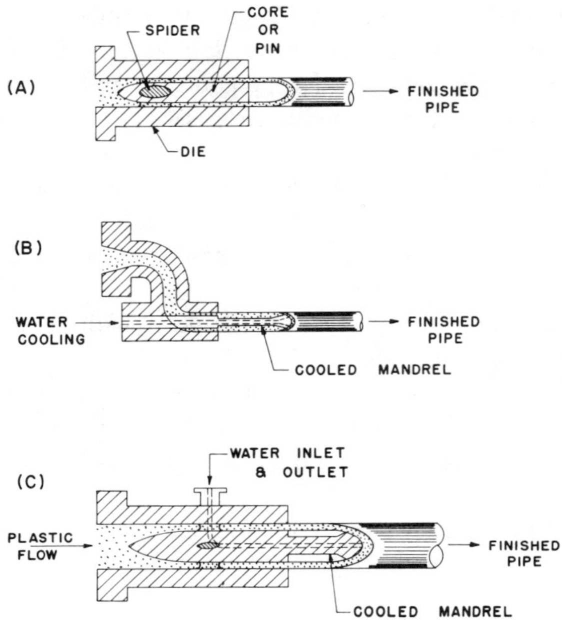

Figure 3.3. Different pipe dies. The straight-through type (A) is the most streamlined and also the most common. The offset design (B) has been used for polyethylene in combination with an extended, water-cooled sizing mandrel, but is becoming obsolete because of design (C), which allows water cooling in a streamlined, straight-through form. (*Illustration by author*)

support it. Water can be supplied easily to the extended cooling mandrel often used with such a die; this mandrel is a continuation of the core.

Straight-through dies (Figures 3.4 and 3.5) are now standard in the field. For PVC, offset dies cannot be used, because the PVC would stagnate and decompose in the elbow or other "dead spots." Also, the mandrel cooling system is seldom used with PVC, and the passage of water through the core is not practiced. For polyethylene, many offset dies were originally used (and some are still performing well) because polyethylene is more stable and less likely to decompose in the

Figure 3.4. Construction and assembly of a straight-through pipe die. *Top:* The three components of this die —from left to right, the *die base* with *spider arms,* the *die* itself, and the core or *pin. Middle:* The core has been screwed into the die base at the left, and the die is being placed over the core. *Bottom:* The finished die is ready to be bolted onto the extruder. The annulus out of which the plastic will extrude can be seen at the lower right. (*Courtesy of Union Carbide and the author*)

Figure 3.5. Large die for 12-inch (30-cm) PVC pipe. (*Courtesy of Prodex Division, Koehring Co.*)

Figure 3.6. Straight-through die for polyethylene pipe with water-cooled extended mandrel. (*Courtesy of Prodex Division, Koehring Co.*)

die, and also because it is more in need of internal water-cooling. However, straight-through dies with water-cooled mandrels (Figure 3.6) were introduced about five years ago, and are well-known now.

Pipe dies are externally heated with cylindrical resistance bands or blocks, as seen in Figures 3.5 and 3.6. The core is sometimes also heated to ensure a smooth inner surface, especially with polyolefins. For PVC, the core is sometimes cooled with water or air, especially when a high melt temperature is used, to avoid localized decomposition on the core surface.

The die parts are normally machined to form a perfect circular annulus, designed to give a uniform wall thickness. Adjustment is sometimes needed during extrusion, and most dies have a few bolts which displace the cylinder with relation to the core. The actual dimensions of the annulus depend on the sizing technique that is used, as well as the size of pipe desired. This is discussed later in this chapter when the various sizing techniques are compared.

Pipe dies are made of tool steel and are sometimes chromium-plated to give a smoother pipe surface. For PVC they are sometimes plated to prevent corrosion, although unless the plating is very thick and well cared for, the plated dies tend to wear down, scratch or chip, after a short period of use.

Sizing Equipment. Four sizing techniques are common: the vacuum trough, the sizing sleeve, the internal mandrel, and the set of sizing plates. They are illustrated in Figure 3.7.

The *vacuum trough* is the newest and now the most popular method in the United States. The pipe is led into one end of a long, closed trough and out of the other end. The trough is filled with water, which surrounds the pipe in direct contact, providing excellent cooling. The pipe passes through fixed metal collars or rings as it moves through the trough, which give it its final size. The first collar, at the entrance to the trough, is the most important. For polyolefins, a little stream of water is directed at the pipe just *before* the entrance, as a lubricant. Inside the trough, a vacuum is drawn over the water, which reduces the pressure in the entire trough and causes the still-soft pipe to expand against the collars or rings, preventing collapse and ensuring good round sizing of the *outside* diameter. This expansive force is generated by the pressure differential between the outside (under vacuum) and the inside (open to the atmosphere).

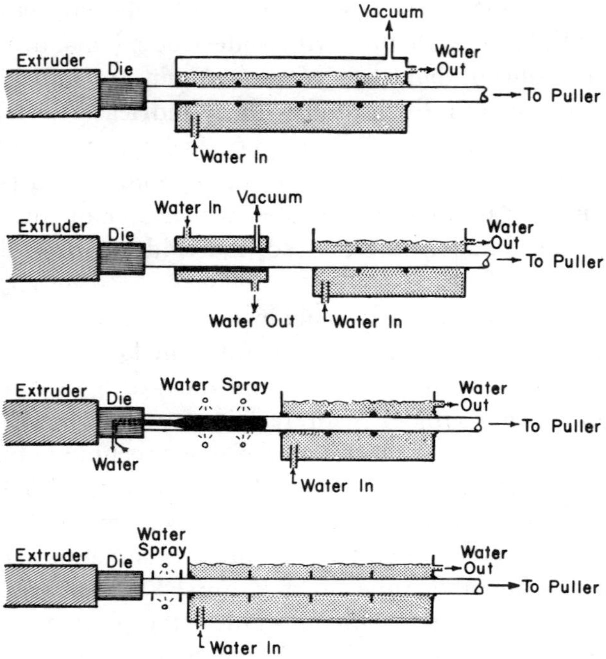

Figure 3.7. Four different ways of sizing and cooling pipe: *Top:* The vacuum-trough method, most popular today in the United States. *Next:* The sizing-sleeve method, used particularly in Europe, but also elsewhere. *Next to bottom:* The extended-mandrel method, used particularly for polyolefins and other plastics which need much cooling, as well as close control of internal diameter. *Bottom:* The sizing-plate method, a less-expensive but efficient cooling system. All these methods are described more fully in this chapter. (*Illustration by author*)

The number and locations of the rings are adjustable, and they become slightly smaller as the pipe passes through the trough to account for shrinkage, especially for crystalline plastics.

This method gives excellent cooling and is virtually unlimited in output rate, as there is very little frictional drag acting on the pipe. Of course, for very high outputs a very long trough must be used to provide enough cooling.

For large pipe, 100 mm (4 in.) or more in diameter, it may be difficult to keep the hollow pipe down below the surface of the water without distorting or marring it. In such cases, the trough is not filled at all, but water is sprayed and cascaded all around the pipe as it passes through, and the entire inside of the trough is held under vacuum.

In the *sizing-sleeve method,* the *outside* diameter of the pipe is fixed as it is hardened by contact with a water-cooled metal sleeve (usually brass). This contact is achieved by air pressure inside the pipe, or by drawing vacuum through the perforated inside surface of the sleeve. The sizing-sleeve method has been used almost exclusively in Europe, but is not common in the United States. The sleeve is a separate unit from 20 to 50 cm (8 to 20 in.) long. When vacuum is used, there is a gap of from 5 to 20 cm (2 to 8 in.) between the sleeve and die face. The interior surface of the sleeve is roughened to prevent seizing (sometimes a lubricant is applied during extrusion), and the passage is slightly tapered to allow for shrinkage, especially for polyolefins. The exit end is thus the smaller end, but it must still be slightly larger than the desired final outside diameter of the pipe, as the plastic will shrink further even after being sized (about 2 per cent for PVC and ABS; 5 per cent for polyolefins). If internal air pressure is used rather than vacuum to force the plastic into contact with the sleeve, the sleeve must be set very near the die face, and is often attached to the die through an insulating gasket. This prevents the pipe from expanding too much between the die and the sleeve. After emerging from the sleeve the pipe receives further cooling in a water trough, but it is already effectively sized when it leaves the sleeve. Several sizing plates (discs with holes) may be included in the trough to maintain roundness. If there is no trough, streams of water called cascades are directed at the pipe as it comes out of the sleeve.

The sizing-sleeve technique can produce a very glossy, smooth, closely controlled outer surface, as it does not contact water until after it is hardened. This avoids the water marks and ripples that are possible with other methods which cool the outer surface directly with water. However, the absence of direct water contact also means a lower cooling rate and limited output, as the heat must be transferred out through the walls of the metal sleeve. One system alleviates this problem by injecting a little water onto the pipe surface as it passes through the sleeve; the vaporization aids cooling and also acts as a lubricant. With all sleeves, however, additional cooling cannot be gained by indiscriminately increasing the sleeve length, because this increases contact area, which in turn increases frictional resistance to the pulling of the pipe through the sleeve. Other disadvantages of the sleeve include the strains put into the outside surface of the pipe by pulling it along a chilled inside of the sleeve, and the larger crystal size in polyolefins (which may weaken the pipe) produced by the slower cooling.

The *extended-mandrel method* (see Figures 3.3 and 3.6) is best suited to polyolefins as it gives additional internal cooling and internal support. The shrinkage of the plastic as it cools helps to keep a tight contact with the metal mandrel which extends from 15 to 30 cm (6 to 12 in.) beyond the die. Mandrels as long as 60 cm (24 in.) have been tried but mechanical difficulties have prevented their widespread use. Mandrels are made of brass, aluminum or stainless steel, and in all cases the surface is roughened to prevent the plastic from sticking and seizing as it passes over the surface. The mandrel is tapered with the larger end nearer to the die; this end is about 6 per cent larger in diameter than the smaller end, regardless of mandrel length, for most polyolefins. This taper permits shrinking as the plastic cools while moving over the mandrel. The end of the mandrel is slightly larger than the final desired pipe size, to allow for further shrinkage. As with the sizing sleeve, 5 per cent is recommended for polyolefins, and 2 per cent for polystyrene, ABS, and PVC.

Inside the mandrel, the passage for cooling water must be uniform and streamlined to ensure constant temperature all around. Cooling is further aided by water cascades over the pipe, both over and after the mandrel. A small split ring is set around (but not touching) the pipe just after it leaves the die; its purpose is to prevent the oscillation of the near edge of the water cascade, which might otherwise cause water marking. After leaving the mandrel, the pipe is further cooled to room temperature—partly by more cascades, and ultimately by passing through a trough with sizing plates to maintain roundness, much as with the sleeve method.

The cooled mandrel and the cooling medium (water) must be thermally insulated from the rest of the die. In the past this insulation problem was the reason for using an offset die, as in a straight-through die the water must enter through one or all of the spider arms. This could cool these arms too much, inhibiting the complete rewelding of the split streams, and producing lines of weakness in the pipe. Now, however, straight-through dies are made in which the path of the water through the spider arms is insulated with air or fluorocarbon plastic, thus leaving the arms relatively hot, and discouraging poor welds. The offset die is therefore made unnecessary, which eliminates its problems of leakage, nonuniform pipe walls, heat loss, and difficulty in cleaning.

The mandrel method yields a smooth and closely controlled inside surface and requires no air pressure. Also, it has the significant advantage of high output rates, as it cools from both outside and inside surfaces. It has the disadvantage of less control over the outside surface, and where outside gloss is important the sizing sleeve is preferred. External heat-polishing is possible with mandrel-sized pipe, to improve gloss, but it is difficult—especially at high speeds—and it is not ordinarily done.

The *sizing-plate method* (Figure 3.8) was quite popular in the United States before the development of the vacuum trough. The pipe takes its size as it is drawn through a series of brass or stainless steel plates 3 to 6 mm ($1/8$ to $1/4$ in.) in thickness, much as metal wire is drawn from metal rod. A positive air pressure is kept in the pipe to force it against the plates.

Figure 3.8. At left, an assembled series of sizing plates for 1-inch (2½-cm) pipe. At right, several more plates of various sizes. (*Courtesy of Union Carbide*)

The plates have entrances beveled to 45° and their hole sizes become progressively smaller as the pipe cools and contracts while passing through them. For medium- and high-density polyethylene, the first plate is about 10 to 20 per cent larger than the final pipe diameter, the second plate is 4 per cent larger, the third is 2 per cent larger, and all successive plates perhaps 1 per cent oversize. The first three plates are commonly 7 to 10 cm (3 to 4 in.) apart. Additional plates will be needed, their number depending on the linear speed and the plastic being extruded. High speeds and hard-to-cool polymers like polyolefins mean more plates. The exact number and placement of all the plates is still found by trial and/or experience, but the first plate is usually 7 to 10 cm (3 to 4 in.) from the die face. A slight trickle of water or a stream of air may be needed at the entrance to the first plate to lubricate the pipe surface and reduce pock marks (water marks).

The first plate can be the entrance to a water trough, or else one or more plates can be set ahead of the trough—3 or 4 are common—in which case cold water is cascaded around the pipe between these plates and the trough entrance. The latter system gives faster cooling of the pipe, but care must be taken that the entire surface of the pipe is covered with water; otherwise water markings and ripples will occur. Within the water trough, the plates are set perhaps 30 cm (12 in.) apart. These maintain roundness during final cooling, as with the other sizing methods.

The sizing-plate method yields controlled outside diameter, like the vacuum trough and sleeve methods. Sizing plates are rather inexpensive, and any kind of die can be used with them. This method offers high output because water contacts the pipe directly, providing fast cooling independent of mandrel or sleeve friction. However, it does require internal air pressure to help the pipe make firm all-around contact with the first plates that it enters, and it requires care in the manufacture and maintenance of the plates. Also, not all plastics will draw down uniformly as required, and because of the direct water contact, marks on the outside surface are always a potential problem. For large pipe (above 5 cm = 2 in.) the sizing-plate method presents collapse problems, and other methods may be preferred.

The success of the vacuum-trough method can be understood because it has similar advantages to sizing plates, yet it eliminates the problems of internal air pressure. This makes it easier to start up and operate. The vacuum trough can, in fact, be considered as an improved form of the sizing plate method.

Sizing plates can be modified in other ways, too. For example, one system uses a series of hollow plates through which vacuum is drawn. In effect, they are little vacuum sleeves, but since water can be cascaded between them, the rate of cooling can be high. They also eliminate the need for internal air pressure to ensure full contact with the plates, although friction is increased.

It is possible to combine the sizing plates or the vacuum trough with an extended mandrel in one system, with the pipe passing off the mandrel into the set of plates. Thus, some positive control is gained over both outside and inside surfaces.

For a given size of pipe, the proper die opening will vary depending on the sizing method used. With sleeves, the die diameter is usually a bit undersize, and the annular opening is just a little thicker than the final pipe thickness. With a mandrel, the core of the die must be larger than the mandrel—7 per cent is reasonable for medium-density polyethylene. This avoids catching and fouling where the pipe first contacts the mandrel. If the core is too large, though, too much drawdown is needed, and not all the mandrel length will be contacted and used for cooling. The annular opening in the die for mandrel sizing of polyethylene is commonly about 20 per cent larger than the final pipe thickness. With sizing plates, more drawdown is used than with the other methods, particularly with polyolefins. A die diameter about 50 per cent greater than the final pipe size is common, and sometimes even larger dies—up to twice the final diameter and 2½ times the final thickness—are used. Such highly oversize dies are not recommended, however, as the high drawdown required introduces stresses that may weaken the pipe.

In all pipe die design, the swell at the die and the ultimate shrinkage must be accounted for. Polyolefins show more of both swell and shrinkage than either PVC, polystyrene or ABS, but these values can be varied by operating conditions and die geometry, too. Swelling, in particular, can be reduced by higher melt temperatures, lower linear speeds, more streamlined die interior, and longer die lands. The figures in the preceding paragraph should be used as a rough guide only, with experience, test data, and suppliers' recommendations as more specific sources of information.

Internal Air Pressure. It is necessary with some sizing methods to keep a positive air pressure inside the pipe. The air can be led through a spider arm—or, in offset dies, through the back of the core—and controls are often needed to maintain an absolutely constant pressure. Of course, something must keep the air from going out of the open end of the pipe. For small pipe coiled on big reels, the end is sealed by crimping or capping at start-up, and the same air stays in the pipe all the time. As the pipe coil grows, only enough air is added to fill the inside of the new pipe and to replace any leakage at the sealed end. For big pipe cut in lengths, an internal plug is attached to the end of the die by a wire from 1 to 3 meters (3 to 10 ft) long, as shown in Figure 3.9. This plug has a soft sealing surface

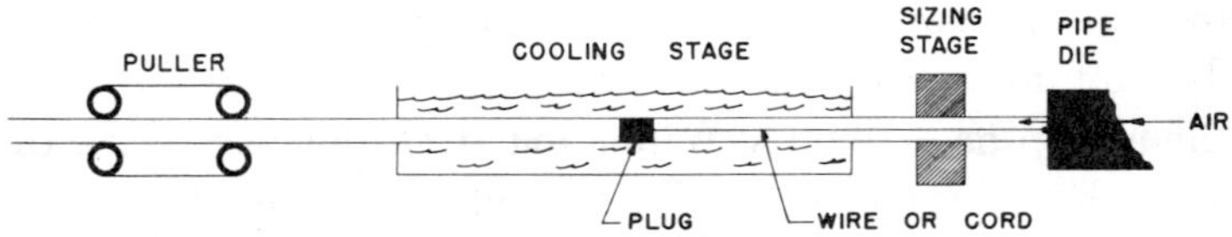

Figure 3.9. Air pressure can be maintained inside a pipe during sizing by this method. A plug is attached to the die with a flexible wire or cord. The pipe passes around the plug, and any air that leaks past the plug is replaced through the die. (*Illustration by author*)

such as leather, which allows the pipe to slide over it while still keeping most of the air between it and the die. Air is added through the core to replace any air that leaks past the plug. This device is tricky to start up, and is not often used now. One novel modification does away with the wire and holds the plug in place with a strong magnet placed around the outside of the pipe. This plug, of course, must have an iron center. All plugs are more or less streamlined in shape; a spherical plug is easiest to align and start up, but offers only line contact with the pipe interior, and is thus more likely to have air leaking past it. A cylinder has the most contact area, but is harder to align. A common compromise is a short torpedo, barrel or oval shape.

Water Troughs. Water troughs are almost always used with all sizing methods. These troughs are much longer for polyolefins than for PVC or ABS because heat of crystallization of the polyolefins must be removed. The higher the density of a polyethylene, the greater the cooling requirements, for this reason. Trough lengths range from 6 to 12 meters (20 to 40 ft) for polyolefins, and perhaps half of these values for PVC and ABS; they all vary in cross section from 10 to 40 cm (4 to 16 in.) square, depending on size and rate of pipe. The pipe enters and leaves through holes in the ends of these troughs and some device is needed to keep the water from flowing out around it. This device might be an iris diaphragm, a rubber sheet with a large hole or an X cut in it, or just rags stuffed around an oversize hole in the metal end plate.

Pulling Machines. The pulling force that draws the pipe through the sizing and cooling system is provided by machines especially designed for this purpose, as illustrated in Figures 3.10 and 3.11. The pulling machines are important because an extremely uniform pull must be maintained to give uniform pipe size. Also, the force needed to pull the pipe through the sizing apparatus is great, and the puller must grip tightly and evenly without slipping.

Pullers are either of the caterpillar-tread or the gripping-wheel type. A caterpillar puller, as in Figure 3.10, consists of two, three, or more caterpillar treads parallel to the pipe and distributed equally around it. They press with considerable force into the pipe, while being driven. A two-tread caterpillar sometimes has concave treads to assist in gripping. One manufacturer uses pairs of small rubber tubes set parallel to the pipe instead of continuous treads, thus providing four points of contact around the pipe instead of two. Another machine has six caterpillar treads with three of them retractable; the other three are used for smaller pipe and all six are used for big pipe, where there is space and need for them. Still another novel system uses little suction cups mounted on two caterpillar treads; vacuum is drawn through the cups, which gives them a firm hold on the pipe. One more design uses two ordinary automobile tires as the "caterpillar treads." When partially inflated, they provide a firm and strong grip.

Gripping wheels are used in sets of two, three, or four pairs of wheels, one above and one below the pipe, each with concave or beveled contact surface for better gripping (see Figure 3.11). They provide a more positive pull than most caterpillar types, but apply this over a much smaller area. The caterpillars are usually a little better and more expensive, and are also more common now.

A pipe puller has stepless speed variation over about a 5 to 1 range. Typical commercial pipe speeds are 2 to 20 meters (6 to 65 ft) per minute over a diameter range of $1\frac{1}{2}$ to 10 cm ($\frac{1}{2}$ to 4 in.). Some large and heavy-walled pipe, 15 cm (6 in.) and over, may run as slow as 0.6 meter (2 ft) per minute.

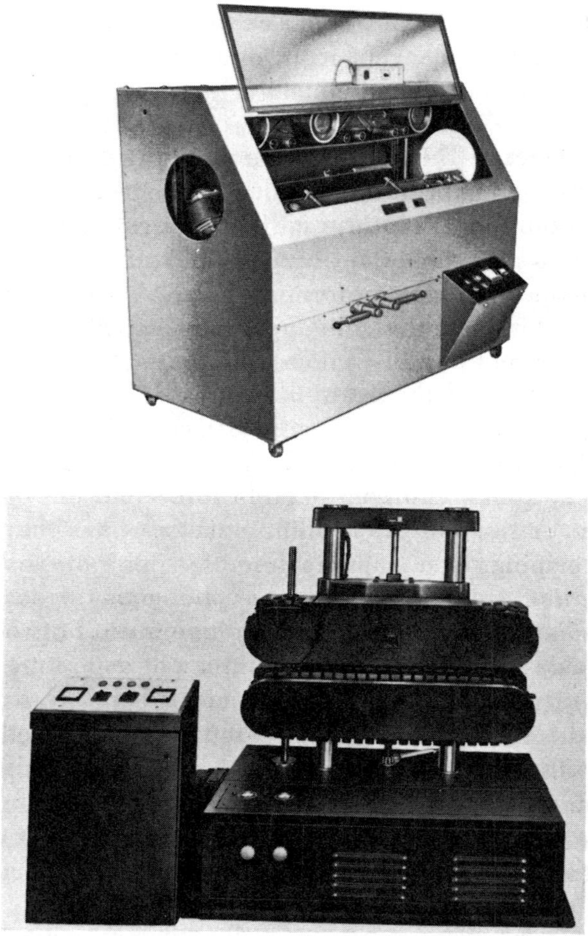

Figure 3.10. Caterpillar-type pipe pullers. *Above:* The three-track types give a strong and non-slipping pull. (*Courtesy of Gatto Machinery Development Corp.*) *Below:* Two-track caterpillars are also used (see also Figure 3.12), and are even more popular as they are less expensive. (*Courtesy of U.S. Anger*)

Figure 3.11. Gripping-wheel type of pipe puller. Pipe enters from left. (*Photograph by author*)

It is desirable to control the puller speed from a point near the die and sizing device, so that the operator can make adjustments in puller speed while looking at the pipe as it is being sized at the die. Electric, electronic, and magnetic speed variators lend themselves to such control, and are therefore popular.

Controls. Automatic measurement and control of pipe size is possible, and is used on a few of the most modern and well-equipped lines. The best-known system measures the capacitance through the pipe and relates this to the thickness, based on calibration in advance with the specific plastic. The sensing head can move around the pipe, showing when the pipe is out of round as well as when it is generally too heavy or too light. Photoelectric and mechanical sensing are also possible. All these systems can be made to feed back to control the puller or the extruder drive.

Cut-Offs. Some operations making large diameters at slow speeds still use a man with a saw to cut the pipe into specified lengths. General practice, however, is to use an automatic cut-off which travels with the pipe as it cuts, and then goes back to its original position and waits for the next signal to cut (see Figure 3.12). The signal usually

Figure 3.12. Cutting apparatus for rigid PVC pipe. This machinery not only cuts pipe into lengths (big circular blade at left) but also has an extra attachment which slits the pipe longitudinally if desired to make rain gutters for construction use. (*Courtesy of Technical Machinery Corp. and Amut SpA*)

comes from a microswitch that is struck by the pipe when it reaches a certain desired length. These saws are circular, with blades specially designed for cutting plastics.

Coilers. Only the very smallest diameters of rigid PVC and ABS pipe can be coiled, and these only on big (2-meter = 6-ft diameter) reels. Larger sizes, comprising most of the PVC, ABS, and polystyrene markets, are sawed to length. However, polyethylene up to 5 cm (2 in.) is commonly coiled, and this includes most polyethylene pipe. Very small pipe (1.2 cm = ½ in.) is sometimes coiled in 1-meter (3.3 ft) coils; 2-meter (6½ ft) coils are necessary with 2½- and 5-cm (1 and 2 in.) pipe. Often the pipe could be forced around a smaller diameter, but then a permanent set would result, causing later difficulty in unrolling and application. Coiling is done on a large center-wound reel, as shown in Figure 3.13. Its drive need not be too elaborate, as it is the puller that provides the uniform pull on the pipe. However, the coiler must wind the pipe neatly, evenly, and tightly. A level-wind device is often employed. Occasionally, the

Figure 3.13. Coiling machine for flexible pipe. The four arms are adjustable to give variable coil diameter. (*Courtesy of John Royle & Sons*)

coiler is merely a wheel driven with belts from the puller, and in the simplest installations the pipe is just coiled by a hand crank.

OPERATION OF THE PIPE EXTRUSION LINE

Predrying. Many pipe compounds, especially polyethylene, are colored black to improve long-term weather resistance. The pigment (carbon black) absorbs water easily, and if this water is not removed before or during extrusion, rough pipe surfaces will result. The drying is normally done in a hopper-dryer right at the extruder. A vented extruder will accomplish the same result.

Mixing. Good mixing is always needed in the barrel to ensure delivery of a uniform melt to the die, and it is sometimes of extra importance—as when PVC powder blend is the feed, or when two or more plastics are being blended in the extruder to obtain a combination of properties, or when color is being added in the extruder. Carbon black, however, is never added directly to the feed; instead, precompounded black concentrate pellets are used.

Most extruders of PVC pipe prepare their own hot-processed compounds from resins, stabilizers, colorants, etc., mixed according to their own private recipes. Many types of machines are used for such compounding, including roll mills, Banbury mixers, and specially designed extruders. On the European continent everyone "mixes his own"; but in Great Britain and the United States, some mixtures are sold ready to extrude, in either powder or compounded pellet form, made according to standard tested formulas by the major PVC suppliers. This costs a bit more, but eliminates the precompounding step as well as the need to experiment and develop a private formula. It is thus appealing to the small producers, but such producers make up a fairly small and shrinking portion of the industry, and the trend is away from this compound buying and toward more and more mixing right in the pipe plant.

Some PVC pipe is made by direct extrusion of a mixture of the components, without any intermediate hot-processing step. This was formerly thought to produce weaker pipe, but with good equipment, entirely satisfactory pipe can now be made directly from powder. Twin-screw extruders are used, particularly in Europe, and vented single-screw extruders are also possible. Output rates can actually be higher when powder feed is used, because there is less heat history and the compound can be run hotter (lower viscosity, hence faster) with no greater danger of decomposition. In addition, there is a saving due to the elimination of one step. Powder extrusion does require more expensive equipment and introduces possible feed and mixing problems; nevertheless, these are often overcome, and it is estimated that almost 50 per cent of the PVC pipe in the United States is now extruded directly from powder, and this figure is still growing.

Twin-screw extruders can give better mixing than single-screw extruders at lower temperatures, and they are widely used for PVC pipe (especially large pipe) in both Europe and the U.S.A. The makers of single-screw machines point out the limits to bearing size, higher equipment cost per unit output, and increased maintenance as serious objections. Apparently, salable pipe is being made in large quantities on both types of extruder. See pages 10 and 279 for further comparison.

Temperatures. Pipe is extruded at relatively low temperatures; among the various extruded products, only tubular film is run substantially cooler. There are several reasons for using low temperatures for pipe. First, for rigid PVC, low temperatures mean less chance of degradation. This can be translated into less stabilizer (cheaper compound), faster rates, or just kept as a safety factor. For all materials, but especially for polyethylene, fast and uniform cooling is important and is the usual limit to production rate. A low extrudate temperature gives the cooling system less of a job to do, thus increasing potential output. Low temperature also means a high viscosity at the die, which helps accurate sizing; in fact, if the plastic gets very hot, it becomes very difficult to size at all.

The preceding paragraph would seem to present a good case for very low extrusion temperatures. This was indeed the rule several years ago, and it is still followed in many plants. But superimposed on this is the growing desire for faster extrusion speeds, which in turn require more efficient mixing. These in turn raise extrusion temperature, especially in single-screw machines. Therefore, pipe producers have tried to get around the need for low temperature, and some have succeeded: they have improved their PVC formulations to give greater thermal stability, they have improved cooling devices and lengthened water troughs to obtain more and faster cooling, and they have improved sizing techniques to allow lower viscosities at faster linear speeds. In such installations typical extrusion temperatures might be 190 to 210°C for rigid PVC, 205°C for medium-density polyethylene and polystyrene, 215°C for ABS, and 225°C for high-density polyethylene. Temperatures from 10 to 25°C lower are still in common use with older equipment, or in very critical applications, or with twin-screw machines where good mixing is relatively independent of high melt temperature.

Start-up. The various sizing methods require different start-up techniques. The extended-mandrel system is easiest to start, as the pipe is positively supported until relatively cool. The sizing-sleeve method with vacuum and the vacuum trough are almost as easy; however, if pressure is used with a sleeve instead of vacuum, the pipe must be plugged at the end, or a floating internal plug is needed. This plug must be attached to the die before start-up and carefully held in position during threading of the pipe through the take-off; considerable skill is required to start up quickly. With all systems, a threading aid—such as a long rope, rod, or tube—is very useful if the puller is a great distance from the die, or with large pipe where the linear speed is very low. Such an aid is tied to the first extrudate and provides a steady pull on the plastic from the very start.

With sizing plates, the hot pipe end is first sealed and then threaded through the plates without sizing. Air is then let into the pipe, which expands to fill the first sizing plate, and then the others. Quick start-up of a sizing-plate system requires a skilled and experienced operator, as the plates receive a hot unformed pipe and must size and cool it in a very short distance. With such experience and skill, this system can be started almost as fast as the others.

Stacking and Storage. This is a real problem in a pipe plant, because all pipe is bulky, coils are inconvenient to move about, and most lengths are too long to be easily handled. Coils are sometimes wrapped with paper or plastic film (special machines are sold for this purpose) and stored horizontally—often on pallets. Cut lengths are usually left unwrapped, but may be prewrapped or boxed if they have small diameters. Long pipes look neat in horizontal racks, but are difficult to get at unless the entire length of the rack is accessible, and there is room to remove the pipe, wrap it, and get it to the shipping location. Therefore, they are often stored outdoors. In hot and sunny climates they should be covered; this protects them from degradation by the sun's ultraviolet light, which might become significant with all except black or specially stabilized materials after several weeks or months. A more immediate danger, however, is the softening and consequent distortion (and in extreme cases, sticking together) caused by the heat of the sun. This is most serious with black pipe, which—although it will withstand ultraviolet rays for years—is most in need of shade as it can heat up from absorbed radiant energy to a temperature well above that of the surrounding air. Pipes piled high on each other may aggravate this situation because of the pressure exerted on the pipes at the bottom.

PROBLEMS AND SOLUTIONS

Common problems in pipe extrusion, along with their causes and proposed solutions, are listed below:

Problem	*Solution*
Indented pock marks with direct water cooling	Caused by air bubbles on the pipe surface under water. Correct with large-volume, low-velocity water cascade.

Problem	*Solution*
Raised pock marks	Caused by water droplets on surface in air cooling zone. Eliminate any splashing by putting up a shield to protect hot pipe. Use water cascades instead of sprays.
Surface defects appearing at die	(a) Moisture can cause surface marks, especially with black or filled compounds. Drying prior to extrusion will help. (b) Unmixed resin or additive particles are possible, and require more mixing in the extruder. (c) Foreign contamination—can be reduced by new or finer screen pack. Check system for handling regrind, inspect virgin resin for sources of contamination. (d) Particles may also come from decomposition in stagnant areas within extruder, especially with PVC.
Rough surfaces inside or outside	(a) Can be caused by moisture, or low melt temperature, or dirty metal surfaces. Sticking and seizing on mandrel or sleeve surfaces can cause roughness. (b) Very high linear extrusion speeds can cause melt fracture at the die, which is cured by higher temperature, longer die lands, and smaller internal entrance angles into the die-land section. (c) When roughness is observed, try to locate the exact point of its origin. A strip of roughness on the underside is caused by uneven water cascading; direct another cascade from bottom upward, or cascade all around the pipe. Strip roughness on the outside can also be caused by a misaligned sleeve or some foreign matter caught in the sleeve or sizing plate. (d) A rippled strip of roughness can be caused by a difference in wall thickness at that point. (e) Roughness all over, especially in little V-shaped marks pointing away from the die, arises from insufficient mixing, and occasionally also from moisture

Problem	*Solution*
	in the melt. Better mixing is achieved by higher melt pressure, cooler screw, slower speed, or better screw design.
Discolored material	Decomposition of compound, usually PVC. Reduce temperatures, or use a different formulation (more stable or less viscous), or improve streamlining of extruder and die interiors.
Splitty pipe	(a) Look for visible weld marks, and check if failure is along weld lines. Extrusion may be too cold or too fast. Raise melt temperature or improve die design by increasing land length to develop more pressure. (b) Excessive drawdown can produce splitty pipe even with proper rewelding of split streams. Check pipe for unbalanced orientation in extrusion direction.
Circumferential ridges	Caused by periodic drag on mandrel, vibration of take-off equipment, or uneven take-off speed. This is remedied by finding proper mandrel temperature or fixing the take-off. The puller may be operating at the low end of its range, which can cause irregular pull.
Circumferential waviness	Caused by material surging, or nonuniform water cascade, or leaking of water thru the sizing plates. Use smaller first sizing plate, or a more uniform flow of water around the pipe. Check plate for wear and nicks. For surging remedies, see "better mixing," top line of this page.
Unbalanced wall thickness	Adjust at die. Make sure the elements of the take-off are all properly aligned and that there are no hot or cold spots in the die.
Pipe out of round	Sizing devices inadequate or out of shape, especially in water trough; or else, pipe is too warm when it reaches the puller.
Lack of gloss	Same problems and remedies as rough surface, discussed above. Raising melt temperature is most likely remedy. Some die

Problem	*Solution*
	cores are electrically heated to give better internal gloss. The exterior can be flame-polished if linear speed is slow. Sizing sleeves can give best outside surface. Formulation can be improved to give better gloss; e.g., lubricants in PVC.
Poor strength	(a) See "Splitty pipe" above. Also, indented printing can weaken pipe, as can too much drawdown at too low temperatures. Look carefully for nicks and grooves that could concentrate stress. (b) Plastic may be degraded by thermal decomposition (PVC), ultraviolet light (long storage in sunlight), or use of too much scrap in the feed. Mold a plaque from the weak pipe and compare its basic physical properties with those expected from new compound. (c) Poor mixing can cause weak pipe even if appearance is acceptable, particularly with PVC. Remedies are given above under "rough surfaces." Strength is very much dependent upon material formulation.

STANDARDS AND CODES

Buyers of pipe are among the most cautious purchasers of plastics, as their choices often have to last many years, sometimes buried and/or under pressure. It takes many years to collect test data and experience to prove a plastic worthy of inclusion in a standardized code, and only lately have there been significant breakthroughs (notably in drain, waste and vent uses, often called DWV pipe). Still, most pipe used in the United States today is for non-standardized applications, and the potential when more code approvals are gained is truly vast.

Most major countries have issued standards for plastic pipe, and these are discussed further in references listed in the Bibliography at the end of this chapter, particularly the article by Reinhart entitled "Plastic Pipe Standards." Some of these standards are given in Table 3.1.

TABLE 3.1. Pipe Standards

UNITED STATES

CS–256–63, PVC pipe.
CS–254–63, ABC pipe.
CS–207–60, rigid PVC pipe.
CS–197–60, flexible polyethylene pipe.
CS–255–63, Polyethylene pipe.
CS–228–61, Modified polystyrene sewer and drain pipe and fittings.
CS–206–57, Solvent-welded cellulose acetate/butyrate pipe.
CS–237–61, Dimensions and tolerances for light rigid PVC pipe.
CS–220–59, Dimensions and tolerances for light rigid ABS pipe.
CS–219–59, Dimensions and tolerances for solvent-welded rigid ABS pipe.
CS–272–65, ABS drain/waste/vent pipe.
CS–270–65, PVC drain/waste/vent pipe.
CS–218–59, Rigid ABS pipe, iron-pipe sizes.
ASTM D2464–65T, threaded PVC schedule 80, fittings.
ASTM D2465–65T, threaded ABS Schedule 80, fittings.
ASTM D2466–65T, socket PVC Schedule 40, fittings.
ASTM D2467–65T, socket PVC Schedule 80, fittings.
ASTM D2468–65T, socket ABS Schedule 40, fittings.
ASTM D2469–65T, socket ABS Schedule 80, fittings.

GREAT BRITAIN

BS–1972, Polyethylene tube for cold water transmission.
BS–1973, Polyethylene tube, low density, general purpose.
BS–3284, Polyethylene pipe for cold water transmission.
BS–3505, Rigid PVC pipe for cold water transmission.
BS–3506, Rigid PVC pipe for industrial uses.

WEST GERMANY

DIN 8062, Rigid PVC dimensions.
DIN 8072, Flexible polyethylene dimensions.
DIN 8073, Flexible polyethylene properties.
DIN 8074, High-density polyethylene dimensions.
DIN 8075, High-density polyethylene properties.
DIN 8061, Rigid PVC properties.
DIN 16928, Rigid PVC pipe, application (fittings and sealing, etc.).
DIN 16929, Rigid PVC pipe, chemical resistance.

In the United States, members of the plastics industry association (Society of the Plastics Industry, the "SPI") have been working for almost 15 years to promote proper application of plastic pipe of standardized quality. In 1951, the National Sanitation Foundation (NSF) started work on its certification program, under SPI sponsorship. The NSF, a private, government-supported institution, grants its seal of approval only to products meeting its specifications: the supplier must disclose the formulation in confidence, the pipe must be labeled with the safe working pressure, and it must not contain any reworked or scrap materials. To enable identification of approved materials in the field, harmless chemical tracers are sometimes added to the formulation and are discernible by analytical techniques. The NSF program has been directed at applications where the pipe is to convey drinking water to people or animals. Approvals and purchases of this nature are locally regulated in the United States, and there is still much controversy as to what additives and materials are safe. In Europe and Japan, rigid PVC is widely used for potable water; in the United States, polyethylene has been favored, but PVC is finding increasing use. The NSF program is currently being expanded to include DWV pipe as well as pipe for potable water.

The pipe division of the SPI is known as the Plastic Pipe Institute (PPI). It has supported the writing of the commercial standards listed in Table 3.1 as "CS," which are issued by the Department of Commerce, and are, in effect, voluntary cooperative agreements among the pipe producers. Not all producers are involved and there is no legal requirement to produce to these standards. However, more local and industrial requirements and codes are being based on these commercial standards, and it is becoming correspondingly more profitable to produce to their specifications.

Two important accomplishments of the PPI are the Model Plumbing Code and the SDR–PR rating system. The Model Code is a suggested section for inclusion in local plumbing codes which do not yet have satisfactory provision for the use of plastic pipe. Many big cities, including New York, St. Louis, Philadelphia, and Chicago, have changed their codes to include some plastic pipe, but to varying degrees and only after much debate. There are many localities still with outdated codes, and the PPI continues to work toward their responsible change.

The SDR–PR system provides both a standardized pressure-rating system for all plastic pipe (based on 10,000-hr pressure tests with water at room temperature), and a standardized system of dimensions more suited to plastics than the iron-pipe or arbitrary dimensions previously used and still prevalent. The SDR–PR system (Standard Dimension Ratio, Pressure Rated) includes a special nomenclature, as shown in Table 3.2, which lists major plastic pipe materials by the new system.

TABLE 3.2. Pipe Material Nomenclature

Key: First digit specifies type and second digit specifies grade, according to ASTM classifications. Last two digits give hydrostatic stress rating in hundreds of pounds per square inch.

Resin	Type	Grade	Stress Rating * (psi)	Nomenclature
Polyethylene	2	3	500	PE-2305
	2	3	630	PE-2306
	3	2	630	PE-3206
	3	3	630	PE-3306
Polyvinyl chloride (PVC)	1	1	2000	PVC-1120
	1	2	2000	PVC-1220
	2	1	1000	PVC-2110
	4	1	1600	PVC-4116
ABS polymers	1	1	630	ABS-1106
	1	2	1000	ABS-1210
	2	1	1250	ABS-2112
	1	3	1600	ABS-1316

* Stress ratings are at 23°C, based on tests with water, according to ASTM test D1598-63T.

Other organizations which have done standardization work in the United States on plastic pipe include the ASTM (American Society for Testing and Materials), the American Gas Association, the American Petroleum Institute, the American Water Works Association, and the American Standards Association—as well as several government and military agencies which have specialized uses. The PPI is co-operating with the U.S. Public Health Service in the planned revision of the American Standards Association National Plumbing Code (ASA A40.8-1955, an important guide for local authorities) to give greater emphasis to plastics.

Much pipe, including half of all polyethylene extruded in the United States, does *not* meet U.S. Commercial Standards nor does it qualify for NSF approval. It is sold largely for private, rural applications, and often contains reworked and scrap plastic. Some of this pipe invariably finds its way into pressure applications, and its occasional failures have hindered the efforts of the industry to extend the use of properly rated plastic pipe. The British have minimized this problem by subsidizing the purchase of polyethylene pipe by farmers—but only pipe that meets the British Standard. This has made it cheaper to buy the standardized pipe than the "offgrade" pipe, and has driven much of the offgrade from the market.

Although progress is being made toward uniform diameters and wall thicknesses, the lengths of plastic pipes are still not satisfactorily standardized. Pipe is often cut to customer specifications, which delays delivery and sometimes increases cost. Also, plastic pipe is sometimes ordered in the short lengths customary for metal pipe; this takes away one of plastic's advantages—its light weight allows it to be handled in much longer lengths than metal, which speeds up installation and requires fewer fittings.

TESTS

Burst Strength. This is a very common and important test, and is run for both short and long terms and at many temperatures (ASTM test numbers D1599–62T and D1598–63T). For short-term testing, called instantaneous burst, the pipe is first brought to the proper temperature and the pressure is then rapidly increased until the pipe fails. In long-term testing, the pipe is held at a known temperature and pressure, and the time to failure is measured (Figure 3.14). The curve of pressure in the pipe versus time to failure can then be plotted, and often is extrapolated to longer periods of time to find the maximum working pressure safe for actual service. Such curves are shown in Figure 3.15. The hoop stress in the pipe is the critical stress

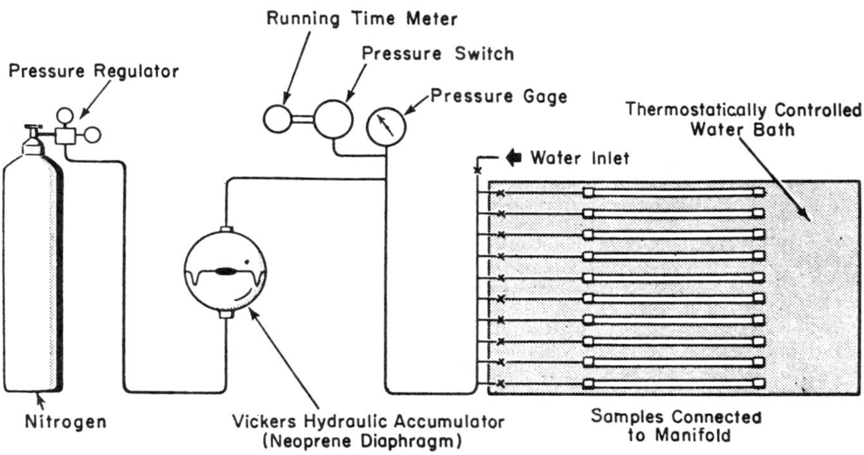

Figure 3.14. Typical test apparatus for long-term hydrostatic pipe testing. (*Courtesy of ASTM*)

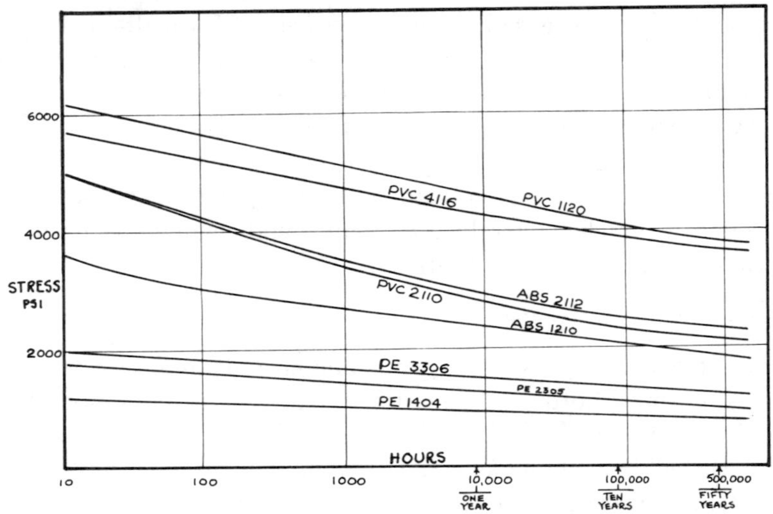

Figure 3.15. Typical long-term strength data for eight common pipe materials, tested with water at 23°C. The material numbers refer to the standardized nomenclature, as given in Table 3.2. The "stress" refers to the hoop stress in the pipe, which depends on both internal pressure and pipe dimensions. Each material can withstand highest pressures at the shortest time periods, but the behavior under long-term testing is a better indication of performance in actual use. Current recommended practice is to test at least up to 10,000 hours, extrapolate to 100,000 hours (over 10 years), and take one-half of this extrapolated value as a safe long-term design stress. (*Data courtesy of the Society of Plastics Engineers and F. W. Reinhart; illustration courtesy of Edison Technical Services, Inc.*)

value and depends on the pipe dimensions as well as on the pressure. It is now calculated from the formula:

$$\text{Hoop stress} = S = P\,(D-t)/2t$$

where

D = outside diameter of the pipe,
P = line pressure inside pipe,
t = minimum pipe wall thickness, and
S = hoop stress.

Long-term burst strength is the most important measure of actual pipe performance, and is not necessarily predictable from short-term (instantaneous) burst data. The short-term tests are still used sometimes for comparison because they are much easier to run; unfortunately, experience speaks against such practice. The short-term tests do have some value for those applications where the pipe is subject to "water hammer" or other temporary high pressures.

Two types of failure are possible in a burst test: brittle and ductile. The brittle failure is common with most rigid materials, especially at lower temperatures. The ductile failure, in which a blister first forms, is most common with tough, high-elongation materials such as the polyethylenes. With some polyethylenes, however, brittle failures are often encountered at high temperatures (e.g., 60°C) after aging. The type of failure is a useful tool in analyzing the cause of failure.

Impact Tests. Izod impact strength is determined for the basic materials, such as ABS and PVC, that are used in rigid pipe. The most important distinction has been between Type I and Type II PVC pipe, both of which are widely used. Type I is low in impact strength but more resistant to chemicals because no impact modifier has been added. Type II is much tougher but less chemically resistant. Recently, a dual-purpose PVC has been introduced which passes the ASTM specifications for both Type I and Type II.

Impact strength is also frequently tested by falling-ball tests (see Figure 4.11, page 122). For example, a Type II PVC pipe with a nominal size of 2½ cm (1 in.) can withstand at least three drops of a 5-kg steel ball from a height of 3 meters (or a 10-lb ball from 10 ft). Formulations even tougher than this can be made if necessary, and many impact modifiers have recently been introduced (e.g., chlorinated polyethylenes, acrylics, ABS types).

Weathering Tests. Much pipe is used outdoors. With polyethylene and ABS, such pipe is colored black, thus gaining excellent resistance to ultraviolet light. For PVC, gray is more common. Accelerated testing of weather resistance is usually limited to tests of the basic material in a "Weatherometer" or similar machine. Also, some large end-users conduct actual service tests on pipe, which may include weathering. Some PVC pipes have an ultraviolet-absorbing additive added for outdoor use, but these additives are expensive and not universally employed. In hot and sunny regions, this problem should receive special attention.

Creep. Pipe must often contain liquids under pressure for long periods, and this puts it under continuous loading. Hence, resistance to creep (slow deformation under load) is important. A special test has been devised to test plastic creep, using molded or extruded bars, and is shown in Figure 3.16. Creep depends on temperature, environment, and formulation, as well as on the stress.

Figure 3.16. Carey-Lander apparatus used to test creep of plastic bars. A specimen is seen clamped between the two black holders. The entire device is about 12 inches (30 cm) long. A weight is hung from the grooved end (extreme upper right), and the distance that the specimen has stretched is measured at regular time intervals. See U.S. Patent 3,196,670. (*Courtesy of Union Carbide*)

Chemical Tests. Pipe materials undergo many chemical tests in connection with their varied applications, and the literature of pipe makers and material suppliers is a rich source of such data. In particular, Type I PVC is noted for chemical resistance and much of its sales success is based on this advantage over ferrous pipe. The laboratory chemical tests are usually reported as "per cent weight change," or "per cent change in flexural strength, tensile strength, or elongation," or simply as "attacked" or "resistant." Temperature is a very significant variable—some substances will attack a plastic when hot but will be harmless at room temperature.

Environmental stress-cracking is a particular kind of chemical attack which must be avoided with polyethylene pipe. Certain chemicals cause premature failure of some polyethylenes when under stress. Polyethylenes vary widely in their resistance to this attack; the lowest melt indices are the best, and additives such as butyl rubber can be of great help. The plastic can be tested for stress-cracking tendency as pipe or in bar form; the latter is more common. The tests, described and illustrated in the chapter on Wire and Cable Covering, all require immersion under stress in a known cracking agent. The time to develop cracks is measured at definite temperatures and agent concentrations.

With ABS and PVC, pipe can be tested for internal stresses by immersion in a solvent (acetic acid for ABS, acetone for PVC). The tests are standardized as ASTM D1939–62T and D2152–63T, respectively.

SPECIAL TECHNIQUES

Lining and Coating of Metal. Plastic-lined metal pipe is commonly used for corrosive liquids and gases, combining the strength and rigidity of metal with the chemical resistance of plastics. The plastic is chosen with regard to both cost and chemical properties. Epoxies and vinyl plastisols are applied as liquids, and extruded solid thermoplastics are also used. They are extruded into tubes somewhat smaller than the metal pipe, inserted, and expanded under heat and pressure. Hot air can be used but hot liquids are most effective. An adhesive may be precoated on the plastic or in the pipe before expanding, if necessary. Metal pipe can also be lined with any thermoplastic by the fluidized-bed principle. Saran, plasticized vinyl, and polyolefins are among the common thermoplastic lining materials.

A related technique is the extrusion-coating of the *outside* of steel pipe. This is an adaptation of the method used to sheath large electrical cables, and involves passing the pipe through a cross-head or offset die after precoating it with a butyl rubber sealant/adhesive. High-density polyethylene is extruded around the pipe as a tube, and is it drawn and shrunk around the pipe as it cools. These coated pipes can be buried indefinitely in all types of soils with no danger of corrosion. They are competitive with pipes wrapped with extruded polyethylene tape, or coated with bituminous materials.

Occasionally, two-layer plastic pipe is made; i.e., one plastic is used to line the other. One available combination joins an exterior of black polyethylene with an interior of natural uncolored polyethylene. The black offers weather resistance, while the natural plastic may be lower in cost and attractive in appearance. Such pipe can be made by separate extrusion and expansion (like lining metal pipe), or by simultaneous extrusion from two extruders through one die.

On-site Extrusion. In special cases, it is more economical to make plastic pipe right in the area where it is to be used. An extruder is mounted on a truck with a suitable power supply, and the truck travels along a prepared trench, extruding a continuous length of plastic (usually polyethylene) into it, with no fittings or other joining needed.

Vertical Extrusion. Although most pipe is extruded horizontally, there are a few installations which extrude down into a water bath. This gives exceptionally good cooling, but the pipe must be flexible enough to be bent around so that it can come out of the bath. Care must be taken to avoid flattening of the pipe caused by bending it while it is still warm enough to be distorted. This technique resembles the method used to make some tubular films, notably saran. One of the major technical problems is the maintenance and circulation of the water *inside* the pipe.

Biaxial Orientation. In 1959, a method was developed in West Germany for extruding polyethylene pipe oriented in both axial and circumferential directions. This process is shown in Figure 3.17. The pipe is extruded much thicker and smaller than the final dimensions, and is sized and cooled with a sleeve, as usual (other sizing methods could also be used). Then the pipe passes through a long, closed reheating bath where it is brought up to orienting temperature. This temperature is controlled to within 1°C. As the pipe emerges from the bath it is expanded by internal air pressure and re-sized in another sleeve to take its final dimensions. Liquid pressure around the pipe prevents expansion while the pipe is still in the preheating section. The pipe puller is set to pull faster than the extrusion rate at the exit from the reheater; thus the pipe receives axial orientation, too. The reheating is necessary because the entire thickness must be exactly

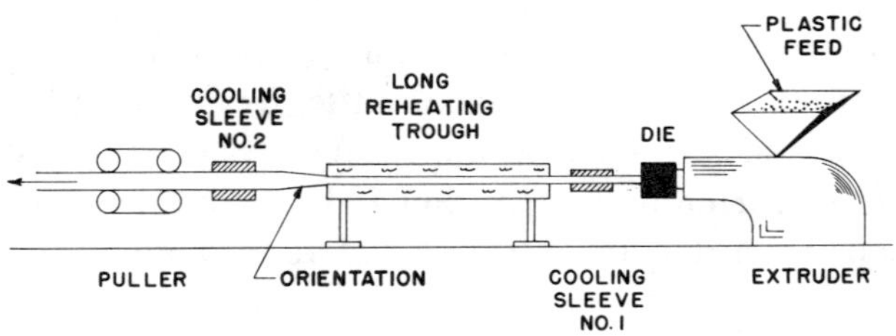

Figure 3.17. Plant for biaxially orienting pipe. Pipe is extruded and cooled, and then slowly *reheated* to exact orientation temperature. It is expanded by internal air pressure upon leaving the pressurized reheating chamber. (*Illustration by author*)

at the right temperature to get and control the proper amount of orientation. Cooling down from extrusion temperature cannot easily be done to such precision, so the pipe is initially overcooled and slowly reheated. This oriented pipe has higher strength than ordinary pipe of the same dimensions; conversely, a thinner pipe (less material cost, lighter weight) may be used to hold the same pressure. Use of such pipe has been hindered in the United States because specifications and standards are set for unoriented pipe, and the manufacturer of the oriented product is unable to alter his dimensions to compensate for his higher production costs. Potential buyers have also been worried about long-term deorientation and weakening of the pipe, especially at high temperatures and pressures. Shotgun shells have been the main application for this technique in the United States.

High-molecular-weight Polyethylene Pipe. In 1958, an American firm introduced a high-density polyethylene of such high melt viscosity (low melt index) that it could not be extruded in normal pipe extrusion equipment. It has high gloss and exceptional resistance to stress-cracking, while remaining coilable in pipe sizes up to 5 cm (2 in.). Fittings and welding were problems, but this pipe was able to sell on the basis of the above advantages. Material supply was at first limited to a few extruders, but now there are more suppliers and these materials are generally available. They are made in several viscosity grades and are not all alike. To extrude this pipe, the original supplier developed special equipment, including a gear-pump assembly which distributes the softened extrudate to many separate passages, making as many as 24 pipes at once. This assembly is needed because with only a single or double die the resistance would be too great to push the plastic through at a reasonable output rate. With the gear pumps, the resistance is reduced and linear speed of each pipe is slowed down to perhaps ½ meter (20 in.) per minute, allowing cooling in the extended die parts. There is one pump for each two pipes—little extruders can also be used as pumps. They act as pressure boosters to help force the polyethylene out of the dies. Once they are out, the pipes are already rigid and cool enough to handle. No further sizing or cooling is done, and the pipes are led directly to their coiling stations. Since the original development, new extrusion techniques and screw designs have been devised to handle high-molecular-weight polyethylene on more conventional equipment, nevertheless several of the original special systems are still in active and profitable operation.

Cross-linked Pipe. Very high strength and increased resistance to heat and chemicals can be obtained in polyethylene pipe by cross-linking the molecules, either (a) on the surface, where it counts most, or (b) throughout the entire pipe. Method (a) involves passing the pipe through a bath containing a cross-linking agent. Unfortunately, this links the outside surface, while it is the *inside* surface that is often more important, as it may be carrying the corrosive chemicals, and it bears the highest stress when the pipe is under pressure. With method (b) the cross-linking agent is already mixed into the plastic, and reacts with the molecules just after leaving the die. Plastic temperatures must be very carefully controlled throughout the process. Organic peroxides such as dicumyl peroxide are the usual cross-linking agents, and often a large percentage of carbon black (e.g., 50 per cent) is incorporated. One such line of pipe is now offered in the United States and has found some success in the chemical industry. Long-term testing is in progress, and standards are being written. The cost of this pipe has been set too high to make any sizable penetration into the biggest pipe markets, but when testing is further along, perhaps this price will come down, and then this product may compete with metal in hot-water applications heretofore denied to plastics.

Irradiation is another possible method of cross-linking, but it is still rare because of the cost of the irradiation equipment and its operation.

Printing. It is often necessary to print information on the pipe about the size, pressure rating, material, extruder's name or trademark, etc. This is normally done right in line with the extruder, after cooling and before coiling or cutting. Marking machines, usually using a small offset wheel, are specially made for this purpose. Care should be taken to avoid deeply engraved impressions in the pipe, because these can act as stress concentrators and promote early failure under pressure; raised printing is preferred, although it is more subject to abrasion. Pretreatment of the surface may be employed with polyethylene but is not common.

Joining. Much work has been done on techniques for joining plastic pipe, and some of the more important references are included in the Bibliography. Joinings can be generally classified into permanent (welds) and removable (most fittings). Some methods are shown in Figures 3.18, 3.19, 3.20, and 2.1, and there are many

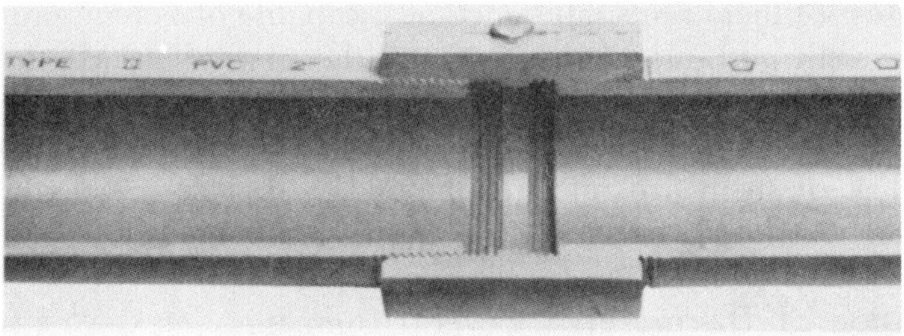

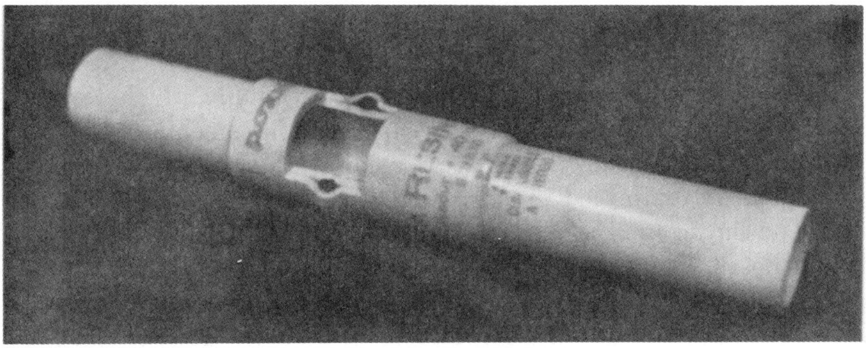

Figures 3.18, 3.19, 3.20. Some of the ways of joining plastic pipe. *Top:* A threaded molded fitting joins two lengths of threaded 2-inch (5-cm) PVC pipe. (*Courtesy of A. M. Byers Co.*) *Center:* A solvent adhesive and a smooth molded fitting are used to join two lengths of plain 2-inch (5-cm) PVC pipe. Note that this pipe can be thinner than the threaded pipe of the same pressure rating, because there are no threads to act as points of maximum weakness. (*Courtesy of A. M. Byers Co.*) *Bottom:* An elastomeric ring seal connects 2-inch (5-cm) pipe. The joint is slightly flexible, it can take vibration and expansion, and still remains water- and air-tight; see also Figure 2.1. (*Photograph by author*)

more. Most of them are simple methods, capable of being applied by one man with portable equipment, and the ease of installation is a major advantage of all plastic pipe. Fittings comprise a large market in themselves—approximately 10 to 15 per cent of the weight of the pipe. Most of them are injection-molded, although some are made from short extruded lengths, as in Figures 3.20 and 2.1.

Extrusion of Thermosets. Several companies make "extruded" pipe from thermosetting materials. One technique involves curing of a length of phenolic pipe in a standard molding press with an open-end mold, and advancing the pipe out of the mold so that it forms one end of the cavity. The next shot cures in contact with this end and attaches to it, and so forth.

Glass-reinforced epoxy and polyester pipe is also being made continuously, with and without filament-winding. One type is sold with an extruded PVC lining, which may act as a support for the thermosetting exterior during its cure. Equipment for making continuous glass-reinforced thermosetting pipe is commercially available in West Germany and perhaps elsewhere.

Other Reinforced Pipe. Machines are available which form a continuous helix from steel wire or other reinforcement. This helix can be fed directly into the pipe extrusion die, enabling continuous production of reinforced pipe. Woven or wound fibers can also be used, and have found a market for many years in the better quality flexible hoses. Two or more materials can be combined in a reinforced pipe in many ways; one British product has a polyethylene exterior, a PVC interior, and a fabric reinforcement between the two. Another product, originating in Greece, has a spiral of rigid PVC embedded in a thick tube of plasticized PVC, and is made with two extruders and a common die.

Corrugated Pipe and Tubing. This product is made by extruding a thin PVC tube into an endless chain of split mold segments, ridged to produce corresponding ridges (corrugations) on the entire surface of the tube. Much greater rigidity is thus obtained per unit weight, the result is flexible yet strong. The corrugated product finds its primary use as electrical conduit, and the machinery was first developed in Germany for this use.

REFERENCES

General References

"Plastic pipe 1965," small book by Dr. C. L. Mantell, published by Noyes Development Corp. and sold for $15. Reviewed in *Plastics Technology*, October 1965. Discusses fabrication methods, markets, materials.

"A new look at plastics pipe," three-part staff-written article on the plastics pipe industry, *Modern Plastics*, May-June-July 1965. Discusses standards, economics, new materials, listings of makers of pipe and fittings.

"Plastics pipe symposium," *ANTEC* 1956. Representatives of du Pont, Goodrich, Eastman and U.S. Rubber contributed to this 51-page survey of uses, properties, and fabrication technology. Much is outdated, but it is still a good introductory lesson, and the butyrate information is largely still valid.

Applications of Plastic Pipe

"Abwasser-Kanäle aus Hart-Polyäthylen" (Drainage piping from high-density polyethylene), W. Imhoff, E. Rottner and E. Gaube (Hoechst), *Kunststoffe*, January and February 1967. How to put in ground, proper dimensions, transmission capacity, applications, fittings and sealing.

"The development of drainage techniques using plastics in the U.S.A.," T. W. Edminster and J. L. Fouss, *Poliplasti,* December 1966.

"Einsatz von PVC-Rohren in Gasrohrleitungsbau der DDR" (Installation of PVC pipe in gas transmission lines in the DDR-East Germany), W. Altmann, *Plaste und Kautschuk,* July 1966. Report of experience with PVC in this area.

"Polypropylene pipes," G. Diedrich *et al.* (Hoechst), *Kunststoffe,* April 1966. Describes applications and strength tests.

"Plastic pipe for water and gas transmission," abstracts of papers presented at symposium in Paris, January 1966. *Plastiques Modernes et Elastomeres,* February 1966.

"Rigid PVC water pipes in East Germany," W. Schrader (Central Institute for Welding Technology, East Germany), *British Plastics,* April 1965. Survey of 20 years' experience with water piping, including drains and waste lines. Discusses joining methods, also.

Survey of plastics pipe use in gas transmission. *Gas,* March 1965. Reports who uses and why; 7000 miles in use as of early 1965.

"Plastic pipe in the gas industry," E. Simonescu *et al., Materiale Plastice* (Rumanian), January 1965. Reports authors' experience.

"Study of applications of rigid PVC pipe for petroleum products," M. Petrescu, *Materiale Plastice* (Rumanian), January 1965. Reports author's experience.

"Plastic pipe in gas distribution," I. S. Webster (Public Service Electric & Gas Co.), *SPE Regional Technical Conference Papers, Newark Section,* November 5–6, 1964. Thorough and competent review of use of PVC and ABS plastics in this area.

"Hart-Polyäthylen-Rohre für Abflussleitungen und Beregnungsanlagen" (High-density polyethylene pipe for drainage and industrial disposal systems), K. Richard *et al.* (Hoechst), *Kunststoffe,* March 1963. Tests and results, coupling systems, emphasis on oriented pipe.

"Rodents and plastics," J. Labryn, *Plastica,* January 1963. Describes damage to wiring and pipe, discusses repellents and their testing.

"Plastics pipe for drinking water," short notes on allowable lead content in PVC for potable water, and requirements for polyethylene pipe for similar use. *Kunststoffe,* November 1962.

"Einsatz und Verarbeitung von Hart-PVC-Rohren" (Installation and manufacture of rigid PVC pipe), A. Schneider, *Kunststoff-Rundschau,* February 1962. How to use rigid PVC, with applications and performance data.

"Materie plastiche nella irrigazione," E. Ferraris, *Materie Plastiche,* February, March and May, 1961. Properties, specifications, and uses of plastics in irrigation. Dozens of excellent photographs, many in color.

"The use of low-density polyethylene tube in United Kingdom water supply," D. A. Gill (Water Research Association), *British Plastics,* May 1959. Discussion of problems: results of questionnaire sent to users asking about details of installation, causes of any failures, etc.

General Information on Individual Plastics as Pipe Materials

"Verarbeitung von PVC hart auf einem Doppel schnecken-Extruder" (Processing of rigid PVC in a twin-screw extruder), G. Menges and P. Klenk, *Plastverarbeiter,* December 1966. Temperature and pressure studied in pipe extrusion.

"Formulating and processing rigid PVC powder blend pipe compounds," R. Jalbert (Rohm & Haas), *SPE ANTEC* 1966. Economics of powder blending, using acrylic modifiers. Properties given. Brief comment on processing requirements.

"How to extrude rigid PVC powder," R. M. Krainz (Diamond Alkali), *Plastics Design and Processing,* September 1965. Extrusion in a vented extruder into pipe, discusses equipment and operation including screw design, temperature control, power needs. More information on this topic is in *SPE Journal,* February 1965, article by J. A. Rolls *et al.,* also of Diamond Alkali.

"Extrusion of rigid PVC dryblend on single screw extruders," F. Reifenhäuser (Reifenhäuser), *SPE Regional Conference Papers,* Newark Section, November 6, 1964. Explains problems but gives neither design nor formulation data.

"Emulsion versus suspension PVC," A. Rodeyns (Solvay), *British Plastics,* August 1962. Comparison of the two, as pipe materials, concludes no major difference between them.

"Development of ziegler polyethylene pipes," K. Richard *et al.* (Hoechst), *Kunststoffe,* June 1962. Discusses test methods, orientation, creep fittings.

"Equipment and processing recommendations for extrusion of polypropylene pipe," A. S. Ahmad *et al.* (Avisun), *SPE Journal,* April 1963. Screw and die design in some detail, power consumption, brief mention of sizing and takeoff equipment. Similar article in *Plastics World,* January 1962, same authors.

"Properties and applications of ultra-high-molecular-weight linear polyethylene," J. H. Lowery (Allied), *Plastics Design and Processing,* April 1963. Physical properties of Allied's special pipe compound, as well as processing information. Earlier articles on same material and process are in *Plastics Design and Processing,* October 1961 (by S. A. Mruk of Allied) and in *SPE ANTEC* papers, 1962 (by S. A. Mruk, discusses testing and properties).

"Polybutene-1 as a material for pipes," Dr. J. Plenikowski and Dr. O. Umminger (Hüls), *Kunststoffe,* November 1965. Structural characteristics and creep rupture strength.

"Polyvinyl dichloride pipe," R. D. Savage (B. F. Goodrich), *SPE Regional Conference Papers, Newark Section,* November 5, 1964. Properties, processing conditions and equipment (including screw design) and application information.

"Delrin acetal resin pipe," J. F. Cogdell (du Pont), *SPE Regional Conference Papers, Newark Section,* November 5, 1964. Properties, advantages, applications, costs, fittings.

"High-duty nylon piping," *Plastics,* March 1957. Description, with five photographs, of British plant to make large nylon-6 pipe. No conditions given.

Equipment and Operation: Extrusion, Sizing and Cooling

"Die Fertigung von Rohren" (The manufacture of pipe), A. Rothenpieler, *Kunststoff und Gummi,* August 1966. Covers extruders and auxiliary equipment for PVC and polyolefin pipe.

"Rundown on pipe sizing methods," R. B. Doherty (Conair), *Plastics Technology,* June 1966, based on a paper given at *SPE ANTEC,* 1966, as well as article in *SPE Journal,* October 1964. Evaluation of the various sizing methods commonly used, particularly the vacuum trough.

"Methods for calibrating pipe from low-density polyethylene and polypropylene," A. P. Lichobabenko *et al.* (U.S.S.R.), *Plasticheskie Massy,* June 1966.

"Critique of calibration systems for high-density polyethylene pipe," V. N. Ryabchenkov and A. P. Lichobabenko (U.S.S.R.), *Plasticheskie Massy,*

April 1966. How to avoid surface problems; effects on appearance and mechanical properties.

"Influence of type of heating and cooling on mechanical properties of PVC pipe," D. Grootenboer, *Plastica,* February 1966. Slow cooling shown to be better.

"Methods of sizing extruded plastic pipe and tubing," C. Gatto (Gatto Machinery), *Plastics World,* December 1964. Brief description of various methods.

"Vacuum sizing of plastic pipe," J. L. Vermillion (Eastman), *Modern Plastics,* October 1964. Brief description of vacuum-trough sizing technique.

"PVC pipe technology," A. A. Van der Wal (Wavin, N.V.), *Plastics,* October 1964. Describes technique of annealing at 100°C, continuously, to eliminate or reduce high benzene absorption, and strengthen pipe. This is important in gas transmission.

"Die Erzeugung von Kunststoff-Rohren aus der Sicht des Maschine herstellers" (The production of plastic pipe from the point of view of the machinery maker), *Der Plastverarbeiter,* October 1964. Description of Kestermann equipment, including dies.

"Calculation of the length of calibrating units," G. Schenkel (Troester, now at Stuttgart Technical College), *Kunststoffe,* March 1961. Theoretical approach to design of a vacuum sizing sleeve.

Equipment and Operation: Measurement and Control

"Measurement and automatic correction of extrusion eccentricity," staff written, *British Plastics,* February 1966. Short description and photograph of Swiss equipment (Fischer) which rotates a transducer around the pipe.

"Measurement and control of plastics tubing or pipe wall thickness," A. Halegua (West Instrument, division of Gulton Industries, and licensee of Haynes and Haynes in England), *SPE Regional Conference Papers, Newark Section,* November 5, 1964. Illustrated description of equipment and operation; the Haynes system measures capacitance, converts to wall thickness, records and feeds back to control the extruder if desired. Similar discussions are found in German, in *Kunststoff und Gummi,* November 1963 (by A. C. R. Haynes); in French, in *Revue Belge des Matières Plastiques,* January 1965 (also by Mr. Haynes); in *International Plastics Engineering,* April 1961 (staff written); and in *SPE Journal,* December 1961 (by A. Willer, of West Instrument Corporation).

"Economics and technology of close tolerance extrusion," E. A. O. Mange (Stewarts and Lloyds), *British Plastics,* August 1963. Plea for money, effort and intelligence; use of automatic weigh feed, Haynes meter, etc.

Equipment and Operation: Pulling, Coiling, Cutting, Fittings

"Flaring and stamping of extruded pipes of rigid PVC," V. Decans, *L'Officiel des Matières Plastiques,* May and November 1964, reviewed in *SPE Journal,* November 1964.

"Coiling machines for plastics profile and tubing," U. Finck, *Kunststoff und Gummi,* August 1964.

"Apparatus for continuous making of slits in profiles and pipes," A. Weiss (Frieseke & Hoepfner), *Kunststoffe,* June 1964. Short illustrated review, product used in agriculture for soil drainage.

Standards and Tests

NOTE: Further information on creep and stress-cracking testing is found in Chapter 7 (Wire and Cable Covering) and in the bibliography at the end of Chapter 7.

"The testing of pipes and fittings," P. F. V. Lloyd (Durapipe Technical Services, Ltd.), *Plastics,* May 1967. Test techniques, apparatus and data for plastic pipe and fittings, mainly ABS and PVC, but also nylon, polypropylene, polyphenylene oxide (PPO) and ethylene/vinyl acetate copolymer (EVA). This author also has a pertinent letter to the editor in *British Plastics,* May 1967, referring to the following article by Cann.

"Plastics pressure pipe—design criteria and performance assessments," J. M. Cann (British Resin Products), *British Plastics,* February 1967. Failure mechanisms; testing; German, Dutch and British standards.

"Long-term hydrostatic strength characteristics of thermoplastic pipe," F. W. Reinhart (consultant, technical director of Plastic Pipe Institute; formerly with National Bureau of Standards), *Polymer Engineering and Science (SPE),* October 1966. Test data on PVC, polyolefins, ABS, cellulosics. Further data and extensive description of standards and tests for pipe in article by F. W. Reinhart: "Plastic pipe standards," *SPE Regional Conference, Newark Section,* November 5, 1964; earlier discussions of standards/tests, also by F. W. Reinhart, are in *Modern Plastics,* September 1964 and *SPE Journal,* February 1961.

"Zeitstandfestigkeit von Rohren aus harten Polyäthylen und Polypropylen unter dem Einfluss von Chemikalien . . ." (Long-term chemical resistance of high-density polyethylene and polypropylene), E. Gaube *et al.* (Hoechst), *Kunststoffe,* October 1966.

"Recent tests on PVC pipe under pressure," *Plastica,* April 1966. Can use mix of suspension and emulsion polymers.

"Die Druckprüfung von Polyäthylen-Rohren" (Pressure testing of polyethylene pipe), S. Maegaard, *Kunststoff-Rundschau,* May 1965. Discussion of hydraulic test method.

"Use properties of plastic pipe," D. F. Kagan and L. A. Kantor, *Mechanika Polimerov* (Riga, Latvian S.S.R.), March 1965. Comment on test methods, etc., and durability of polyethylene under load.

"Impact resistance of plastic pipe," J. D. Rourk (U.S. Rubber), *SPE Regional Conference, Newark Section,* November 5, 1964. Effects of process variables on impact strength of ABS pipe, tested by proposed ASTM method. Test method and results discussed.

"Short-term stress-rupture testing of thermoplastic pipe," P.F.V. Lloyd (Durapipe), *Plastics,* August 1964. Description of apparatus and method, and results obtained from the test, developed by Durapipe and Fittings Ltd. Further information on Durapipe equipment for both short and long-term testing is in *International Plastics Engineering,* January 1963.

"The assessment of the impact resistance of rigid PVC pipe," J. M. Cann (British Resin Products), *British Plastics,* September and October 1963. Description of standards, apparatus, test methods (including falling weight), types of failure; discussion of variables and results for several plastics.

"Ductile failure of polyethylene pipe," S. Mruk (Allied), *SPE Journal,* January 1963, and *SPE ANTEC* 1962. Photos, graphs, tests; explanation of how neck forms, workhardening, and failure.

"Propriétés et essais de tubes géants en PVC rigide par Wavin" (Properties and tests on large rigid PVC pipes by Wavin), G. Cherubin, *Industrie des Plastiques Modernes,* July 1962.

"Determining the impact strength of PVC pipes," K. Richard *et al.* (Hoechst), *Kunststoffe,* August 1961. Two different methods described. Other early works on pipe standards and testing, some of which contain extensive data, are found in: *Plastics,* September 1960 (European standards); *Modern Plastics,* May 1960 (predicting long-term performance from short-term tests); *Kunststoffe,* April 1960 (German standards): *Kunststoffe,* September 1959 (requirements for drinking water transmission, for polyethylene): *Plastics,* April 1959 (testing of polyethylene and PVC); and *Plastteknik* papers, Stockholm 1958 (short- and long-term tests and data, including cycling conditions).

Special Techniques

"Extruding plastic sheath over steel pipe," *Chemical Engineering,* December 23, 1963. Describes coating of pipe with polyethylene at 30 meters/minute.

"Strengthened pipes from Ziegler polyethylene," K. Richard *et al.* (Hoechst), *Plastics,* December 1961. Describes biaxial orientation, advantages and method. Further information about same process in *Kunststoffe,* July 1960; and in *Modern Plastics,* November 1960.

"Drossbach tube corrugating system," *International Plastics Engineering,* October 1965. Short illustrated description of German equipment used to make conduit, drainage pipe, air blower ducting, etc.

"Extrusion of cross-linked polyethylene tubing," C. J. Benning (Grace), *Modern Plastics,* December 1966, and *SPE ANTEC* 1966. Continuous chemical cross-linking, peroxide agents, downward extrusion, all densities of polyethylene.

"Fittings for plastic pipe," M. Usab (International Pipe and Ceramics), *SPE Regional Conference, Newark Section,* November 5, 1964. Not in collected conference papers; contact author or SPE.

"Kalibrierlose Klebeverbindung von PVC Rohren" (Uncalibrated adhesive fittings for PVC pipe), *Der Plastverarbeiter,* July 1964.

"The joining of Ziegler-polyethylene pipes," K. Richard *et al.* (Hoechst), *Plastics,* January 1961. Similar article in German, in *Kunststoffe,* June 1960. Many illustrations, descriptions of various joining methods.

"Continuous helix-generating machine for reinforcing extruded tube," *International Plastics Engineering,* April 1961. Machine makes a continuous coil from wire and feeds it into an extruder die to make reinforced tubing.

"Behälter und Röhrenbau aus Hostalen nach dem Wickelverfahren" [Containers and pipe from Hostalen (polyethylene) by the winding method], *Kunststoffe,* January 1958. Extruded tape is wrapped around a rotating mandrel perpendicular to the extruder and slowly moving along its axis. Diameter can go up to 3 meters (10 feet).

FOUR
sheet

PRINCIPLE

Sheet is made by forcing molten thermoplastic through a long horizontal slit. The hot web thus made is carried around polished metal cooling rolls and is then cut up or rolled up as desired. The process is illustrated in Figures 4.1 and 4.2.

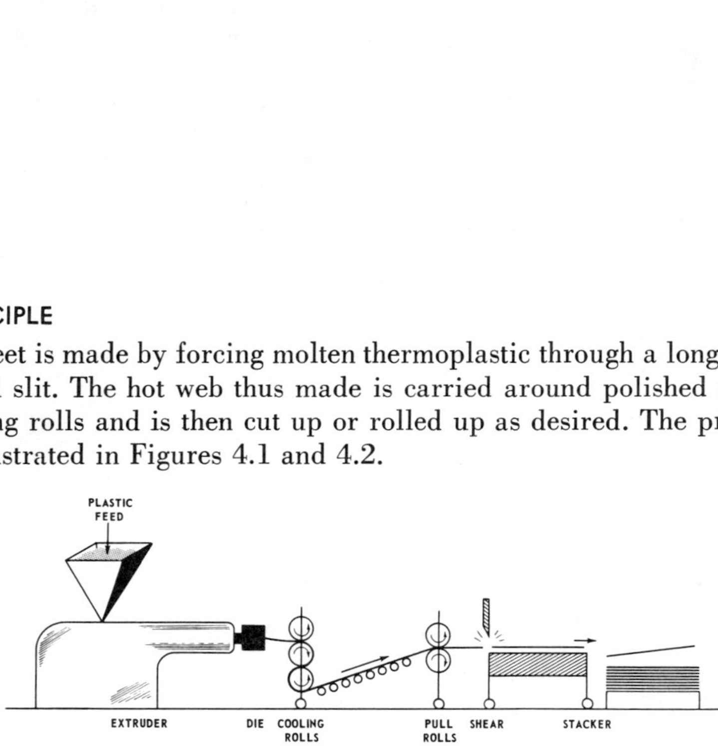

Figure 4.1. Sheet extrusion line. *Left to right:* Extruder feeds the hot sheet into a stack of metal cooling rolls. Sheet cools further in air, is drawn through rubber pull rolls, and is cut to size. For this flexible sheet, a wind-up is used in place of the shear and stacker. (*Illustration by author*)

MATERIALS

High-impact polystyrene is the most important sheet material. High-density polyethylene, PVC, and ABS are also extruded into sheet in

Figure 4.2. Photographs of sheet extrusion lines. At top, a medium-sized line in operation. (*Courtesy of Prodex Division, Koehring Co.*). At bottom, a high-speed line for thin polystyrene sheet, rated at over 1800 lb/hr (820 kg/hr). (*Courtesy of NRM Corporation*)

significant amounts. Sheet extrusion grades usually have a high melt viscosity, because: (1) die resistance is relatively low, (2) the high viscosity discourages sagging during thermoforming, and (3) the physical properties are better (higher molecular weight).

DEFINITIONS

Extruded sheet is often distinguished from film at 1/4 mm (.010 in.) thickness—anything less is film. However, people still use the type of manufacturing equipment as the basis for calling a product sheet or film, regardless of thickness. Thus, "sheet" includes high-impact polystyrene and cellulosics down to 0.1 mm (0.004 in.) because sheet equipment is used in manufacture, while "film" may include polyethylene as thick as 1/2 mm (.020 in.) if blown on film equipment.

PRODUCT DIMENSIONS AND EQUIPMENT SIZE

Sheet as thick as 3 mm (1/8 in.) is very common, and 6-mm sheet is readily available. Examples of the latter are the high-impact polystyrene and PVC used to form entire refrigerator inner liners. The available width ranges from narrow strip only a few centimeters wide to sheets almost 2 meters across. Wider extruded sheet is possible, but rarely seen.

For production economy, sheet is usually run wide—90 to 150 cm (36 to 60 in.); if the end products are narrow, they are slit from wider sheet. Thus, large machines can be used, ranging from 40 to 150 hp and 90 to 150 mm (3 1/2 to 6 in.) in barrel diameter. The most common system is a 120-mm (4 1/2-in.) extruder with a 100-hp drive. Some older extruders of this size are still in satisfactory commercial operation with 50- or 75-hp drives.

EQUIPMENT

Dies. Sheet dies are nearly all center-fed, as shown in Figure 4.3. A few end-fed dies are found in special cases but center-feeding predominates because wide sheet is easier to make (the plastic travels only half as far to get to the ends of the die), sheet properties are more symmetrical about the center, and the whole system can run in a straight line, parallel to the extruder. End-fed dies can be run parallel to the extruder too, using a right-angled elbow adapter, but they have never found wide use for sheet. The entrance to the die is an opening of circular or oval cross section, which flares out into the

Figure 4.3. Close-up of extruder die and sheet entering rolls. The sheet can be seen as a band of white going around the lower roll. (*Courtesy of Prodex Division, Koehring Co.*)

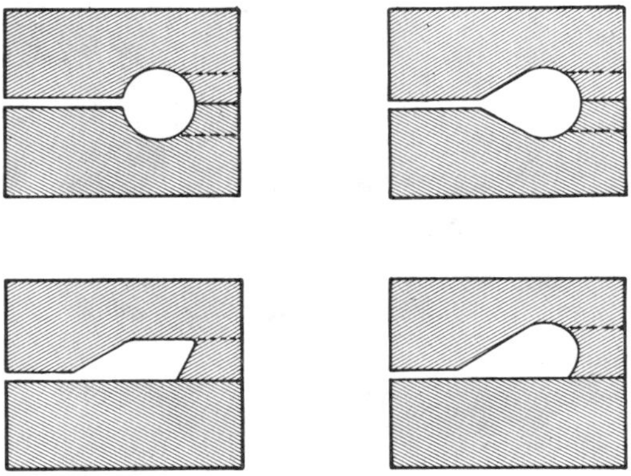

Figure 4.4. Various cross-sections of flat sheeting dies: circular, teardrop, angular and flat-sided teardrop. (*Illustration by author*)

manifold, a long passage that distributes the plastic along the width of the die. Several cross sections of manifold are used (see Figure 4.4). A circular manifold is cheapest to make, while a tear-drop is the most streamlined. A good compromise between cost and streamlining is a flat-sided tear-drop, shown in Figures 4.5 and 4.6. Streamlining is important because the plastic moves slowly in the die, and

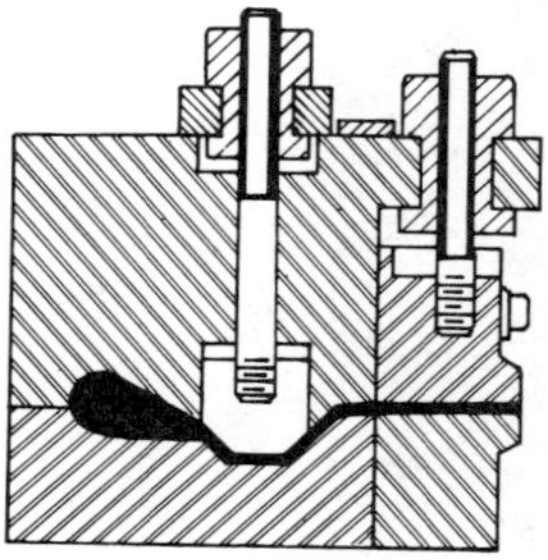

Figure 4.5. Cross section of a typical sheet die. This one has a teardrop manifold; some older dies have circular manifolds.

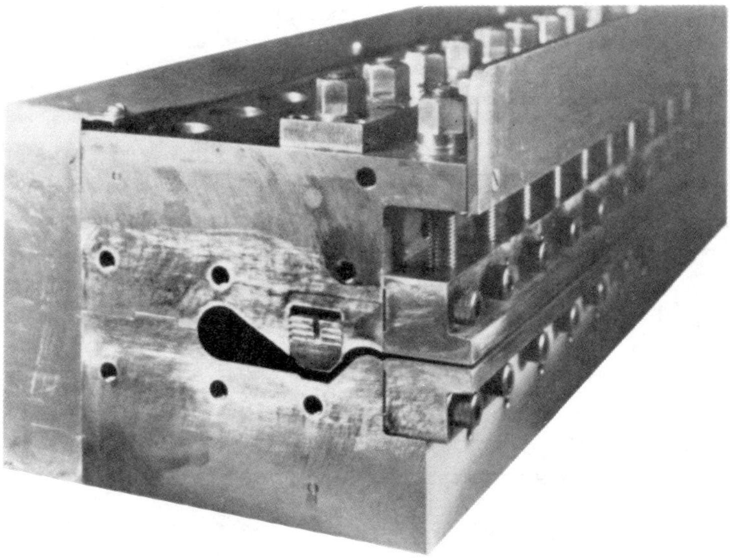

Figure 4.6. A die similar to that in Figure 4.5, with end plate removed to show manifold (teardrop section in center), choker bar (adjustable bar just to the right of the manifold), and adjustable lips (at right edge of die). The choker bar is raised higher than the normal production position, which may be as low as within 1 or 2 millimeters (40-80 mils) of its seating surface. (*Courtesy of Prodex Division, Koehring Co.*)

sharp bends or crevices are likely to hold pockets of stagnating plastic, which will yield discolored or contaminated sheet, and sometimes even decomposed sheet. To prevent this stagnation at the manifold ends, bleeder holes are often drilled in the end plates. They are opened by turning a bolt in the end plate which releases a small flow of plastic through the hole. Occasionally, they are left open all the time. This technique helps quicken color or material changes, since the ends are the last parts of the sheet reached by the new material.

The manifold can be in a straight line (straight-through construction) or bent to form a "Y." These are shown in Figure 4.7. The bent form is often called a "fishtail" or "coat-hanger," and may vary from a slight bend to a pronounced fan shape. It is more costly to machine, but is more streamlined, which is desirable for polystyrenes and polyolefins, and imperative for PVC. Another advantage of the fishtail die is that when the plastic reaches the ends of the manifold, it has less distance to travel to get to the lips than the plastic in the center. This compensates for the shorter path for the central material, and promotes a more uniform flow out of the die. Some of the best dies have a manifold that tapers to a very small cross section at each end, thus minimizing residence time and possible stagnation. Still another advantage of a more streamlined manifold is the ease of cleaning and the tendency to self-clean. This is especially important with sheet dies, which may weigh up to a half ton, and are very time-consuming to disassemble and clean. New designs have recently become available, especially for PVC, that are very streamlined, and which disassemble very quickly.

These principles of streamlining and manifold design are also valid for flat-film and extrusion-coating dies.

The extrusion of rigid PVC sheet has received much publicity, and has been the object of much work all over the world. In all successful cases, the die has been constructed to give maximum streamlining, making stagnation less likely. A coat-hanger die is usually used, and sometimes a special screw tip or head assembly is used to prevent stagnation directly in front of the screw (controlling the screw-tip temperature with hot oil is also recommended).

One of the most interesting devices to avoid stagnation is a driven screw (or two screws) which fills the manifold, threaded so as to distribute the plastic toward the ends of the die. Such machines are already in commercial use in Europe, America, and Japan. They

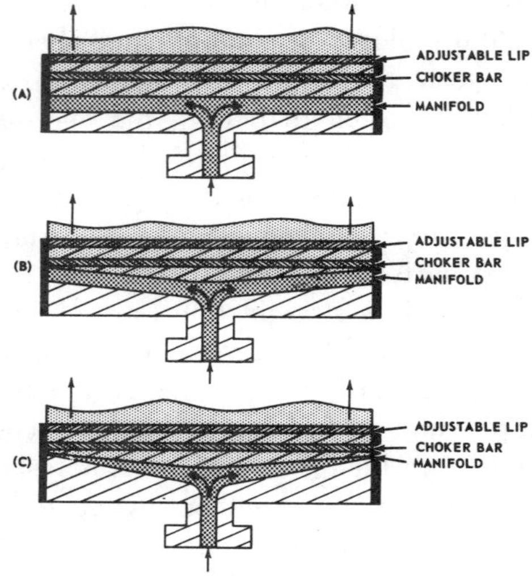

Figure 4.7. Sectional views of different sheet dies, showing three different types of manifold. Design (A) is a conventional one, with a straight-through manifold of constant cross section. Design (B) is improved, as its manifold path is more streamlined, but its cross section is still constant. Design (C) is most streamlined (and most expensive), with tapered manifolds getting smaller toward the ends, as well as angled manifold halves as in design (B). Both (B) and (C) are called "coathanger" or "fishtail" dies. A true "fishtail," not shown, has been built without any distinct manifold, but rather with the plastic stream gradually widening and flattening out as it passes through the die. This is an expensive construction and is rarely seen. (*Illustration by author*)

eliminate the need for a coat-hanger manifold, substantially reduce the effective die resistance, and give an additional control over thickness uniformity. The main problems are the bearing regions at the manifold ends where the screw enters and where it seats. These screw dies were originally developed for thick sheet, 1 mm (.040 in.) and over; thinner sheet required closer die lip spacing, which developed very high pressure inside the manifold, causing leakage around the bearings. Now, however, one manufacturer of this equipment is advertising its use for thin rigid PVC down to .05 mm (.002 in.), and for even thinner gauges of plasticized film.

An adjustable choker bar (see Figure 4.6) is found on most sheet dies. It is a relatively thin metal bar running across the die, and can be flexed by the many bolts connecting it to the die body. It is set in a special groove designed to minimize stagnation, and acts as a kind of valve; in fact, it is the usual means of regulating thickness across the sheet. The operator normally tries to keep thickness constant across the sheet, although sometimes he deliberately runs the ends thin, if they are to end up as thermoforming trim. The choker bar should not be thought of as a regulator of melt pressure in the extruder. Its effect on melt pressure is small (about 30 atm at most); a valve in the extruder head does a much better job, and has less effect on the thickness profile—i.e., the pattern of thickness across the sheet. The bolts that adjust the choker bar should enter from the top. A few dies have been built with the adjusting bolts on the underside, to balance the lip-adjusting bolts and give more uniform heating, but this has created undue inconvenience to the operators and correspondingly poor thickness control. Some dies have a fixed choker bar—i.e., a permanent restriction built into the die body.

The die lips themselves are the major factors in determining the final sheet thickness. They are cut square to give maximum rigidity and protection from nicking. Their opening is adjusted by bolts similar to those on the choker bar, but these should not be used to adjust thickness during operation. Rather, they are set before start-up, usually at or near the desired sheet thickness. Older systems use alternating bolts, where every other bolt pushes down on the lip and the remaining ones pull up. Modern dies now use closely spaced identical bolts with fine threads which both pull up and push down the die lip, depending on the direction in which they are turned. Both lips have usually been separate strips of steel, but the latest systems favor a flexible upper lip, integral with the die body.

The choker bar combines with the die lips to give enough resistance to force the plastic out to the ends of the manifold before it can all come through the center of the die. Since thick sheet and wide sheet offer less resistance to flow, these operations will require longer die lands and more "choking" to ensure uniform and full distribution of material across the die. The die lands are the internal faces of the lips, usually parallel or nearly parallel to each other, and are about

3 to 5 cm (1.2 to 2 in.) long. Resistance to flow, hence melt pressure in the manifold, varies directly with the length of the lands and inversely with the *cube* of the distance between the lands. This distance, of course, can be varied and is the actual die lip opening.

Exceptions to the need for long land lengths are dies for extruding slabs of foamed polystyrene or polyethylene. Their lands are short to give small foam cells and a smooth sheet surface. Thin foamed polystyrene sheet is normally extruded on equipment similar to that used for blown tubular film (see Figure 4.12).

Many widths can be made from one die with polyolefins and polystyrene. Since these materials do not spread decomposition quickly out of a stagnated area (like PVC does), a simple brass dam can be placed over the die face, blocking off just as much of the width as desired (see Figure 4.8). The plastic will stagnate behind the dams, but this only means that purging and possible die cleaning will be needed when wider sheet is desired. Occasional bleeding out of plastic through the end plates will prevent much of the stagnation behind

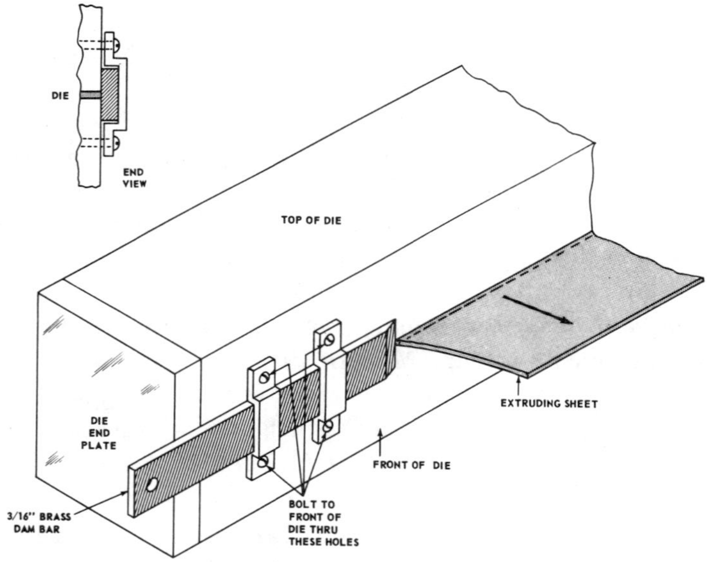

Figure 4.8. Illustration of the use of brass "dams" on the ends of the die face to give variation of width. The brass dam is held firm by the two brackets, and the inner end is beveled to ensure a clean sheet edge. (*Illustration by author*)

the dams. These external dams are much easier to use than the complex internal plugs that were once preferred, and external dams are now generally accepted. Neither type is satisfactory for PVC, where any stagnation at all will spread out through the sheet. For this reason, a PVC die is always run full width, which may have a significant effect on the economy of production.

There are four basic types of heaters for sheet dies:

System	*Advantages*	*Disadvantages*
Band heaters	Lowest cost. Allows solid die body, which means a smaller die, more uniform heating.	Exposed terminals. Leakage or warping may cause poor contact. Insulation more difficult. Hard to disassemble.
Block (cast) heaters	Like bands, but better contact and easier assembly. Less maintenance.	Exposed terminals, hard to insulate, a little more expensive than bands.
Cartridge heaters	Easiest to assemble and replace. Simple to insulate.	Die body has internal holes, requiring a bulkier die with more internal temperature gradients. Heaters sometimes stick in holes. More maintenance than block heaters.
Induction heaters	Fast starting, very uniform heat, air gap is inherent insulation, long heater life claimed.	Expensive, bulky, may be hard to disassemble.

Bands, blocks and cartridges hold most of the market, and are used about equally. Occasionally, a die is heated with circulating oil. This gives uniform temperature control, and can make a compact, simplified die (no wires). However, oil heating is never as clean and easy to run and control as electric heating. With conventional operating techniques, oil cannot give adequate zone control on very wide dies, as only one or two control circuits are conveniently possible. More circuits would increase the die cost, and would require a great amount of plumbing all over the die. If the die is well insulated, it can be run at a single temperature, with perhaps a second circuit controlling the lips, and oil heating will then work very well. All thickness adjustment must then be made by turning bolts.

Sheet dies are often left uninsulated—probably for easier access to bolts and heaters. However, the die radiates much heat, which is both inefficient and uncomfortable, and full insulation is recommended. Asbestos-board covers and enclosures are occasionally used, custom-designed around the particular die.

Sheet dies are made from tool steel, often with chrome-plated internal surfaces—i.e., those that contact the plastic. The lips are always plated or hardened.

The die is divided into several heating zones—usually three or five. The temperatures of these zones are measured in the metal and actuate standard control devices. Some dies have pressure gauges and stock thermocouples (immersed in the plastic stream) at several points in the manifold, to give a better idea of material behavior in the die.

A sheet die can be very heavy—½ ton or more—therefore, it is commonly supported by a little cart rolling on the same rails as the take-off. A hoist is often mounted above the die and is used to open the head and change dies. It can also be used as the die support during operation, instead of the cart.

Sheet Take-offs. As the sheet leaves the die it is normally in a horizontal plane, and is picked up by the middle roll of a vertical stack of three polished metal cooling rolls (either chrome-plated or special alloy). All these rolls are driven from a common source. A close-up view of the sheet entering the roll stack is seen in Figure 4.3. The sheet can either travel "down the stack" (over middle roll, around and under bottom roll), or "up the stack" (under middle roll, around and over top roll). The former method is far more common, minimizing sag problems at the die, and making laminating easier. In the past, some firms made "up the stack" units, and claimed that this method kept the most important surface—that taken from the middle roll—face up, and less liable to scratch on the subsequent conveying equipment. However, such scratching can be (and has been) avoided by proper design of equipment, no matter how the sheet is taken around the rolls.

The sheet passes from the cooling rolls through 1 to 3 meters (3.3 to 10 ft) of air space, often supported by small idling rollers. One manufacturer uses roller-skate wheels instead of the rollers. The sheet is drawn through this space by a pair of rubber pull rolls. For extra cooling, fans can be placed above and below the sheet between the cooling rolls and the pull rolls. In special cases (see Figure 4.2) a water trough and/or one or two more cooling rolls are put in this space.

All the rolls are driven at similar surface speeds either by a worm gear system or a chain drive. Separate drives run the metal cooling set and the rubber pulling set, and their surface speeds are often synchronized by mechanical, electrical or electronic means. The pull rolls usually run a little faster than the cooling rolls. The difference between the two speeds is important and controls the tension in the sheet between the roll sets. Typical linear speeds of sheet on modern equipment are 2 to 3 meters/min (6 to 9 ft) for 2½-mm sheet (0.100 in.), and up to 15 to 20 meters (50 to 65 ft) per minute for thinner sheet of ¼-mm (0.010 in.). A good versatile take-off will run from ½ to 20 meters/min, and for this it may need more than one speed range. Most take-offs in operation today actually cover a narrower range.

All three metal rolls are usually the same diameter (30 cm = 12 in. is common), but some units have two 30-cm rolls and one roll at 20 cm (8 inches) as the top one in a "down the stack" system. Compared to three 30-cm rolls, the 20–30–30 system is less costly, and there is no loss in cooling capacity as the sheet is wrapped around the two 30-cm rolls only. This system also allows the die to be set closer to the roll nip, reducing sag problems. Larger roll diameters than 30 cm are occasionally used for high output (230 kg = 500 lb/hr or more), especially for polyolefins, which need more cooling than polystyrene or PVC. The rubber pull rolls are usually 15 to 20 cm (6 to 8 in.) in diameter.

The internal roll design must provide excellent heat transfer so that the entire roll surface remains at the same temperature during operation. Also, it should allow easy cleaning. The roll temperatures are controlled by closed-circulation heat exchangers, usually containing water. Oil is also used occasionally; it is more expensive, but allows higher temperatures and is therefore useful for polyolefins. Almost all systems have one exchanger for each metal roll. Some water systems are not sealed, and fresh cold water is drawn into each circuit to cool it. This is less expensive than a heat exchanger and offers quicker response, but hastens internal fouling of the rolls, especially in hard-water areas.

Inadequate roll cooling is a frequent cause of problems, and this is not always obvious. Rolls may become internally coated or fouled, or may be of poor design, or the water-pumping capacity may be too small. Any of these conditions may produce temperature differences from one side of the sheet to the other, sticking of the sheet to the rolls and similar faults which generally reduce permissible output rate.

The whole take-off system rides on angle-iron tracks which have stops attached to them in order to prevent the take-off from rolling into the die. Some simple mechanical device is often included to push the heavy take-off along the tracks. Sometimes the heat exchangers are mounted separately, rolling on wheels and connected to the rolls through flexible hoses.

A new concept in sheet take-offs has already received limited trial. Instead of rolls, the hot plastic sheet fails on a flat, stainless- or chromed-steel conveyor belt traveling around two large hollow metal drums. The extrudate falls on the belt as it passes over one drum which is oil-heated to a temperature approaching extrusion temperature. The plastic then cools slowly and eventually passes over the other drum, which is chilled. This system can produce sheet with little or no internal stresses, because no tension is applied to pull the plastic into and through the cooling apparatus. Also, very high gloss can be produced which remains after thermoforming, because the polished belt is very hot when it first contacts the plastic. A related technique, rubber-belt take-off, does not have as many advantages, but is in commercial use in many countries. It is a simple system, and is used to make sheet for continuous thermoforming into containers.

The gloss of polyethylene, ABS, and some vinyls is adequate when passed through the conventional system of polished metal rolls. However, high-impact polystyrene is naturally rather dull, and the gloss imparted by a shiny roll will disappear almost entirely in a thermoformer. To counter this lack of gloss, a thin film of biaxially oriented, clear, unmodified polystyrene can be laminated to the sheet during extrusion (see Figure 4.9). This film can be preprinted to produce a decorative effect as well as the high gloss. Another popular gloss-producing technique involves passing the sheet under a bank of intense radiant heaters which melt and glaze the surface. The heaters are placed between the die and the metal cooling rolls, or sometimes just after these rolls. Both of the gloss-producing systems have disadvantages: the laminate costs extra money and weakens the surface in flexure and impact strength, and the heat glazer also weakens the surface and may even discolor it. A carbon-dioxide or nitrogen

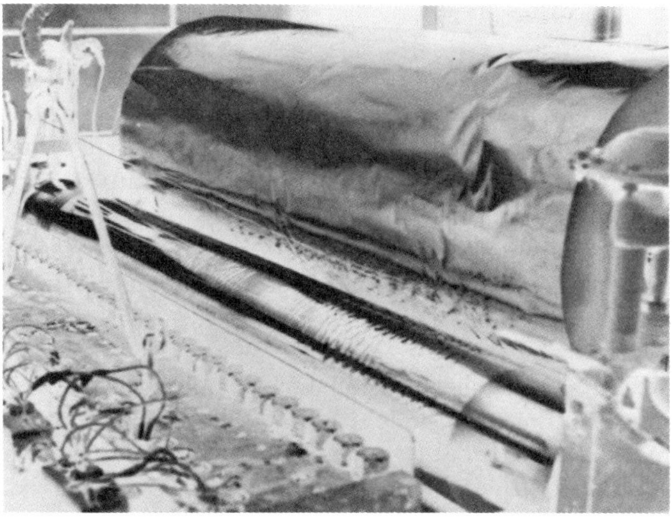

Figure 4.9. View of a laminating system from above. The sheet die (lower left) is extruding hot sheet into the bite of the polished cooling rolls. The top roll is clearly visible. The thin, transparent laminating film (large roll, upper right) feeds over and around this top roll, which preheats it. The film then enters the bite to be pressed into the hot extruding sheet, forming a permanent bond. (*Courtesy of Union Carbide*)

atmosphere greatly reduces the harmful effects of glazing but is inconvenient to use and no specialized equipment is commercially available. As a general rule, independent of all these techniques, a higher extrusion temperature at the die lips yields a more permanent gloss, and the raw materials themselves vary in inherent gloss.

After being drawn through the rubber pull rolls, the sheet is either rolled up on conventional film-handling equipment, or else cut into individual sheets by a timed shear. Electric eyes or microswitches, activated by the moving sheet, are used to trip the shear at the right times.

Trimmers are often used to cut sheet to exact widths. Razor-blade cutters are popular for thin sheet, hardened metal shearing wheels cut medium thicknesses, and circular saws are needed for the thickest products. Trimming is sometimes omitted, as width can be held to 5 mm (0.2 in.) with dams; moreover, the edges are often cut off after forming and do not become part of the finished product.

OPERATION

Drying. A pre-dried feed is a necessity for impact polystyrene sheet over ½ mm (0.02 in.) thick. Below this thickness, the marks on the surface caused by moisture tend to disappear. Drying is also necessary for nylon, cellulosics, and styrene-acrylonitrile copolymer, and for any other polymer with hygroscopic additives like fillers, carbon blacks, and pigments. Predrying in a hopper-dryer is the usual technique. Vented extruders are used too, and some plastics such as nylon are shipped sealed and moisture-free.

Formulation. Most sheet compounds, including those serving the big appliance market, are bought already formulated, compounded and colored, directly from the raw-material suppliers; however, PVC extruders often mix their own compounds. There is a small amount of coloring of other materials done in the extrusion plants, using dry powdered pigments or color concentrates in the required plastic. With polyolefins, the normal pipe, wire coating, and monofilament grades are commonly used for sheet. These are usually rather low in melt index. PVC pipe compounds have been successfully used for sheet, but sheet extrusion is more critical and a special formula is generally preferred. More stabilizer may be needed, viscosity must not be too high, nontoxic additives must be used in sheet for food packaging, and absorbers of ultraviolet light must be added for outdoor building components.

Pre-blending of different grades of polystyrene is well known, and super-high-impact intermediates are sold for this purpose. The pre-blending is done by merely tumbling the different grades of pellets together or running them through a continuous proportioning device, and then extruding them under reasonably high pressures (e.g., 200 atm or more). No hot compounding in advance is needed. A wide range of properties is thus available from just two components, often at substantial cost savings.

Output Limits. In a well-designed system, extruder power and cooling capacity of the take-off are balanced to yield a predictable and desired maximum output. However, many producers are further limited by problems such as air bubbles (low melt pressure, wrong type of screw); improper gearing and insufficient available horsepower (too little reduction of motor speed); inability to hold constant thickness (light weight die, broken bolts, surging in extruder); high compound temperature (take-off cannot cool sheet fast enough); or inability to run at fast take-off speeds for thin sheet. The question: "why don't you turn the screw faster?" often leads to interesting answers. A lot of horsepower and investment goes unused in sheet extrusion because the above factors remain uncorrected.

Mixing. Good mixing of the melt in the extruder is important for any sheet. With either a blend or a single material, poor mixing yields temperature variations which cause ridges, waves and streaks in the sheet. High pressure in the extruder is the recommended cure, and extruders and screws should be designed with this in mind.

Start-up. A length of paper, cloth, or plastic is used as a starting aid, and is threaded through the rolls before the extruder is turned on. When the first plastic comes out of the die, it falls on the starting aid and sticks to it. Then the take-off is started and the sheet is pulled steadily from then on. The roll temperatures should be set quite low to avoid the possibility of sticking, until the operator gains enough experience to predict proper roll temperatures in advance. The shear is set to cut narrow strips, suitable for easy regrinding, until good sheet of uniform thickness is being produced. This may take anywhere from 15 min to 3 hr, depending on the skill of the operator, the stability of the equipment, the size and speed of the equipment, and the required tolerances.

Roll Control. Accurate control of roll temperatures is important to ensure best possible surface. This is especially true for polyethylene, where a change of only a few degrees can turn a smooth glossy sheet into a dull and wrinkled one. The rolls should be run as hot as possible, just safely short of the sticking point. If they get too cold, a warped and curled sheet results, and spots and pocks appear on the surface. Polyethylene should be molten and translucent or transparent all over the first roll that it contacts, crystallizing to a more opaque solid just as it peels off this roll. Sometimes a curl can be produced in the sheet *if the last roll is too hot*—the sheet is then left warm enough to be distorted by the pull of the rubber pull rolls, and the extreme edges then curl in because they cool off the fastest.

Special Note on Polyethylene. The spots and smaller wrinkles caused by cool rolls *will disappear* in most thermoforming operations. The spotted appearance of the sheet and its failure to lie flat and straight are certainly unwelcome factors, but they may not cost much and may be worth accepting in order to make a salable formed product at the highest possible rate.

The one roll without surface contact (the top roll in a "down the stack" system) is not used at all when absolute minimum internal stress is desired. It otherwise helps achieve more uniform thickness; this is a minor effect, and some of the gain in uniformity is lost again during thermoforming because of a recovery or "plastic memory" effect. The top roll is necessary to apply pressure when a surface film or thin sheet is being laminated, or when the extrudate is being embossed. Ordinarily, the temperature of this roll is less critical than the other two; however, when it is used to heat the laminating film, it must be very closely controlled—often to $\pm$ 2°C or better.

Die Control. Adjusting the choker bolts of a sheet die to give consistent and even thickness is a matter of "getting the feel of the die." An operator must learn just how much each bolt must turn for a given thickness change. Markings on the bolts can be added to serve as a rough guide, but final adjustment is still a matter of skill.

Some average commercial tolerances are:

Nominal Thickness	*Tolerance Range*
0.25 mm = 0.010 in.	0.23 to 0.28 mm = 0.009 to 0.011 in.
0.75 mm = 0.030 in.	0.7 to 0.8 mm = 0.028 to 0.032 in.
2.50 mm = 0.100 in.	2.4 to 2.6 mm = 0.096 to 0.104 in.

A controversy exists as to how much the die heaters should be varied to assist in getting uniform thickness. Varying the temperature across the die produces varying exit temperatures across the extruding sheet. These are unadvisable for polyethylene because of the narrow range of cooling-roll temperatures that is required. The variation is acceptable within limits for most other materials. It affects the stress levels in the sheet, which may have an effect on thermoformed parts, and this effect can be reduced by higher extrusion and/or thermoforming temperatures. If these stress variations are considered harmful, then the die must be set to give a constant-temperature extrudate (*settings may not all be the same*—it is the melt temperature that counts), and all adjustments to thickness are then made with the choker-bar bolts. Where the stress variations can be tolerated, or are insignificant, it is much easier to work with the heater settings first, adjusting them until the output per heating zone is the same all across the die. Further adjustments would then be made with choker-bar bolts.

The buildup of hot material in the bite of the cooling rolls is called a "bank" or "bead." It must be carefully watched, and kept as small as possible. A uniform rolling bank, barely visible across the sheet, is a good indication of uniform sheet thickness and properties.

A system is available in which bolts are turned by a powered device, actuated by the operator. Although this is a more reliable and steady turning force, it is not automatic control, as the operator must still decide which bolts to turn and how much to turn them. A truly automatic control system is still not generally available, and would be a welcome development. It already has been described in a U.S. patent.

Low-level Orientation. These are the stresses introduced in the sheet by the pulling and compression of the take-off rolls. Usually this type of orientation is undesirable as it makes one direction strong at the expense of the other; it also makes printing go out of register during thermoforming. These problems are prevented by high melt temperature, low tension between the two sets of rolls, low degree of drawdown (ratio of die opening to final desired thickness), little or no pressure applied by the top (laminating) roll, and high temperature on the middle roll (which prevents the sheet from slipping and stretching over it). Sometimes, low-level orientation is deliberately introduced—e.g., to increase shrinkage and thus to reduce sag in thermoforming high-density polyethylene. Control of this type of orientation is difficult, and it is best to try to avoid any need for it. This can often be done by proper design of the formed piece and its mold, and proper specification of the thermoforming procedure and equipment.

This low-level orientation should not be confused with deliberate orientation of film and thin sheet to get greatly improved strength. This is high-level orientation, and is usually done both parallel and transverse to the direction of extrusion (biaxial orientation). It is discussed further on page 161.

Trim and Scrap. Reusing the thermoforming trim and off-size sheets is a necessary part of the economics of sheet extrusion. This reworked material may occasionally run as high as 75 per cent of the total sheet weight! A more common range is 15 to 35 per cent, and this figure is very much dependent on the design and size of the formed object and its mold. All thermoplastics can be reworked, but a few things must be remembered:

(1) Trim and scrap must be ground up and returned to an extruder, which may be in the same building as the thermoformer, or may be in some other company's plant. This handling can become expensive. Therefore, in a captive or proprietary operation, the reground material is almost always worked off internally. In custom work, where a sheet extruder sells his sheet to an individual thermoformer, that thermoformer may sell his trim (preferably already ground up) back to the original extruder, or else to a reprocessor or scrap dealer—in all these cases, the price will be well below that of the original material.

(2) The grinding and extra handling are potential sources of contamination. Money will be saved and quality improved if scrap sheet is kept off the floor, covered, and ground up and put into containers as soon as possible. Plastics develop large charges of static electricity which will attract dust and dirt if left exposed in the open. Also, metal contamination has occasionally been traced to the regrind, either due to careless handling or to a chip from a granulator blade.

(3) With some plastics, especially vinyls, the reground material may be less thermally stable than the original material. When one is reworking this material back into the same product, some extra stabilizer may be needed. In extreme cases, the trim cannot be properly extruded at all, and must be sold for less critical applications such as floor tile.

(4) Reground sheet may be slightly different in color. When a high percentage (25 per cent or more) is being added to new material, this color variation may show up in the final sheet.

(5) Reground high-density polyethylene is sometimes *lower* in melt index than the original material. This can be a help to thermoforming, as a stiffer melt gives less sag.

(6) In extrusion-blending operations, addition of some reground trim makes the blending a little easier, as the trim is already blended.

PROBLEMS AND SOLUTIONS

Some common sheet extrusion problems are noted below along with their suggested solutions:

Problem	*Solution*
Straight lines in extrusion direction	
(a) die lines	Inspect die for nicks, clean die lips.
(b) moisture	Pre-dry material or use vented extruder.
(c) scored roll	Resurface the roll.
(d) film laminate too cold	Raise temperature of laminating roll.
Straight lines in cross direction	
(a) sticking to roll	Lower roll temperature, or stock temperature, or blow air between sheet and roll.
(b) chatter in take-off or die	Repair take-off, improve die support.
Curved lines	
(a) poor mixing of melt	Use different screw, or higher pressure in extruder, or lower screw speed, or lower temperatures.
(b) big rolling bank in bite of rolls	Reduce pressure between rolls, make thickness more uniform leaving the die, speed up take-off.
(c) film laminate too hot (small curved lines near one edge)	Lower temperature of laminating roll; check roll surface for variation; clean inside of roll if necessary.
Spots or pock marks:	
(a) poor mixing	Use better screw, higher pressure, lower temperatures, lower screw speed.
(b) moisture	Pre-dry material or use vented extruder.
(c) cold rolls	Raise roll temperatures.
(d) contamination	Use finer screens, clean die, improve handling of material.
(e) trapped air	Use different screw, vented extruder, higher pressure, lower rear-barrel temperature, lower screw speed.
(f) dirt on rolls	Clean with damp or dry rag and nonabrasive kitchen cleanser.
Warping	Raise middle-roll temperature, raise or lower bottom-roll temperature, reduce or increase tension, eliminate drafts, put external heaters on edges.

Problem	*Solution*
Inability to hold constant thickness	Check extruder for constancy of output (lack of surging), check operator's skill in die adjustment, use heavier die, check for broken adjustment bolts, check uniformity of material, keep records of variation of temperature, pressure, and output to indicate possible causes.
Dull surface:	
(a) "roll plating" with high-density polyethylene	Lower compound temperature or lower roll temperature; clean roll with wet rag.
(b) poor gloss	Raise melt temperature at die exit, or use an auxiliary glazing technique.

STANDARDS AND TESTS

Most testing of sheet is practical, and is done by the end-user in connection with his particular end-use. Also, several countries have standards for extruded sheet, particularly high-impact polystyrene (American Commercial Standard TS–5652, British Standard 3290). These standards cover properties such as impact strength, orientation, and dimensional tolerances. Some of these properties are further discussed in the following paragraphs:

Impact Strength. This is commonly measured by the notched Izod test (ASTM D–256), and United States high-impact polystyrene materials are generally classified and priced by this test. When interpreting Izod values, it is important to observe the nature and size of the test specimens. Values obtained on thin (3 mm = 1/8 in.) injection-molded specimens can be deceptively high, because of the linear orientation induced by the molding procedure. Extruders should only compare values measured on unoriented, thick extruded sheet or compression-molded plaques. Despite the more traditional status of the Izod test, the falling-ball or falling-dart tests have become very popular (see Figure 4.10), and are considered more closely representative of actual service, especially for thick sheet. They have the advantage of being applicable to many finished formed products, as well as the sheet itself. People seem to be able to grasp the significance of this kind of test more easily than the Izod. Unfortunately, there is still no uniform standardization for falling-ball or falling-dart tests, although there are a few important systems such as that developed by General Electric.

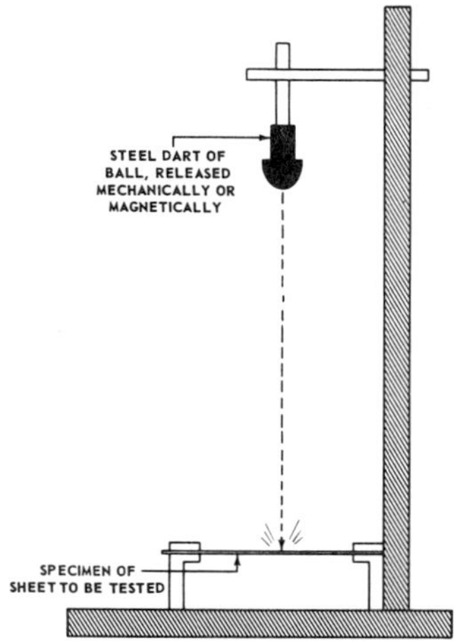

Figure 4.10. Falling-dart test, as used to measure impact resistance of plastic sheet. Typical height is 30 inches (75 cm) and typical weights are from 1 to 5 pounds (450 to 2300 grams). (*Illustration by author*)

Orientation. These tests usually apply to low-level orientation, but the principles are entirely relevant to high-level, biaxial orientation, too. In such tests, the sheet is brought to a high temperature, where the stresses introduced during extrusion and take-off can be relieved without completely melting the sample. The sample is allowed to shrink, and the dimensional changes are measured. The best temperature and time vary with the material and sample dimensions, but as a rule, polystyrenes and polyethylenes can be tested for 15 min and 150°C. An oil bath or an oven may be used.

The relationship of this test to the end-use should be realistically determined before setting the test specifications and limits. It will cost money to demand stress-free sheet where it is not needed, but it means lost sales to ignore this property where it affects product performance.

Tolerances. Tolerances that are realistic depend on the steadiness of take-off equipment and the ability of the extruder to produce melt at a uniform rate, and the thickness further depends on the die adjustments. The micrometer and tape measure are standard equipment for the extruder operator and the quality inspector. Some typical tolerances are given earlier in this chapter.

Appearance. Sheet users sometimes set standards for gloss, often measured with a Gardner Glossmeter. The requirement varies greatly with the material and end-use. A most important thing to remember, especially with high-impact polystyrene, is that the gloss should be measured *after* thermoforming, if the sheet is to be thermoformed. It is easy to make a sheet look good before forming, but the only gloss that counts is that remaining after forming.

Thin-sheet Toughness. This is important for high-impact polystyrene for containers. A number of empirical tests and testing machines have been developed, involving folding, twisting or bending the sheet a number of times until it cracks or otherwise fails. It is important to run such a test in both the extrusion direction and the transverse direction; otherwise, if the sheet contains some orientation, deceptively low or high values will result.

Heat Stability. This is usually measured by ASTM D–648, which heats up a number of small specimens under flexural load until they bend and fail. Polystyrenes are run at about 18 atm, and polyethylenes at about 5 atmospheres load. In Great Britain, the Vicat softening-point test is used, based on the pressure of a special tool into the sheet, which is heated until it gets soft enough to allow indentation. Some special applications have their own tests; e.g., sheet for coffee cups is formed into a test cup, filled with boiling water while supported only at the rim, allowed to cool, and visually compared with standards.

Creep, Fatigue, and Other Mechanical Tests. These are often run where the sheet will have to support a load for a long time, such as in a refrigerator door liner or in building applications. Specific tests have been developed (see Figure 3.16, page 80), and are described in the technical literature. Creep and fatigue testing is tedious and costly, and many end-users compromise by selecting only the tests with the most significance to them. An example is the door-slam test run by many refrigerator manufacturers, in which a finished door is assembled and slammed repeatedly in a special machine. At intervals, the plastic liner is inspected for cracks.

Taste, Odor, and Toxicity. Much sheet is thermoformed into containers for food or into items that come in contact with food. Such objects must not have an unpleasant odor, nor transfer any taste to the food. The chief "offenders" here are styrene monomer (for styrene-based plastics) and oxidized polyethylene (for polyethylenes). These can be detected by panel tests in which either the sheet or the formed item is enclosed with some food such as sweet butter in an air-tight jar; the butter is tasted the next day by the members of a panel. Alternatively, the plastic is placed by itself in the jar, and the jar is later opened and smelled by the panel members. For applications where food is actually packaged in plastic, most countries have government bureaus which set regulations and approve all ingredients in the formula.*

Ease of Thermoforming. Sheet materials can be tested for their ease in achieving deep draws. Small thermoforming machines are used, and a few have been specifically developed and designed for this purpose. These special machines not only test depth of draw, but can be used to find the best forming times and temperatures for various materials.

SPECIAL TECHNIQUES

Continuous Extrusion/Thermoforming. Machines are sold that can take a hot sheet from the extruder and thermoform it right in line. Adaptations of linear thermoforming machines are most common, and are often used for forming disposable cups and containers from polystyrene. A rotary-drum machine is simpler than the linear type, but it is not widely available, it involves multiple molds, and it is restricted to the manufacture of rather small, uncomplicated items. The continuous operation from pellets to formed products has the advantages of less labor, faster forming time, and less need for storage space—but it sacrifices versatility and independence in scheduling and maintaining the machine components. Because of this, many container manufacturers (probably a majority) prefer not to run a completely continuous operation, but use separate extrusion and forming lines, set near one another.

* This is particularly important with PVC, which is expected to have a bright future in the United States for food packaging, if the formulators can meet government (FDA) rules and still sell at competitive prices. To date, high sheet prices and the many middlemen between extruder and food packer have kept PVC sheet usage low for food packaging. This is especially true in the United States, where costs of material, equipment, operation, and development cannot themselves justify the high sheet prices—and the only explanations left are either processing difficulties and inefficiencies of a major nature, or quick-profit pricing. It is still more acceptable to management that a product does not sell because of high price, compared to making and selling the same product at a low profit (even if this low-profit stage may be temporary).

Continuous Thickness Measurement. Because extruded sheet moves slowly, there is a time lag between adjustment at the die and measurement of the finished, cooled sheet. This makes adjustment very time-consuming, as the operator must wait several minutes sometimes before he can be sure of the effect of each adjustment that he makes. Furthermore, if the sheet is being rolled on a core, it cannot be cut to check thickness without spoiling the roll at some point. Therefore, many machines have been developed to make a continuous, non-destructive thickness measurement of the moving sheet.

Nuclear-energy gauges are the best known of these devices; they work by placing a radiation source on one side of the sheet and measuring the amount of radiation that passes through, which will be a function of the thickness. These nuclear gauges are very accurate and very expensive, but their manufacturers claim that they will quickly pay for themselves in terms of better tolerances and lower aim thicknesses, as well as less scrap and rejected sheet. Actually, these gauges are only as good as the operators who must read them, interpret their readings, and adjust the die bolts. The gauge measurements are far more accurate than any operator could hope to follow with his die adjustments. Also, many extruders are paid by the pound for their products; in such a case, closer tolerances will improve product quality and may give a corresponding sales advantage, but no actual material cost will be saved. Nuclear gauges will really be able to pay for themselves when they are connected to a control mechanism linked to the die bolts, or to little cartridge heaters in the manifold, thus automatically regulating the thickness at all points across the sheet. Already, some units are *controlling* variation, as well as measuring and recording it, in the machine (extrusion) direction. One system turns the bolts by itself, but only when the operator makes it do so. This is not true automatic control at all, but it is only one step away.

Other less expensive devices are also sold to do this continuous measurement. One of the newest is a magnetic device in which two magnets "float" on a controlled cushion of air, one on either side of the sheet; as the thickness changes, the magnets move closer together or further apart, and this is reflected in their magnetic field, which is measured and recorded. Mechanical gauges are even cheaper, but less accurate. They must be built very rigidly and carefully in order to avoid any variation of the reading due to the flexing of any of the components of the device. Optical gauges are also available, in which increasing thickness blocks a light beam aimed at a photoelectric cell, and cell and beam adjust to positions that are dependent on the thickness.

Extra-thick Sheet. Anything much over 3½ mm (0.14 in.) is unusual. Thicknesses up to twice this value can be made on most conventional equipment, provided that the take-off will operate at low enough speed with adequate horsepower. This problem is compounded at low speeds because the sheet is thicker and requires more force to give enough pulling tension. The die must be constructed to fill out completely and uniformly, even with the large lip openings, which tend to reduce the pressure inside the die. Large, high-output machines are helpful, of course, as they allow a higher linear speed.

For sheet over 7 mm (0.28 in.), special problems begin to appear. The most famous application of this kind is the 25-mm (1-in.) polyethylene sheet used as a radiation barrier in nuclear-powered ships. Special take-offs have been built for this extrusion which cool the sheet very slowly to avoid internal shrinkage voids. In addition, a huge die and extruder are required to get complete fill-out of the die at a reasonable linear rate. This sheet is made from regular and boron-containing polyethylene according to United States Military Specification MIL-P-19336C(Ships).

Static Elimination. Static charges will accumulate as the sheet passes through the take-off—especially if it is rolled—and these charges attract dust. They can be minimized by contacting grounded metal wires or chains over the sheet surface as it is rolled or stacked. There are also bars that ionize the air in the immediate vicinity of the surface, thus preventing buildup of a charge. Another popular technique is to run the sheet through a bath of detergent solution during take-off; this forms a static-dissipating film. Alternatively, the solution can be sprayed on the finished sheet or formed product. The bath is used for disposable containers and the spray for appliance parts. Finally, some extrusion compounds are made with an antistatic agent built right into the formulation.

High Slip. It is important that nested containers do not stick too firmly together. This depends on the coefficient of friction of the surface, which in turn depends on the material formulation. Take-off conditions have an effect also, but this has yet to be definitively studied and reported.

Printing. Much sheet is printed—either for direct use, or else deliberately distorted for subsequent thermoforming; during the thermoforming the distorted pattern is stretched into the proper form. Polystyrenes and vinyls are easy to print; polyolefins are harder, and require a surface oxidation treatment which can be given in line with the extruder. The formulation has much to do with the ease of printing of the sheet, as certain additives such as fatty lubricants inhibit ink adhesion. Extrusion conditions have effects too; very high temperatures produce oxidized surfaces, which are more receptive to most inks. A frequent requirement is the reduction of sheet stresses (low-level orientation) to an absolute minimum. This allows good registration of the printed area when it is thermoformed. The stresses are minimized by having little or no drawdown at the die (die opening set at approximate final sheet thickness, and die very close to first roll), avoiding any squeezing between pairs of cooling rolls, and operating with as little tension as possible between cooling rolls and pull rolls. An ingenious compromise, often used for thermoformed lids, involves preprinting a sheet with no distortion—but only printing in those areas which remain flat during thermoforming.

Laminating. Another method of putting a special surface on sheet is to laminate a film or thin sheet of the same material to the hot extruding sheet, right on the take-off roll, as illustrated in Figure 4.9. The film is preheated by passing it around one roll before pressing it into the hot plastic, and no adhesive is needed. The preheating temperature is most critical for biaxially oriented films such as polystyrene, because such films will shrink and pucker if too hot, and fail to adhere if too cold. The combination of polystyrene film with a high-impact polystyrene base is the most common laminate. The film is oriented to permit handling and to permit printing without breaking. However, the laminated surface does not have the original strength of the oriented film; this is lost as the film is first preheated and then directly heated by contact with the extrudate. Unoriented polystyrene film has been used as a laminate, but it is not easy to print, and must be handled very carefully because of its extreme brittleness. When printed film is used, the printed side is usually placed next to the hot sheet; thus, the print is kept below the final surface, protected with the thickness of the film. Another important application of this process is the lamination of colored, embossed, or UV-absorbing films or thin sheets to a less expensive base (for example, ¼-mm = 0.010 in. surface on 2-mm = 0.080 in. base).

This process has been expanded to include lamination of hard-surfaced printed polyester or flexible and rigid PVC films to high-impact polystyrene. The film is precoated with an adhesive, rolled up, and then unrolled during lamination at which time the heat of the extruding sheet cures the adhesive. Finally, thermoplastic fabrics like nylon and some acrylics can be laminated to thermoplastic sheet in conventional equipment with no adhesive; the porous, fibrous fabric forms a mechanical bond with the hot, soft sheet. High extrusion temperatures are recommended for this application, to get better bonding. A solvent-type adhesive is sometimes used for even better adhesion. All of the preceding combinations provide more ways to utilize the inexpensive rigidity and easy processability of high-impact polystyrene.

Embossing. If the middle take-off roll is patterned with the reverse of the desired surface, the sheet will take that surface during extrusion. Some roll pressure must be maintained in the bite, and the embossing-roll temperature must be kept as high as possible short of sticking. Polyethylene and ABS are the most frequently embossed sheet plastics. The same effects can be achieved by patterning the thermoforming molds. This latter method has not been too popular in the past; with normal molds it is not good for deep embossing as the sheet is unable to fill into the deepest cracks. However, porous metal molds allow any texture to be reproduced faithfully on the mold. Other considerations are: (1) pre-embossed sheet tends to "wash out" in areas of deep draw, and its pattern distorts in any draw at all; (2) if a patterned mold is female, pre-embossed sheet would need a male mold to make the same product, and vice versa; and (3) pre-embossed sheet can be used for many products, while a patterned mold is limited to that one product.

Tubular Extrusion of Sheet. For flexible materials such as polyethylene, thin sheet (0.1 to 0.5 mm = 0.004 to 0.020 in.) can be made by ordinary vertical tubular extrusion such as that used for film and described in detail in Chapter 5. The more rigid plastics are harder to flatten and collapse as a tube. Rigid sheet of greater thickness is sometimes made by horizontal extrusion of a heavy-walled tube (as in pipe extrusion); the tube is slit near the die and carefully opened up and flattened out as it is conveyed through the take-off. This method is used in a few plants to make rigid PVC sheet for subsequent corrugation, and has the advantage of a streamlined, uniform, rather compact die. It is also used on a smaller scale (tube diameter 35 mm = 1.4 in.; wall thickness ¾ mm = 0.03 in.) in a strip-fed continuous thermoformer for containers.

The most common use of tubular extrusion for sheet is for expanded (foamed) polystyrene (Figure 4.11), where the increase in tube diameter smooths out the wrinkles that would ordinarily form at the die from the fast volumetric expansion there. Extruded polystyrene foam is sold for packaging and insulation. It has a fine, barely visible cell structure, is easily printable, and is quite flexible and tough. It has been made in conventional tubular or cast film machinery by direct extrusion of polystyrene beads that contain a hydrocarbon

Figure 4.11. Tubular extrusion of foamed polystyrene sheet. Under about ½ mm (20 mils), sheet can be blown directly, as in extrusion of tubular polyethylene film. Above this thickness, it is usually produced by drawing the extruded tube over a cooling/sizing mandrel. (*Courtesy of Chemische Werke Hüls AG*)

(usually pentane) plus a small amount of a nucleating agent. The beads are similar to those for steam-molding. Some foaming agent is lost to the atmosphere upon extrusion, and care must be taken to ventilate the area to avoid explosion or vapor inhalation. More pentane diffuses out of the cells upon standing, so this care must also be extended to storage areas. The final film or sheet is quite odor free.

Lately, several processes have been developed which allow the use of unmodified, clear polystyrene pellets as the feed to the extruder. The foaming agent (hydrocarbon or fluorohydrocarbon) is injected into the barrel during extrusion, while any other additives are mixed with the polystyrene just before extrusion. The economic situation greatly favors this process, and most foamed polystyrene sheet is already being made this way. One system uses a very long vented extruder; the foaming agent is injected into the vent, and the remainder of the barrel is used to mix and cool the product. Another system in commercial use has two extruders, one feeding the other. The first one is a vented machine in which foaming agent is injected and mixed, and the second is a larger, slower-turning machine just for cooling. Still another system employs a twin-screw machine. The extrudate must come out quite cool (120–140°C) because the melt viscosity is too low at higher temperatures, and because its good insulating properties make further cooling difficult. The take-off is still basically similar to polyethylene, but a horizontal bubble is usually used, the flattener is lined with fluorocarbon or simulated velvet to prevent sticking, the cooling section is very long, and winding the light, bulky sheet requires some modifications to standard machinery. At least one producer solves these problems by thermoforming right at the extruder.

Post-forming in Line. Flat sheet can be bent into a variety of continuous shapes while still warm. Polystyrene, for example, can be extruded flat and then bent into a U-shape for lighting fixtures. Metal jigs, sometimes coated with fluorocarbon plastics, are used to form the sheet (see Figure 4.12). A typical postforming application is the corrugated heavy PVC paneling that competes with polyester, aluminum, and other materials in construction markets. A postforming die for longitudinal corrugations is shown in Figure 4.13. Both longitudinal and transverse corrugation are done, but each requires different forming equipment.

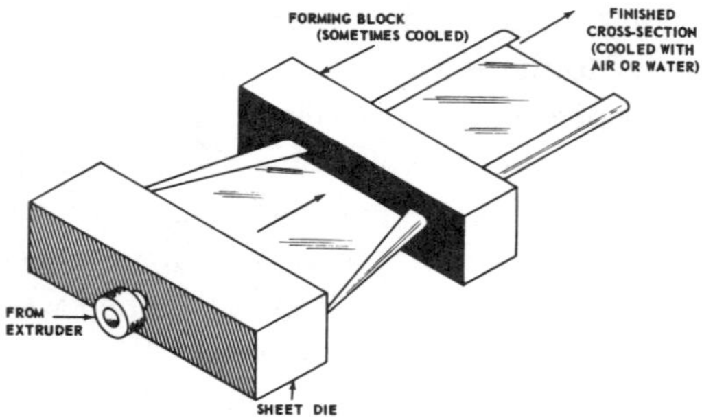

Figure 4.12. Postforming. Flat sheet can be postformed in specially built dies to make a variety of different shapes. (*Illustration by author*)

Figure 4.13. Forming apparatus for the manufacture of corrugated panels from PVC or other rigid thermoplastic. There are several systems for making such products. (*Courtesy of U.S. Anger, Inc.*)

Welding and Fabrication. Construction from flat sheet is still a major method of making complex structures, especially when only a few are needed. All thermoplastic sheet can be welded with rod of the same material; furthermore, sheet can be heated and bent over wooden, plastic or metal forms. Fabricated acrylic and cellulose acetate/butyrate sheet finds much use in both outdoor and indoor signs and displays, and polyethylene and PVC are also commonly fabricated. Many books and articles are available which describe the various fabrication operations in detail.

Special Designs. Plastic sheet is used in a vast number of specialized structures. Since it can be thermoformed into three-dimensional shapes, design possibilities are immense, and have hardly been tapped. Some of the possibilities include: double-wall sandwich construction with or without foam inside; resilient packaging material made by forming flexible sheet over hundreds of little bumps; various patterns, grids, and corrugations to increase rigidity of a sheet; honeycomb structures; and combination foam-sheet laminates.

Continuous Forming of Hollow Articles from Sheet. Hollow containers and ducts are being made by extruding two sheets, or by folding over a single sheet, and clamping the resulting double layer between two female molds. A heat seal is made around the clamping line and a vacuum is drawn through the molds causing the sheets to separate, fill the molds, and form a hollow article. This method is used for making small bottles from PVC, and is applied on a larger scale to make complex air ducts for automobiles from high-density polyethylene. Hollow articles can also be made by conventional thermoforming techniques and machinery, using split or collapsible molds, and at least three such mold-design methods are known. The products are used mainly for dairy packaging.

REFERENCES

Sheet Extrusion Technology for Specific Plastics

Polystyrene: "Influence of extrusion conditions on high-impact polystyrene sheet," M. G. Blumental *et al.* (U.S.S.R.), *Plasticheskie Massy*, May 1966.

PVC: "Extrusion of rigid PVC sheet," D. J. Dowrick (ICI), *Plastics*, October 1965. Discusses equipment, conditions, formulations with suggestions.

ABS: "ABS sheet extrusion," E. J. McDonald, Jr. (U.S. Rubber), *Modern Plastics*, April 1965, adapted from *SPE Regional Conference*, April 1964. Gives screw designs, roll arrangement, die design; discusses instrumentation, drying needs. Similar article by J. R. Kent *et al.* (Monsanto), *Plastics Technology*, November 1964.

Polyolefins: "Production of sheet from polyolefins," H. Oertel, *Der Plastverarbeiter*, February 1966. Problems, conditions, properties. Also "Extrusion and thermoforming of polyethylene sheet," A. L. Griff *et al.* (Union

Carbide), *SPE ANTEC* 1958. Describes equipment and operating techniques for all densities.

Acrylics: Review of MPM's vertical extrusion system for acrylic sheet, in *International Plastics Engineering*, May 1965, and *Modern Plastics*, April 1964. Also, general discussion of extruded acrylics by P. Musso (Montecatini), in *Materie Plastiche*, April 1961.

General Descriptions of Sheeting Lines

"The modern sheeting plant," J. D. Keil (Kleinewefers), *International Plastics Engineering*, September and October 1961. Illustrated description of German equipment, handling all materials. Includes corrugation.

"Description of Norge plant for refrigerator door liners," *Plastics World*, May 1959. Hotpoint plant for same product: *Modern Plastics*, March 1959.

General Operating Technology

"Automatic regulation of sheet thickness," K. Herlitze (Frieseke und Hoepfner), *Kunststoffe*, September 1964. Discussion of various types of nuclear-energy measurement.

"Development of gloss in high-impact polystyrene," F. J. Pokigo and M. J. Roman (Monsanto), *Modern Plastics*, March 1962. Study of separate and on-the-roll glazing with infrared heaters, giving conditions, properties and photomicrographs.

"Techniques for securing gloss on extruded high-impact polystyrene sheet," R. D. Carhart *et al.* (General Electric), *SPE ANTEC* 1962. Six methods tested, infrared glazers concluded as best; discusses mechanisms of gloss production and retention.

"The effect of some extrusion variables on surface finish," S. J. Skinner *et al.* (Monsanto Ltd.), *Plastics*, August 1961. Relation of shear rates to surface smoothness and gloss, effects of temperature, output rate, die opening dimensions, all discussed.

"Sheet extrusion problems and solutions," A. L. Griff (Union Carbide, now a consultant), *Plastics Industry*, September, 1959; *SPE ANTEC* 1959; *Kunststoff-Berater*, April 1961. An illustrated practical guide to solving problems encountered in daily operation, especially with high-impact polystyrene.

Sheeting Dies

"Investigation of extruder heads with wide sheet dies," F. Tomis (Czechoslovakia), *Plasticke Hmoty a Kaučuk* (Czech), January 1967. Study of pressure relationships using models.

"Tapered coathanger manifold extrusion dies," H. N. Stone (Goulding, now division of Crompton and Knowles), *SPE Journal*, July 1966, and *ANTEC* 1966. Basic non-mathematical description and justification of such dies.

"Formgebung eines Plattenspritzkopf nach rheologischen Gesichtspunkten" (Design of sheeting die using rheological principles), F. Röthemeyer, *Kunststoffe,* August 1966.

"Heat transfer in extrusion," W. M. Underwood (Monsanto), *SPE Journal,* February 1963, and *SPE Regional Conference,* 1962. Mathematical study, using sheet and film dies as examples.

Testing of Sheet

"Shrinkage of high-impact polystyrene sheet as a function of extrusion conditions," M. G. Blumental (U.S.S.R.), *Plasticheskie Massy,* August 1966.

"Prediction of ABS extruded sheet surface quality," E. P. Weaver (U.S. Rubber), *SPE ANTEC* 1965. Use of Brabender, Instron and Mooney tests to predict surface, gloss, die lines, ease of flow.

"Extrusion and evaluation of thin-gauge high-impact polystyrene sheet," F. J. Pokigo and M. J. Roman (Monsanto), *SPE ANTEC* 1962. Sheet of 20 mils (½ mm) tested; opening not critical, lower melt temperature and higher orientation improve strength.

"Evaluation of high-impact polystyrene for refrigerator door liners," D. A. Davis *et al.* (General Electric), *SPE Journal,* March 1961, and *ANTEC* 1961. Evaluation procedure and requirements to be met by General Electric's suppliers of raw materials. An extremely thorough program, well described, requiring extensive laboratory.

"Requirements, properties, and testing of extruded high impact polystyrene sheets," R. G. Hochschild (Koppers), *SPE ANTEC* 1960. Tells which tests to run and gives effects of extrusion/forming conditions and specimen preparation upon test results.

Special Techniques

"Enhancing the properties of ABS through lamination of non-ABS films," S. M. McDougle (Marbon), *SPE ANTEC* 1966. Technique, adhesives, properties for combining ABS with films of polycarbonate, oriented and unoriented polyester, polyurethane, and polyvinyl fluoride. Films are applied during extrusion on conventional sheet extrusion equipment.

"Laminated plastics—their applications and technology," V. L. Gliniecki (Dow), *SPE ANTEC* 1965. Refrigerator parts and outdoor fixtures made with PVC and fluoroplastic film laminates over high-impact polystyrene, using adhesive. Technique similar to preceding reference.

"Polystyrene Foam Sheet," series in *Plastics Technology,* August 1966: "Direct-injection extrusion," by L. L. Scheiner (*Plastics Technology* staff). In September 1966: "Extrusion of expandable beads," by T. Martens *et al.* (Koppers). Scheiner's article is useful, but incomplete in this fast-moving, secretive field. Koppers' article explains conditions and properties; bead extrusion is limited by adverse economics.

"Die Verarbeitung von Styropor auf Extrudern" (The processing of Styropor in extruders), W. Schulte (BASF), *Industrie-Anzeiger,* December 1965 (special publication of BASF).

"Polystyrene blown-tubing foam extrusion technology," P. J. Croft (Shell), *Plastics,* October 1964. Good information, some patent references, explains conditions, equipment needs and problems.

"Barand extrusion plant for thin expanded polystyrene sheet," *International Plastics Engineering,* October 1964. Brief review of equipment, which uses expandable beads as starting material.

"Equipment needed for the extrusion of expandable polystyrene beads," V. L. Gliniecki (Dow), *Plastics Design and Processing,* September 1964. Discusses foam sheet, mandrel versus free blow, slitting, rolling, etc. Similar information presented by V. L. Gliniecki at *SPI Foam Conference,* New York, April 24, 1963. Earlier Dow publication in this area: "Controlled-density polystyrene foam extrusion," by F. H. Collins, *SPE ANTEC* 1960 and *SPE Journal,* July 1960. Covers technology, equipment, products.

"Extrusion, fabrication and application of thin expanded polystyrene sheet," D. N. Davies and D. J. A. Lee (BX Plastics), *British Plastics,* August 1963. Properties, comparisons, economics, fabrication, methods and applications.

"Corrugating rigid PVC sheet," G. A. Nisbet (Goodrich) and W. G. Potts (NRM), *Modern Plastics,* July 1961. Description of NRM corrugating equipment, making corrugations parallel to extrusion direction.

"Plastics in roofing; extrusion of corrugated PVC," *British Plastics,* October 1960. Describes market, and mentions acrylic and polyester products. Stresses PVC and describes Reifenhäuser and LMP equipment and processes.

"Extrudieren von Fussbodenbelag" (Extrusion of floor covering), *Kunststoff-Berater,* January 1967.

"The manufacture of floor coverings by extrusion," Dr. A. Romanowski (Müller), *Kunststoffe,* February 1966. Describes a complete Müller production line, including two extruders and annealer oven.

FIVE
film

PRINCIPLE

Most film is extruded by forcing the molten plastic upward through a thin annular opening. A tube is thus made and is expanded further by internal air pressure (an internal mandrel is sometimes used). The tube of film is cooled by blown air and drawn through a pair of rolls to a winding station. For thick film, auxiliary cooling devices may be used. Some film is also made by extruding through a slot opening downward over a chilled roll (casting roll). Both tubular and casting methods are illustrated in Figures 5.1 to 5.3.

MATERIALS

Polyethylene is by far the most popular extruded film, consuming well over a half-billion tons in 1966, in the United States alone. Almost all of this is low- and medium-density high-pressure polymer. Polypropylene film, both unoriented and oriented, is slowly increasing in use. Films are also extruded from polystyrene (always oriented), nylon, and PVC, with the latter rapidly growing in importance.

PRODUCT SIZES

Packaging is the biggest market, where the films are from 0.0007 to 0.002 in. (18 to 50 μ) in thickness. Some films as thin as 0.0003

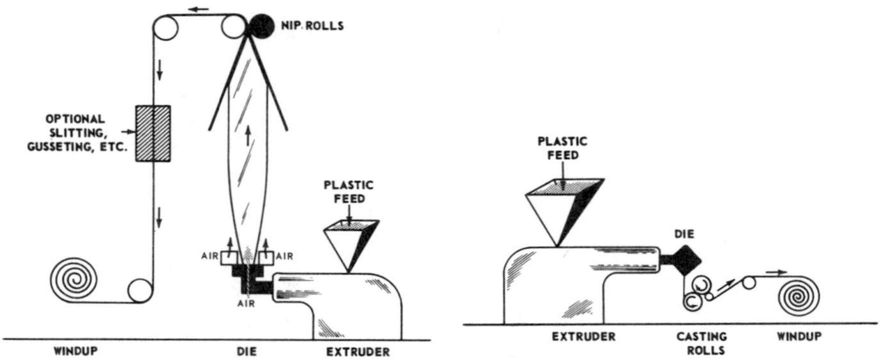

Figure 5.1. Extrusion of plastic film. At left, illustration of tubular extrusion, which is by far the most common. A thin-walled tube is extruded upward and is cooled with air. (Downward or horizontal extrusion is occasionally used.) Air pressure inside the tube expands it as it is drawn up into a flattening frame, through a pair of "nip rolls," and into a winder. At right, the film-casting process is illustrated, using cooling rolls. The film is extruded downward onto the first roll, where it is quickly cooled. Subsequent rolls provide further cooling and direct the film to a winder. (*Illustration by author*)

in. ($7\frac{1}{2}$ μ) are used in garment bags; such thin films (and even thinner) are possible for certain polyethylenes, but most other plastics cannot be drawn that thin either because of tearing or excessive blocking (clinging or sticking together of layers). Equipment commonly used for film can often produce products up to 0.020 in. (0.5 mm) or more; while some sheeting lines (see Chapter 4) are capable of drawing plastic into the film range.* Thus, an overlap occurs in which film producers can supply "sheet" (usually polyethylene) and sheet makers supply "film" (usually polystyrene and cellulosic).

Extruded film is sold in rolls as a seamless tube, or as a flat single layer, or folded in various ways. The tube size is often specified as the "layflat" size, which is the width of the finished roll—half the circumference of the expanded tube. Common layflat sizes are 12 to 40 in. (30 to 100 cm), but polyethylene layflats of 100 in. (250 cm) are made for agricultural and construction uses, and are slit and opened to give 200-in. (508 cm) widths. Even wider films are made

* The division between film and sheet is not clear. With vinyls, the U.S. Tariff Commission calls "film" anything 0.006 in. (0.15 mm) or less. The polystyrene industry usually reserves "film" for products 0.003 in. (75 μ) or less. With polyolefins, "film" ranges up to about 0.007 to 0.008 in. (0.17 to 0.20 mm). However, flat single-layer polyolefin film of any thickness is also often referred to as "sheeting," to distinguish it from tubular film.

Figure 5.2. Large tubular-film extrusion line, showing multiple air rings, guiding rods and windup on an upper floor. (*Courtesy of Barmer Maschinenfabrik AG*)

by a few producers, sometimes up to 480 in. (12 meters) in final (slit and opened) width. Very small tubes can be easily made if needed, but this is not always an economical operation and is not very common.

Flat film, called "cast" film, is usually made in widths of 48 to 72 in., and is then slit as required for specific orders.

EQUIPMENT SIZES AND OUTPUTS

The sizes of film extrusion equipment vary widely. For high-cost materials such as nylon, or where there is a captive or other special

Figure 5.3. Small, "packaged" film extrusion line, designed for users who want to make their own film. The extruder is inside the cabinet, and a narrow tube can be seen emerging from the die (lower left), being cooled by the black air ring, and passing up through the collapsing slats to a winding station. (*Courtesy of Sterling Extruder Corp.*)

use, extruders of 15 to 25 hp (2- to $2\frac{1}{2}$-inch = 45- to 60-mm barrel) are common. For the bulk of the polyethylene film production, 25 to 100 hp ($2\frac{1}{2}$- to $4\frac{1}{2}$-inch = 60- to 120-mm barrel) are used. However, the big markets for polyethylene demand very high output for greatest economy; 150 to 200 hp is now commercial (6-in. = 150-mm barrel), and there is serious discussion of even bigger lines, particularly for wide, heavy films.

Machine size is closely related to product dimensions; if a narrow, thin film is desired it is usually slit from a wider roll made on a large machine. But if for some reason the narrow film must be made by itself, then a small machine is preferred—with a big line, the linear exit speed at the die would be too great, and the film could not be evenly and/or sufficiently cooled. The tubular-film industry often refers to output per unit of die circumference: with good standard equipment, one can expect 7 lb/hr of polyethylene per inch of circumference (1.25 kg/hr/cm). With special accessories such as refrigerated cooling air or a metal cooled mandrel, over 10 lb/hr/in. (1.8 kg/hr/cm) can be achieved, and rates as high as 20 lb/hr/cm

(3.6 kg/hr/cm) are reportedly possible for heavy film with much auxiliary cooling. In all these cases, when the limiting rate is exceeded, it becomes difficult to maintain consistent thickness, and/or to get enough cooling.

When die circumference and total output are thus determined, then extruder horsepower, barrel size, and cooling-tower height follow. For example, if a 40-in. (102-cm) lay-flat is needed, a die of 12 in. (30 cm) in diameter might be chosen. The circumference is about 38 inches (97 cm). At 7 lb/hr/in., this allows a maximum output of 266 lb/hr (120 kg/hr). This would require 40 or 50 hp; a 2½-in. (60-mm) barrel might do, but a 3½-in. (90-mm) barrel is better as it will allow lower extrudate temperatures and thus require less cooling.

Tower height depends on film thickness as well as output. A common size is 35 ft (10.6 meters) for the example in the preceding paragraph. Most towers are between 20 and 60 ft high (6.1 and 18.3 meters), and some 100-ft towers (30.4 meters) have even been built. The required tower height can be reduced by using auxiliary cooling devices (external sleeves, mandrels, refrigerated air).

Good tubular extruder outputs of low-density polyethylene in the United States today, with properly matched dies are:

Machine Horsepower [a]	*Barrel Diameter, in. (mm)*	*Output, lb/hr (kg/hr)* [b]
15–20	2 (40)	50–75 (23–34)
25–40	2½ (60)	100–175 (45–80)
50–75	3½ (90)	250–350 (114–160)
100–125	4½ (120)	400–500 (181–227)
150–200	6 (150)	600–750 (271–340)

[a] Higher horsepowers needed to get lower output figures with very viscous materials, or where much mixing is needed.
[b] Higher output figures possible with higher horsepowers and normal polyethylenes.

Cast-film lines are generally large, with 50 to 150 hp motors driving extruders with 3½- to 6-in. (90- to 150-mm) barrels.

A commercial custom film extruder must serve a great variety of markets; therefore, he usually has a number of different sizes of machines and dies, and part of his efficiency will depend on his skill in scheduling the right jobs for the right machines, utilizing his capacity to greatest advantage.

TUBULAR VS. CAST FILM

In typical tubular extrusion, the melt is extruded upward through a thin annular opening. Air introduced through the center of the die inflates the tube and causes it to expand. A moving bubble is thus formed (Figure 5.4) which is held at a constant size by control of the internal air pressure. The film is flattened as it passes through two converging planes into a pair of pull rolls, and thence to a winder.

Cast film is extruded through a linear "slot" die, usually downward. In the most common cooling method, the flat web strikes a large and moving polished metal roll. It quickly cools, and peels off this roll, passes over one or more auxiliary cooling rolls, then through pull rolls, and finally to a winder. Only about 15 per cent of polyolefin film in the United States is made this way; the remainder is made by the tubular process. With PVC, tubular extrusion is also the leader (especially with plasticized PVC), and its share of the total is about the same as with polyolefins.

Tubular film is generally tougher than cast film, as the tubular

Figure 5.4. Overhead view of a tubular-film die in operation. The expanding bubble is clearly visible. The black hose at the right is feeding cooling air to the air ring surrounding the bubble. (*Courtesy of Union Carbide*)

extrusion process is really a mild form of biaxial orientation. Tubular polyethylene film is stiffer (if starting materials are equal in density), unless a higher-density (and more expensive) starting material is used to compensate, which is the current practice. Tubular extrusion lines are often smaller and cheaper, although advances in cast-film machinery have made it possible to install a large high-speed cast-film line with investment per unit output almost as low as that for a tubular line. A high-output tubular line requires a high tower, and a building big enough to enclose it, while a casting line requires no unusual height. A tubular line produces a product that can be more easily made into bags, edge-trimming can frequently be avoided, and the film width is more easily changed simply by blowing a bigger tube. Tubular lines are generally easier to start up and operate, and since for polyolefins they run at lower temperature the film is less liable to contamination due to oxidation in the system.

Balanced against this array of advantages for tubular film is the most important reason for casting polyolefins: better gloss and clarity. The film is more quickly cooled and the crystalline areas are smaller, yielding better optical properties. (Some of the best new polyolefins do yield *tubular* film which approaches cast film in optical properties, if run under optimum conditions.) Also, thickness is more easily regulated on a cast-film line, and nuclear-energy thickness gauges are more easily adapted.

Cast film is used in packaging (e.g., for bread wrap and overwrapping), where sparkle and clarity are all-important in helping to sell the packaged product. Cast film is currently limited to such markets, and its use is growing more slowly than tubular film. It is often used for nylon, where a tube would be hard to handle because of the rapid viscosity change with temperature. Some rigid-PVC dies were built for casting,* to get better thickness control and avoid decomposition (e.g., with screws in die manifold). Good flat-film PVC dies will continue to be used, but the current trend is toward tubular equipment, even for rigid PVC.

Before the development of adequate chill-roll systems, some clear flat film was made by extruding downward into a water bath. The rapid quenching produces good optical properties, and equipment cost is low. However, it is difficult to get precise control over the water

* The word "casting" here refers to extrusion-casting of a melt on a chill roll. PVC film is also made by solvent-casting, and this product is called "cast film" in the industry. The extrusion-cast PVC, as well as tubular PVC, is simply called "extruded film."

temperature; vibrations and currents can cause little marks on the film; and the water must be kept from carrying over into the finished roll. Liquid-bath systems have been used for tubular film too, notably saran, and this principle has been revived in one current technique for extruding heavy film for sacks at very high output rates.

TUBULAR FILM EQUIPMENT

Dies. Dies for tubular extrusion are shown in Figures 5.5 through 5.8. They are either center-fed or side-fed. Center-fed dies are better, since all points on the lip are equidistant from the feed-entry point; this is desirable for uniform flow, which yields uniform thickness.

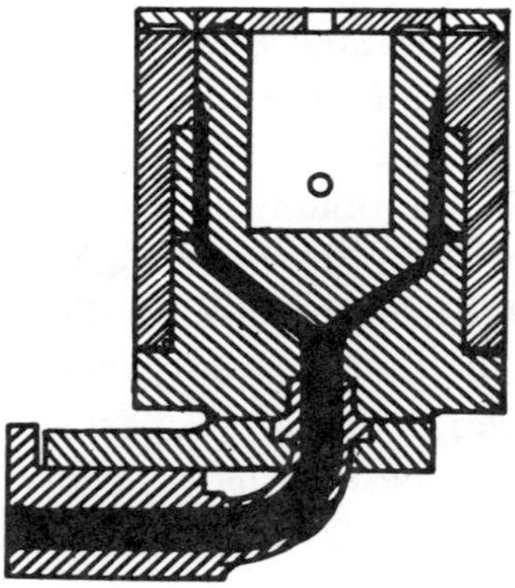

Figure 5.5. Cross-section of a die for tubular film. Plastic enters at lower left from extruder and emerges as a tube at the top.

Figure 5.6. Dies for tubular extrusion. *Left:* Core portion only of unique die in which the melt emerges through two holes 180° apart (one hole visible, arrow) and then passes around in a spiral path as it flows up and out. The last portion of the core—the die land—is smooth. *Right:* A photograph of a simple side-fed die. (*Die at left by F. W. Egan, at right J. & W. Müller; photographs by author*)

Figure 5.7. Large, 60-inch (1500-mm) die for construction and agricultural film, with central entry and branched flow paths to circumference. (*Courtesy of Faeco Division, Koehring Co.*)

Figure 5.8. Rotating 24-inch (60-cm) film die with central entry. The whole die rotates about the adapter neck. (*Courtesy of Sterling Extruder Corp.*)

Center-fed dies are also preferred if the die is to be rotated or oscillated (further described below). Side-fed dies are sometimes less expensive, and were particularly useful for large diameters (12 in. = 30 cm and over), where the internal upward force on the large area of a die like that shown in Figure 5.5 was too great for the supporting "spider" arms holding the core to the outside of the die. Lately, standard-type center-fed dies have been successfully built as wide as 30 in. (75 cm) in diameter, and even wider center-fed dies are possible with special designs (see Figure 5.7). These efforts to avoid side-fed dies were made partly because a side-fed die can only produce a uniform tube over a limited range of melt viscosities; adjustments are possible, but always at some sacrifice (e.g., eccentric core = unequal stresses; raised temperature = more cooling needed).

In center-fed dies, the spider arms always divide the flow into separate paths which must come together and weld completely before leaving the die. If they do not, "weld-lines" are produced, which are lines of weakness. For low-viscosity melts and/or large die gaps, the die resistance is low, and one or more annular restrictions (blisters) may be included (see Figure 5.5). In designing these—and, in fact, in designing the entire die—care must be taken to keep all flow paths as streamlined as possible. Any stagnation will eventually produce bits of decomposed plastic which will then get into the film, causing "fisheyes" or even breaks in the bubble. Also, the bits can lodge in the die lips and cause "die-lines" which weaken the film and mar its appearance.

The die opening is fixed and preset by the operator. A gap of 0.020 in. (½ mm) has been very common for polyethylene (for film of 0.001 to 0.002 in. = 25 to 50 μ)—but for some newer high-drawdown polyethylenes, and for heavy film for sacks and construction uses, gaps up to 0.050 in. (1.25 mm) have been successfully used. Too small a gap increases die resistance and may cause overheating in the extruder and/or reduced output rate; too large a gap may reduce resistance so much that weld lines appear. Also, a large gap requires more drawdown to make a given thickness of film; since polymers are limited in their drawing rate and ability, a large gap may require a slower output rate, or a higher temperature, or, in some cases, the required thickness may be impossible to make.

The operator can adjust several bolts around the die which shift the outer ring in relation to the core, thus altering the gap as needed. He should be careful not to compress the ring from both sides, making its opening slightly oval. This will not usually cure an uneven thickness problem, and may permanently deform the die.

Film dies are usually heated by bands or similar resistance heaters, with a cartridge heater sometimes inserted in the central core. The controlled heating of the core is particularly important for large dies. Induction heating and circulating fluids are also used.

Many tubular dies built today are able to rotate or oscillate around their axes, as illustrated in Figure 5.8. An external drive is fitted to do this, and the adapter between the die and the extruder is suitably designed and sealed to prevent leakage. This seal may be made of fluoroplastic or of brass impregnated with molybdenum sulfide, and it is self-lubricating. Because of the rotation, any thick or thin spots on the film do not build up at the same place on the finished roll. Rather, they are evenly distributed across the roll, which can therefore remain tight, flat, and neat.

In principle, only the center of the die needs to be rotated with respect to the outer ring, or vice versa, to gain the same result. However, rotating or oscillating the whole die around its axis enables the seal area to be much smaller, and reduces the need for precise alignment of the moving parts. In many cases, the air ring above the die is rotated too, to compensate for any inequalities in air flow around this ring (Figure 5.9).

In a few cases the entire windup assembly is oscillated, while the extruder and die remain fixed (Figure 5.10). This arrangement is designed for rigid PVC, where decomposition/stagnation problems make the construction of conventional rotating/oscillating dies difficult. Finally, several firms offer systems where the extruder itself turns around its axis, with the die rigidly attached and coaxial with the extruder (no elbow adapter). These are discussed in more detail on page 163.

Cooling and Sizing. Around and above the die is an air ring, fed by a blower, which directs air onto the outer surface of the emerging film, very close to the die opening (see Figure 5.9). Because a large volume of air is needed, the ring is often large and bulky. The air flow inside the ring is baffled to provide uniform, jet-free flow at all points. The most common type of ring produces a high volume of air at relatively low velocity; at least 40 cu ft/min/in. of die circumference is desirable (= about ½ cu meter/cm of circumference). If velocity is too high, and/or if flow is not uniform, visible bands will be produced on the film, thickness uniformity might be affected, and the stability of the bubble may be impaired.

Air rings are now built so that the gap through which the air flows can be varied in width, thus varying velocity. Annular inserts are provided so that large rings can be used with smaller dies. Also, the

Figure 5.9. Rotating 24-inch (60-cm) air ring for tubular film. Inserts are available to allow use with smaller-diameter films. (*Courtesy of Sterling Extruder Corp.*)

Figure 5.10. Horizontal film takeoff which rotates around its *horizontal* axis to make a tight and even roll of film. Horizontal extrusion is often used for rigid PVC and foamed polystyrene. (*Courtesy of Technical Machinery Corp. and Amut SpA*)

angle of impingement on the hot film may be variable, from 45° to almost parallel to the moving plastic tube (which helps to stabilize the bubble, and improves cooling). The latter (parallel) design, with the air directed upward along the film at high velocity, out of an opening very close to the actual die opening, can be improved even further by enclosing the next few feet of tube in an oversize cylinder topped with a variable iris-type opening. This combination provides a stable bubble at very high linear speeds.

Refrigerated air is sometimes used as a cooling aid, blown through the air ring. This is useful where cooling rate is the limit to output; i.e., where tower height is limited, or where the melt is abnormally hot. Units of 15 to 30 tons in refrigeration capacity are currently being sold for commercial production lines in the United States. For small lines, some firms even use ordinary commercial air conditioners. Such devices are also used to cool the *inside* of the bubble, which keeps the inside temperature more uniform as well as cooler. Both effects allow faster rates.

A novel way of cooling the inside of the bubble is to place a coiled metal tube there, with a fan over it. Cold water is circulated through the coil (which does not touch the film) and the fan blows the cooled air around the enclosed bubble space.

Another cooling technique uses an internally cooled metal mandrel as shown in Figure 5.11(a). A conventional air ring may be placed around the outside of the bubble as usual, for auxiliary cooling. The mandrel precisely fixes final diameter and can do an excellent job of cooling; reports of outputs of 10 to 20 lb/hr/in. of die circumference (1.8—3.6 kg/hr/cm) have been heard. There are disadvantages to the mandrel too: it must be kept absolutely smooth to avoid scratches which will be transferred to the film, and each mandrel is good only for one film size. In contrast, with no mandrel, but with the conventional air-filled bubble, the film size can be quickly changed merely by changing air pressure in the bubble. Thus, such mandrel cooling is suited primarily to high-capacity production of a single size of product for captive use or contract order, rather than for conventional custom extrusion work. Many variations and improvements in the mandrel method have been developed, mainly aimed at maintaining its cooling ability while decreasing the frictional drag of the hot plastic past the mandrel, and avoiding the scratching problem.

A recent development is the adjustable inflatable mandrel shown in Figure 5.11(b). Although it doesn't provide as much internal cooling as a metal mandrel, it does do considerable internal cooling and it increases bubble stability at high speeds, thereby allowing much higher outputs for a given diameter. Unlike a metal mandrel, it can quickly be adjusted for different bubble diameters.

Some tubular film is cooled by extruding downward into a water bath. This causes problems at high speeds, but the method is still used for saran, and is now reportedly in use for heavy polyethylene film for sacks (0.005 to 0.008 in. = 0.125 to 0.20 mm). This film is hard to cool with air because of its thickness; the water provides excellent cooling, and linear speed is not excessive even at very high output rates.

Other auxiliary cooling devices, such as external cooling sleeves and external water-fog sprays, have been known for many years, but have the disadvantages of limiting the range of film sizes (requiring

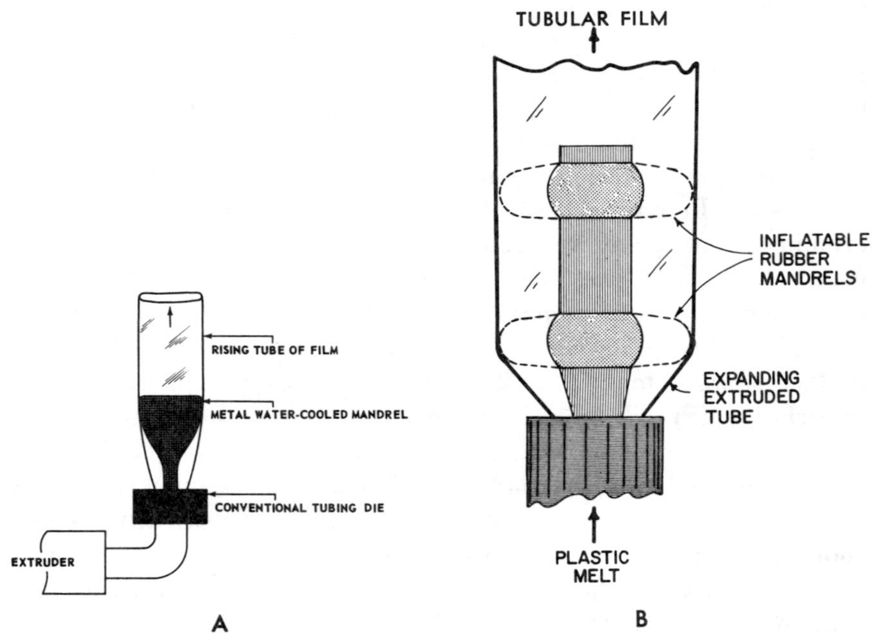

Figure 5.11. Internal mandrels for cooling and sizing tubular film. *A*: A cooled metal fixed-diameter mandrel. *B*: An inflatable mandrel, with rubber "doughnuts" which expand with air to match any tube diameter within its range. C (at right): photograph of "B". (*Courtesy of General Engineering/Radcliffe Ltd.*)

C

Figure 5.11 *cont'd*

a set of devices for different sizes), and also frequently increasing friction and thus limiting rate. They have come back into common use lately for the relatively slow-moving heavy polyethylene film for sacks, where extra cooling is the key to high output rate. Such film is now being made at rates of 600 to 800 lb/hr (270 to 360 kg/hr), and equipment is reportedly being developed now to operate as high as 2000 lb/hr (910 kg/hr), making a bubble 25 ft (7.7 meters) in diameter which would be slit and converted into side-welded sacks!

Take-off. The blown tube passes between two converging planes which flatten it out. For polyethylene, these are commonly made of smooth, hard wooden (maple) slats, 2 to 3 in. wide, set every 4 to 6 in., transverse to the direction of the moving tube. Some lines use composition-board (e.g., "Masonite"), which has a slightly rough surface that prevents sticking. The boards may be perforated to improve cooling; the maple slats are very good in this respect, because air can circulate around them.* Other materials used to make these flattening planes include sand-blasted metal, small long metal rollers, and cloth. The cloth can be coated or impregnated with fluoroplastics; such a slippery surface is essential for foamed polysyrene. A few manufacturers make towers with water-cooled collapsing plates, and some even add wedge-shaped edge plates perpendicular to the nip rolls; the increased cooling is needed for high-speed production of heavy polyolefin film. This system has the disadvantage of unequal treatment of different linear portions of the tube.

The distance between the die and the nip at the apex of the planes is important, as is the angle of the planes. These factors determine the cooling time, and are often adjustable. For polyethylene, the cooling time must be longer than for most other materials, so a polyethylene take-off is longer (higher tower) and has a smaller angle.

The flattening apparatus must be mounted very firmly, as it must remain properly aligned over the die. Misaligned or wobbly flatteners may cause wrinkled film and winding problems. Additional mechanical aids are also frequently used to keep the bubble moving straight and wrinkle-free; for example, an internal "wishbone" can be rigidly connected by a rod to the die, extending upward and spreading the film smoothly just before it passes into the nip of the pull rolls. A common device comprises four horizontal bars whose positions are adjusted to form a square that surrounds the moving bubble perhaps 6 to 10 ft (2 to 3 meters) above the die, and helps to keep it in the proper shape.

At the apex of the flattening planes there is a pair of rolls, one rubber and one steel, which draw the film through the take-off. These are the "nip rolls." They are from 6 to 12 in. (15 to 30 cm) in diameter, with even larger diameters needed for the widest film, to prevent deflection and consequent uneven contact. Most systems have the tube passing around one of the nip rolls and then down one side

* Some lines, particularly for heavy film, have extra cooling air blown through the perforations or other openings in the flattening planes.

of the cooling tower to floor level, where it is wound up. Two variations occasionally seen are:

(1) The "fountain" arrangement, in which the film edges are trimmed off right after the nip rolls. The two resulting layers are separated, and one layer is passed down each side of the take-off. A winding station is thus needed on each side. For efficient trimming, another pair of driven nip rolls is set just above the first pair, and the film is trimmed while firmly held between the two pairs. This system makes the individual flat films more accessible for in-line treating, printing, etc., but makes more work for the operator who must go from one winding station to the other. Also, the thinner single layers must be handled more carefully to get proper winding.

(2) Upper-level winding, in which the winders are up at or near the level of the nip rolls. Often, a two- or three-story building is used, with the extruder on the lowest level and the tube passing up through a hole in the next level(s). The proximity of the winders to the nip makes handling easier.

The winders are independently driven; both surface-winding and center-winding are used. The specification and design of winders, tension controls, etc., are an important and specialized branch of this technology, too complex to be considered here in detail. These components must be selected to match the expected materials, speeds and thicknesses, to wind at proper and variable tension, but not to pull so strongly as to override the nip rolls and affect the drawing and dimensions of the film.

Good operating speeds in recent years have been 50 to 100 ft/min (15 to 30 meters/min) for medium and thick tubular film, and 100 to 300 ft/min (30 to 90 meters/min) for thin film. Lately, however, lines are being designed for speeds at least twice as fast, and a few such high-speed lines capable of over 600 ft (about 180 meters) per minute are in operation now. In the meantime, many older lines are still running successfully at lower speeds.

Along with high speed has come the increased need for automatic roll changing. Most suppliers now offer systems which automatically cut the film and start the following film on a new core. Some units even include a length-measurement device which actuates the automatic changer when a preset length of film has been collected.

Many operations are now done directly in line with the extrusion operation: electrical discharge treatment for printing of one or both sides of the film is very common, and the printing itself is sometimes done in line, as well as sealing and bag-making. With such lines the film is never wound or stored; a plastic compound is put in one end and completed bags are produced at the other.

EQUIPMENT FOR CAST FILM

Dies. Dies for cast film (Figure 5.12) are similar to those used for heavy sheeting or extrusion-coating (see discussions in Chapters 4 and 6). They are usually center-fed, heated by resistance bands, blocks, or cartridges (some now are induction-heated), and are often supported by a separate carriage or overhead cable or frame. They differ from sheeting dies in that they usually extrude downward rather than horizontally, but a few high-speed casting lines have recently been built with horizontal extrusion, eliminating the elbow-adapter.

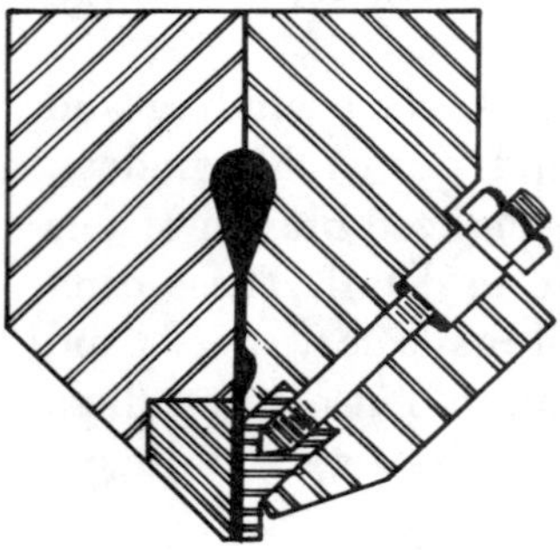

Figure 5.12. Cross-section of die for flat cast film. Bolts on the right adjust thickness along the die. Note the streamlined teardrop-shaped manifold.

A casting die seldom has an adjustable choker bar; instead, the internal flow paths are designed to compensate for the more difficult flow to the ends, and a "coat-hanger" manifold is the result. Lips are adjusted from perhaps 0.010 to 0.060 in. (0.25 to 1.5 mm) by means of a row of bolts pushing and pulling a separate and removable lip bar. Some dies have an integral and flexible adjustable lip, which allows a very smooth, streamlined path. Even in these dies the actual lip edges may be removable and interchangeable for easier maintenance and versatility.

Take-off. Figures 5.1 and 5.13 illustrate the casting process. The film first strikes the main cooling roll (the "chill roll") which is made of polished chromium-plated steel or another hard, corrosion-resistant alloy. Stainless steel is not generally used as it is more difficult to fabricate, has poorer heat transfer ability, and is substantially more expensive. This roll is hollow and baffled to yield better heat transfer to the cooling water, and to maintain an even temperature all over the surface. The roll is held at about 50°C by a heat exchanger. At least one auxiliary roll is usually employed right after the first roll, and from these auxiliary rolls the cooled film passes through pull rolls into a winder.

An "air knife" is now used on the first roll, to force the moving film into close contact with the roll. This prevents air entrapment and improves cooling, both of which help to increase production rate. The pressure of the air knife also prevents "plate-out"—the deposit of certain additives on the roll. In a few cases, an additional air curtain is set on the other side of the descending film of hot plastic—not blowing on the film directly, but rather in a transverse direction so as to create a vacuum between film and roll. This augments the action of the air knife directly blowing on the first side.

The edges of cast film must be trimmed before winding, to get a neat, tight roll of uniform thickness. Razor-blade or equivalent slitters are mounted just before the winder, and cut to exact width. Thus,

Figure 5.13. Takeoff for flat film extrusion. Film falls on lower roll, at extreme right, and then passes through the system to the winder at the left. (*Courtesy of Prodex Division of Koehring Co.*)

width control is perfect, but some trim is always generated—more than is usually made in the tubular process.

Thickness control of cast film (by means of the adjusting bolts) is better than that of tubular film, but high and/or soft spots on the finished roll may still be a problem, even when the edges are trimmed. With tubular film, the die can be rotated to eliminate this problem, but this cannot be done, of course, with a flat die. Considerable improvement can be made by reciprocating the windup from side to side, but even more trim is created than usual.

The production speeds of cast film have risen in recent years from around 350 ft (100 meters) per minute to over 1000 ft (300 meters) per minute in a few cases. Some lines are now being designed for speeds up to 1500 ft (450 meters) per minute!

OPERATION

Formulation. Special compounds have been developed specifically for polyolefin film, and are sold as such by the material suppliers. Many have excellent impact strength and clarity. Additives are usually included to increase slip and prevent blocking (the mechanical cleavage of one film to another). Other possible additives are colorants, antioxidants, and light-stabilizers. When colorants are used in thin film, their dispersion must be very good—both to give a uniformly colored film and to prevent breaks during extrusion caused by undispersed colorant particles.

The density of polyethylene film compounds is important, as it is directly related to stiffness, which affects handling in packaging machinery. Different applications will require different densities. It should be remembered that the *film* density is what counts, and this may differ from the original material density by as much as 0.010 gram/cc. The production method has a significant effect; casting produces less crystallinity (hence lower density) than tubular extrusion. The melt index, too, must be carefully specified; a low melt index (e.g., 1.5) will give a tougher melt, less likely to tear, but a high melt index (e.g., 6.0) may be easier to draw down to very thin film.

With extruded vinyls, the formulations are developed by the extruders, and are mixed by them just prior to extrusion. Much extruded vinyl film is plasticized PVC, and the plasticizer must be carefully chosen to meet performance requirements, which often include nontoxicity. Stabilizers, too, are critical, since extrusion adds considerable heat to the compound and PVC is always susceptible to degradation. Rigid, unplasticized compounds have the most severe stabilizer requirements, as they are most viscous and develop much heat in extrusion. Particularly for these rigid compounds, lubricants are added to make extrusion easier. Also, a small amount of plasticizer (1 to 5 per cent) may be added without impairing stiffness; this must be cautiously done, however, because small amounts of some plasticizers may have an unexpected embrittling effect on PVC.

Start-up. For tubular film, the first extrudate is bunched together and passed up through the nip rolls and into the winder. The "bunch" and the following film cannot be too thick, or it will not go through the nip rolls. As the tube is being formed, air is introduced through the die in small amounts to keep the tube slightly blown. After the threading is complete, more air is blown in to bring the bubble to the desired size. Care is needed to keep the die faces clear of molten resin as this may later be decomposed, become difficult to clean, and cause die lines. A length of paper, cloth, or rope is often first threaded through the system and hooked onto the extrudate at the die. This eliminates the need for pulling the film itself through the system by hand. A common practice is to leave a bubble itself, cooled and stationary, over the die at the end of a previous run; this bubble is then available for quick start-up.

The first minutes of production always yield scrap material, as the system must reach equilibrium. The die bolts may have to be adjusted to get uniform thickness, and the extrusion speed and winder speeds must be balanced to get the desired over-all thickness. Both tubular and casting lines are started quite cool (e.g., 150°C for polyethylene with 2.0 melt index) in order to minimize formation of decomposed material in the system, which could subsequently contaminate the film and/or cause streaks. Once the screw is turning and plastic is coming through the die, the temperatures are raised to normal values (see next paragraph). When the product is one that is familiar, the operator of a small line can be making good tubular film in as little as 5 to 10 min; however, when a new material or different size is involved, it may take an hour or even a day to determine the best conditions, and it will be many days before the operators can run the line at maximum efficiency. In general, a casting line is considered harder to start up and run than a tubular line; some contributing reasons include higher speeds for cast film, and higher temperatures (which make thickness adjustment more difficult, and which require more time to reach from the cool start). Threading a casting line may also be more difficult because the rolls are close together and damage to their surfaces must be scrupulously avoided.

Temperatures. Tubular film is run much cooler than cast film. A low temperature is needed for tubular film (e.g., 170°C for polyethylene of 2.0 melt index); since cooling capacity often limits output, a higher temperature may mean lower output. It also may increase blocking, reduce bubble stability and dimensional uniformity, and promote decomposition in the die with resultant impaired appearance and possible bubble breaks. The higher temperature has the advantages of improved optical properties and sometimes even higher output if extruder power is the limitation. For vinyls a low temperature (e.g., 175°C) is often needed to avoid decomposition; however, recent techniques of high-speed extrusion in vented extruders, direct from powder mix, allow temperatures of 200°C and higher, even for rigid PVC—with corresponding advantages in reduced pressure, higher output, easier drawdown, and more uniform production.

Cast film requires a high temperature; 265°C or more is typical for polyethylene. Cooling is less of a problem because of the large metal-to-plastic contact area, and optical properties (gloss, light transmission, and haze) improve with higher extrudate temperature. Vinyls are seldom run flat; die design to avoid decomposition is a problem, especially for wide film, but several specialized designs have been developed. The temperature requirements are generally the same as for tubular film, as noted in the preceding paragraph. Nylon and polypropylene are cast at about 300°C and 270°C, respectively. Casting is well suited to these two plastics—both have low melt viscosities (particularly nylon) and blowing a bubble is consequently harder, and optical properties are better with the casting process. Polypropylene film is also made by a tubular biaxial orientation process in which it is extruded at low temperature, cooled for solidification, and then reheated and simultaneously blown and oriented.

Blow Ratio and Bubble Adjustment. The blow ratio (or blow-up ratio) is the ratio of bubble diameter to die diameter, and is one of several interrelated factors which determine final film size and properties. A high blow ratio yields the strongest film, as the increased stretching has an orienting effect. A high ratio also means that a smaller and less expensive die is needed for any given film size. But a high blow ratio also encourages bubble instability, requires more drawdown, and magnifies all the imperfections in the die; thus a compromise blow ratio is needed. Bubble instability is a major problem in tubular film extrusion, as it produces wrinkles, thickness variation, and "walking" of the film along the windup roll.

For low-density polyethylene, a blow ratio of about 2.0 is common. Some lines run as low as 1.5 to 1.8, while very good systems and materials can run as high as 3.0. Even higher ratios may be possible with special attachments (e.g., iris collars) which improve bubble stability.

For rigid PVC, the blow ratio is fairly low (2 or less) because of limited drawing ability of the hot melt. Formulations with impact additives and/or processing aids may be blown with higher ratios; this is also the case for the plasticized PVC films.

The blow ratio is determined in advance, when a die is selected to do a given job. Once the extrusion is in progress, variables such as melt temperature, melt pressure, extrusion rate, linear speed, and cooling rate can all be adjusted to give optimum properties. Optical properties must always be compromised with output, as these properties are best when a hot melt is cooled very fast. Mechanical properties are often better with faster cooling, and also depend on the amount of orientation imparted to the film. This is a complex function of all the above variables. If a "square" film (mechanical properties equal in all directions) is desired, the stresses on the film must be balanced in both directions. Too much linear pull, or too high a blow ratio with very little linear pull, will yield unbalanced film, which is desirable in only a few specialized applications. The effects of the extrusion variables on optical and mechanical properties have been studied in detail; many references are listed in the bibliography at the end of this chapter.

Film Casting. In casting, the principles are similar to tubular extrusion, but the variables are somewhat different. They include: melt temperature and pressure, temperature of the chill roll, the distance between the die lip and the roll, and the balance between extrusion rate and linear film speed. These variables determine the time that the plastic travels through the air to reach the chill roll. This in turn affects the amount of neck-in (reduction of width), the surface oxidation, and the ease in drawing to low thicknesses. Some imbalance in mechanical properties is inevitable because all the pull is in one direction. However, if the melt temperature is high and the travel time in air is very short, all drawdown will take place quickly and the tendency to splitting due to one-directional orientation is kept low. The conditions between the die lip and the roll also have a critical effect on optical properties. A 2- to 3-in. gap (5 to $7\frac{1}{2}$ cm) has been recommended for low-density polyethylene, and a smaller gap ($\frac{1}{2}$ in. = 1.27 cm) is preferred for high-density polyethylene. It must be kept in mind, however, that the critical variable is really the time in air, not the distance—and for very fast operation the preceding distances must be increased.

Shutdown. In all processes it is vital to avoid decomposition in the head and die. Such decomposition can be caused by thermal degradation, oxidation of hot plastic in contact with air, or both. For vinyls, thermal degradation is always a danger, and it spreads easily throughout the die, ruining the product. For all materials such degradation and decomposition may produce hardened bits of material, which can break off and lodge in the die lips. They may even be formed in the lips if points of stagnation exist. Such bits cause die lines which are not only unsightly but are also lines of weakness. Also, the bits may break loose and contaminate the film. When a film line is shut down, therefore, material is kept moving as the die cools to about 130°C (low-density polyethylene)—then the extruder is stopped and the die and head are cooled with air as fast as possible, to inhibit decomposition. Polyethylene is frequently left in the extruder *barrel* as well as in the die to prevent air from entering the system and oxidizing any plastic bits left in it. Likewise, before start-up the die is not left hot any longer than absolutely necessary. Frequently the line is shut down with the plastic still in position threaded through the system. This makes start-up easier the next time.

Trim and Scrap. With tubular film no trimming is needed if it is to be sold as a tube. Otherwise the edges of the lay-flat tube are trimmed to make two rolls of single-layer film. With cast film, the edges must be trimmed—both because they are a little thicker than the rest of the film, and because final width requirements are seldom precisely the width of the extruded product. These trimmed edges all constitute an important cost factor and must be reclaimed. Ideally, they should be worked off in some black or other noncritical product.

Sometimes they must be reused in clear film, and must then be handled with extreme care and cleanliness. Equipment is available to draw these edges away continuously either mechanically or by vacuum; they are then ground up and automatically returned to the extruder hopper in a preset proportion. Vacuum systems have the particular disadvantage of drawing in air which may contain possible contaminants. One mechanical system feeds the trim right into the extruder itself. It is also common to chop up the trim and extrude it into pellets for reuse, on extruders fitted to handle such fluffy feed.

It is common among film producers to make both tubular and flat films in widths much larger than actually required by the customers. These wide films are then cut to final widths as needed. This system minimizes edge trim and enables the use and economic advantages of a large, high-output extruder. Even when bags are to be made and a complete tube is desirable, it is sometimes more economical to extrude a very large tube or a cast film, and make the bags with a side weld. When it is known in advance how a wide film is eventually to be divided, the slitting is sometimes done right in line.

PROBLEMS AND SOLUTIONS

The following paragraphs list some of the problems commonly encountered in film extrusion, along with their suggested solutions:

Problem	*Solution*
Poor optical properties	Change compound, or run hotter. If cast film, decrease gap between die and cooling surface. If tubular, increase blow ratio, raise frost line.
Improper slip	More slip agent and/or higher melt temperature increase slip.
Film too weak in one direction (splitty)	If tubular, adjust relation of linear speed and blow ratio to correct balance. If cast, minimize drawdown, run hotter, use shorter gap, smaller lip opening. In both methods, check for die lines which may weaken film.
Film too weak all over	Both methods: change to stronger resin, check for possible degradation of polymer. If tubular, use greater blow ratio *and* greater linear speed, or lower frost line.

Problem	*Solution*
Film discolored or streaked	Check condition and color of feed. Run cooler to inhibit decomposition.
Too many fisheyes (gels or other particles)	These impair appearance and may cause bubble breaks. Review entire operation, looking for sources of contamination. Decomposition inside the system is a major cause. Use more streamlined equipment. Follow recommended start-up/shutdown procedures. In opening bags, avoid pickup of outside dirt or paper fibers and thread. Reduce or eliminate addition of reprocessed trim. Check cleanliness of the air for pneumatic loaders. Keep hopper covered and full. Check plant ventilation, as dusty air must not enter extruder. Clean die often. Check raw material; if fisheye-producing material is there, change to another material, or use fine screens (200 mesh) and/or continuous changer. Use special gel-reducing heads, porous plates, etc. Burnt plastic on die lips may break loose and cause fisheyes and bubble breaks.
Bubble breaks (not from fisheyes)	Drawdown limits may be exceeded: increase temperature, use different compound, get better mixing and more uniform melt. Moisture in hopper, either absorbed into plastic or introduced as contaminant, can cause tiny bubbles and roughness, promoting breaks. Check also for uneven thickness.
Die lines (grooves)	Clean die, follow proper start-up/shutdown procedures, avoid internal decomposition. Check for nicks or burrs on die lips.
Varying thickness	This is evidenced by an asymmetrical bubble and/or heavy spots on the finished roll. Adjust die opening. Check for defective heaters or other hot/cold spots on die. Use uniform, jet-free cooling air. In tubular film, thin bands require increased cooling-air flow and/or *reductions* in gap opening at corresponding positions on die, and vice versa for thick bands. To get neat roll, use rotating/

Problem	*Solution*
	oscillating die or reciprocating slitter. Uneven roll is also possible if wound too hot.
Bubble instability (vibrating or shaky bubble)	Decrease blow ratio (use larger die, or make smaller film). Check temperature and thickness uniformity of melt at the die. If extruder is surging (watch pressure and power fluctuations) reduce screw speed, increase pressure, use different screw, cool screw (last resort, as it reduces output). Check constancy of air pressure inside bubble. Improve air-ring construction, aiming air more in an upward (vertical) direction. Use iris, expandable mandrel, or other stabilizing device. Avoid outside drafts, shield whole system from doors, windows, other air currents.
Wrinkles	These are often caused by bubble instability (see preceding item). Also check for improper alignment of takeoff, skewed rolls, etc. Check for smooth, drag-free passage through flattener and nip rolls. Use auxiliary expander or bowed anti-wrinkle rolls, or wrap a roll with tape in helical pattern diverging from center out to ends.
Blocking	Improve cooling or run at lower temperature. Inflate a secondary bubble just *after* the nip rolls. Use antiblock additives, reduce nip-roll pressure, check for excessive electrical discharge treatment for printing, which can cause blocking. Use nitrogen instead of air in bubble.
Poor cooling	Tubular: Increase height of nip rolls above die, use more air, better baffling inside air ring, better direction of air (smaller angle with vertical), cooling sleeves, fog sprays, metal mandrels, expandable mandrels, extra air rings, refrigerated air. Put cooling section in air-conditioned enclosure. For cast film: use colder cooling water, larger chill roll, air knife. Both methods: run at lower temperatures.
Puckering on cast film	Run chill roll warmer.

STANDARDS AND TESTS

The Society of the Plastics Industry has sponsored Commercial Standard CS–227 issued for polyethylene flexible films. This is a voluntary industry standard issued by the United States Department of Commerce. It classifies films by impact, slip, and density and sets minimum standards for tensile strength, appearance, dimensions, clarity, ink adhesion, sealability, and odor. Another Commercial Standard, CS-238, covers film used in building. There is a U.S. Federal Specification L–P–378 covering polyethylene film .006 inch (.15 mm) or thinner, for packaging of military items. Some others are L–P–504 (cellulose acetate), L–P–375 (soft PVC), and MIL–F–22191/22019/23712 (transparent, flexible thermoplastic). Finally, for food packaging, the U.S. Food and Drug Administration must approve each formulation, submitted to them in confidence, before it can be used.

There are many specialized tests which are important to makers and users of film:

Strength Tests. The simplest one is to grasp the edge of the film firmly in both hands, with about three inches of free film between the hands, and yank the film apart. This test is tried in machine and cross directions. It will show up "splitty" film quite well, but a little practice is needed to perform the test properly. Tensile and elongation tests (ASTM D–882, D–828, D–987) also will indicate this tendency. Highly oriented film has high tensile strength and low elongation in the direction(s) of orientation. Orientation and/or film imbalance is also measured by determining shrinkage in a hot glycerine bath or oven; the greater the orientation in a given direction, the greater the shrinkage in that direction. To test over-all strength, falling-dart tests are used (ASTM D–1709) as are bag drop tests. In the latter, a bag is made from the film, filled with a measured amount of dry sand, and dropped from varying heights. This test also measures the quality of heat seals. Finally there are standardized tear tests (Elmendorf—ASTM D–689) and burst tests (Mullen—ASTM D–774) which are directly applicable to certain uses.

Optical Tests. There are three: haze, transmission and gloss. They are all standardized (ASTM D–1003, D–523, and D–1223), and the proper test instruments are commercially available. These are very common tests and are most important to packaging. For polyethylene, good optical properties are primarily a *surface* phenomenon. This can be dramatically demonstrated by coating a cloudy film with mineral oil on both sides. The oil levels out the surfaces, the refractive indices are about the same, and the film becomes much more transparent.

Chemical Environment Tests. Most important of these is the test for permeability (ASTM E–96/water and D–1434/gases) which measures the transmission rate of water vapor, carbon dioxide, nitrogen or other gas through the plastic film. This is most important in packaging and other barrier applications (e.g., construction). Also, finished film or bags are sometimes tested in contact with fats and oils (ASTM D–722), salts, or other chemicals, to prove their suitability as packaging for particular products. Finally, toxicity tests are performed on laboratory animals in order to prove that certain additive formulations are suitable for food packaging use. Often these tests are avoided by using only ingredients previously sanctioned by the U.S. Food and Drug Administration. These chemical and permeability data are available from the various resin and film suppliers.

Miscellaneous Tests. *Slip* is measured by testing coefficient of friction of the film. *Outdoor life* is tested by means of accelerated ultraviolet exposure, simulating sunlight. The latter test is important for construction and agricultural end-uses. *Sealability* is measured on standard heat-sealing machinery. The MFT (minimum fusion temperature) is measured with a special sealing bar along which the temperature is varied. Finally, *printability* is tested most simply by the "Scotch" tape test. A standard ink is rolled on and let dry. Then a strip of ordinary "Scotch" tape is pressed onto the printing and yanked off; if the plastic retains most of the ink, it has a good surface for printing. This is a common test, but actually a very severe one. More critical tests depend on the behavior of liquids on the films; one test measures the angle at which a water drop slides off the surface, and another—being considered as an ASTM standard—determines the behavior of variable mixtures of solvents on the surface. These tests should also tell when the surface is *over*treated, as an overtreated film may be weaker, block more easily, and seal with more difficulty.

SPECIAL TECHNIQUES

Converting. The manufacture of the film is only the beginning. From then on, it is passed through different processes, many adapted from paper handling, which convert it into its final usable form. These processes are specialized enough to warrant a book for themselves; some are briefly mentioned here:

Form/Fill/Seal Machines. These lines are available and in use for items such as peanuts, dried fruit and beans. Most of these lines make simple "pillow pouches" by parallel sealing of both ends of a tube. Some include simple thermoforming of heavy film into shallow containers, with a sealed-on cover of film drawn from a separate roll. When pouches are made with the seals alternately perpendicular to each other, tetrahedral containers are made; these are used in Europe for household milk, and in Europe, the United States and elsewhere for institutional feeding and single-portion packaging. They use polyethylene-coated paper as well as solid plastic film.

Skin Packaging. In this process the article to be packaged is placed on a precoated or perforated card. Film is heated in a thermoformer and drawn over the article by vacuum drawn through the card. The film adheres to the card and thus encases the article in a "skin." Formerly, cards were coated to effect a heat seal; now some machines can get an adequate seal just by forcing the molten plastic into the fibers, pores, or perforations of the uncoated card. Competing with the skin-packaging technique is curtain-coating machinery which extrudes a molten resin downward in a "curtain"; the article to be sealed is placed on its card and passed right through the molten curtain.

Overwrapping Machines. The machinery now in general use in the United States is capable of handling polyethylene films, although film stiffness ("machinability") is still desirable. This contrasts with the situation several years ago, when special "kits" had to be used to convert cellophane machinery to take polyethylene. Bread, stationery, phonograph records, toys and games are important overwrapped items. Extruded polypropylene and PVC films are successfully competing with cellophane for these markets, and are even replacing some polyethylene.

Metallizing. Film, especially oriented polyester, is vacuum metallized and cut in strips for use as a textile filament (e.g., in auto upholstery). Uncut metallized film is used in space-study balloons, and is now being offered as a "blanket" of high thermal insulating capacity.

Laminating. Film can be laminated to paper, other plastics and even metal (this is done most often with vinyl) to gain the combined properties of both materials. However, when a very thin layer of plastic is used, especially polyethylene, extrusion-coating is usually employed and no film is needed. The two processes compete in such areas as multiwall sacks; in some the inner layer is polyethylene-coated, while in others a thin separate film liner is inserted. There has been much activity lately in these multilayer "structured films"—and many more developments in this area are expected, especially in coextrusion, in which two or more extruders feed a single film die.

Printing. An important part of conversion is printing. It is easy to print vinyl and polystyrene films, but polyethylene—the bulk of the extruded film—requires pretreatment for good ink adhesion. There are three methods: chemical, flame and electrical discharge, with the latter by far the most important. In this clean and effective method, the film surface is treated right in line with the extrusion, before windup. A high-frequency generator or a high-voltage source creates a spark or "corona" electrical discharge. The film is passed over an insulated roll through the path of this discharge, and its surface is oxidized; the oxidation helps attract and retain the ink. The quality of the treatment is impaired by antioxidants and slip agents—this is one reason for treating right on the extrusion line, as close as possible to the nip rolls, before the slip agent fully comes to the surface. Sometimes, even the inside of an extruded tube is treated in line, just after edge-trimming.

Chemical methods of treatment (ozone, chlorine, nitrogen oxides, sulfuric-chromic acids) are the oldest, and are least adapted to safe, high-speed, economical operation. The flame method is used mostly for blown containers, although it can also be used for film if a metal cooling roll is used as a support on the opposite side from the flame.

For all methods, the surface must be protected during the time between treating and printing. The treatment is a surface phenomenon and, if not carefully handled, the oxidized surface can be rubbed off in spots. Machines that treat and print right in line avoid this problem.

Sealing. Another important post-operation is the sealing together of two or more films. Adhesives are rarely used with thermoplastic films; rather, heat is used to melt and weld the sealing surfaces together. The most common method is a simple hot bar, coated with fluorocarbon to prevent sticking. This is most popular with polyethylene. Another way of heating the surfaces is with radio-frequency waves which generate heat *at the interface only;* these are most useful with heavy film, which would take a long time to seal if the heat had to be conducted through all of its thickness. In this way, the heat is generated just where it is needed. This method is used mainly with vinyls, which are often used in greater thicknesses, are more sensitive to heat than polyethylene, and also generate heat more easily at these high frequencies. Another rather specialized method is the "molten-bead" sealer—which is actually a little extruder. It deposits a thin strip (bead) of hot molten plastic between the two surfaces which are then pressed together over the bead. This method is useful for thick films and for continuous work where the intermittent operation of a heated or high-frequency bar is impractical. Finally, ultrasonic sealing devices literally vibrate the molecules at the sealing surface to make intimate contact with each other, effecting a seal with no external heat at all. This method is useful for oriented film, where excessive heat would tend to relieve the orientation and thus weaken the film. It also can seal through most types of contamination, and is generally independent of surface treatments. It finds most use with the more rigid films (polystyrene, polyester, cellulosics, PVC).

Coating. For some extruded film, especially oriented polystyrene, the surface must be coated to acquire certain properties, such as moisture resistance, static electricity dissipation, and anti-fog properties (freedom from moisture condensation on the inside surface of packages of moisture-containing food). This is best done with a solution of the coating agent—with the consequent need to dry the film with air, heat, or both. Sometimes it is done in line with the film extrusion. Another method is extrusion-coating, where the film is a substrate over which hot molten plastic is deposited by an extrusion die. This process is covered in detail in the next chapter.

Gusseting and Other Tube Configurations. Tubular film, instead of being wound as lay-flat film, can be folded in many ways before rolling. A simple gusset is the most common, and can be made during the extrusion operation by mechanical shaping devices between the nip rolls and the die. The nip rolls are usually taped or otherwise altered to ensure equal pressure across the film, with no possibility of air leaking out of the bubble. With such film, a gusseted bag can be made with a simple bar seal across the width of the film. Many variations of this simple gusset are commercially available; the film makers will fold and/or slit the tube in just about any way requested.

Orientation. By stretching the film in one or both directions when rather stiff and cool, the molecules are uncoiled and lined up. As a result, the film becomes tougher. The difference is most remarkable with polystyrene, which is converted from a brittle, useless material into a tough, clear, cellophane-like film. All thermoplastics can be oriented, however, and polyesters, vinyls, and polypropylene—as well as polystyrene—are commercially treated this way.

The simplest orientation method is tubular extrusion, and in any tubular process some strength improvement results. This accounts for the better strength of tubular films compared to cast films. But by far the most significant improvements are achieved with specialized orientation processes which stretch the film at temperatures substantially lower than normal tubular-extrusion temperatures. It is the use of these processes that is normally indicated by the term "oriented film."

There are two basic methods: tubular and flat. *Tubular* orientation is done right in line with the extruder. The film is first cooled, and then is gradually reheated to orienting temperature. Expansion is done with internal air pressure, and the film is restrained from expanding until the proper point is reached. *Flat* film is first extruded (or calendered) and rolled up; then it is oriented longitudinally by drawing between sets of heated rolls running at different speeds, and passes from these rolls into a tentering frame. This frame, adapted from textile-handling machinery, grips the moving film at its edges and stretches it transversely as it passes through an enclosed oven held at closely controlled temperature. The preceding sequence is the most common, but a few lines do the transverse orientation first, and a few others orient in both directions at once. The latter system would be highly desirable, particularly for polypropylene, but a versatile, high-speed, flat, simultaneous system is not yet widely available. (All tubular orientation, of course, can be simultaneous.)

In the tentering frame, tremendous forces are needed to hold and pull the film, and precise heat control is also needed as the degree of orientation depends on the material temperature. This orientation (for both methods) can be measured by shrinkage; typical values of polystyrene are 300 to 500 per cent in each direction; i.e., the oriented length is three to five times the shrunk length. Excessive orientation uncoils the molecules so far that they have no more elasticity, and the film becomes less tough again. The tendency to shrink (relieving strain) makes printing, heat sealing and thermoforming difficult, and special techniques are needed to perform these operations, unless the film is annealed. The shrink properties of unannealed film may be used to advantage, however, for certain packaging applications.

Automatic Thickness Control. Nuclear-energy gauges are available for cast-film lines, similar to those described earlier in the preceding chapter for sheet. The higher speeds and lower thicknesses require changes, but the technology is known, and thickness can be accurately measured in this way at any commercial speed. The use of these measurements to *control* thickness is less common but grow-

Figure 5.14. Pneumatic device to measure thickness of a strip or ring of film cut from a tube. Film is passed under a little roller; an air jet senses the thickness and records it on a circular chart. (*Courtesy of Technigage* Co.)

ing; some lines integrate thickness across the sheet and feed back to the pull rolls or extruder drive to keep average thickness constant, but feedback of measurements in specific areas to specific die positions is still not generally done. Most work with nuclear-energy gauges has been done with cast film; lately they have also been applied to tubular film, where the single layers cannot easily be measured, and where it is more difficult for the operator to make use of the information since die adjustment is less exact. Nuclear gauges are very costly, but less expensive devices are also available. These have more limited use, but may be entirely satisfactory. They include pneumatic devices (Figure 5.14), optical gauges, and magnetic gauges.

Multiple Dies. To keep large, high-output extruders going on small-diameter tubular films, multiple heads have been designed. They require complex die construction and individual valving to insure streamlined and constant flow to all heads. They are tricky to operate and for that reason are used only in special situations, where very large amounts are needed of one or two narrow tubes. It is becoming more common to extrude one big tube, and make smaller tubes and bags by slitting and side-welding.

Vertical Extruders. Several machines are available which have the screw in a vertical position, forcing material upward through a conventional tubular die (see Figure 5.15). The die is simple and is completely streamlined and symmetrical, as no elbow or angular entry is needed. This makes it particularly suitable for PVC, although most activity to date has been with polyethylene. Considerable floor space is also saved. The machine in the illustration slowly oscillates around its vertical axis one revolution every three minutes underneath its stationary collapsing frame, thereby eliminating thick or thin spots on the finished roll (same effect as rotating die). The same effect is claimed by two suppliers of horizontal machines making horizontal bubbles: one oscillates the extruder barrel around its axis, and the other oscillates the entire collapsing frame (Figure 5.10) and winding assembly.

Horizontal Bubbles. In addition to the two machines described in the preceding paragraph, other *non*-rotating machines are known

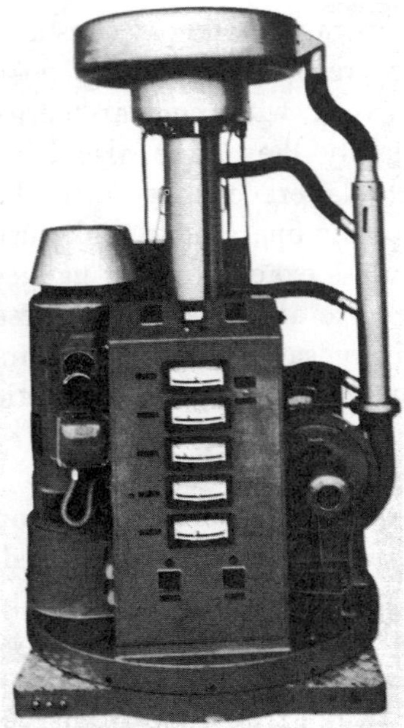

Figure 5.15. A vertical oscillating extruder. The entire machine oscillates slowly around its vertical axis. The metal shaft in the center surrounds the extruder barrel, and the wide, short cylinder at the very top is the air ring, with the die down inside it. The plastic tube is extruded upward through the air ring into a conventional flattening frame, nip rolls and winder. As the film is flattened, a sensing device inside the bubble measures the width and keeps it constant. (*Courtesy of Oerlikon Plastics, Ltd.*)

which produce horizontal bubbles. They have been used for many years for saran, where no right-angle bends can be tolerated because of the danger of decomposition. They are also used for small rigid PVC lines, particularly in Europe. Rigid PVC is subject to the same decomposition problem as saran, but both vertical and horizontal bubbles have been successfully used. The horizontal bubble is also used for some polypropylene orientation lines. Any horizontal bubble process can be considered a horizontal extrusion of thin-wall pipe, blown several times its extruded diameter at the die. The main disadvantage is that rising hot air makes the top half of the bubble cool more slowly than the bottom half. The air ring must compensate for this, or else a different cooling method must be used. PVC is easily run horizontally, as cooling is less critical—less heat must be removed (no heat of crystallization as with polyethylene). Nevertheless, polyethylene also is run horizontally on occasion.

Downward Tubular Extrusion. This requires a normal horizontal extruder and a tubular die pointing downward. The bubble is passed down through a hole in the floor to a winding station on the lower level. This complex arrangement makes it uncommon but its users gain these advantages:

(1) Threading is easier as gravity is a help, not a hindrance.
(2) The die face is more easily kept clean.
(3) Heat from the die rises around the die and *away* from the bubble without hindering the cooling of the bubble.

Another advantage is that this method permits direct liquid cooling of the tubular film, usually with water, which allows a very high output rate. This would be more difficult, if not impossible, with upward extrusion. Mandrel cooling of downward-extruding tubular film is also done.

REFERENCES

Markets for Extruded Film

"Der Kunststoff-Sack" (The plastic sack), *Chemische Industrie,* April 1966. Eight-page discussion of materials, properties and uses.

"Flexible plastic films—chemistry and uses," S. Sacharow (Reynolds Tobacco), *Plastics Design and Processing,* June 1965. Review of each type, with chart and qualitative comparison.

"Economics of mulch," *Modern Plastics,* October 1963. Discusses economics of polyethylene mulch for cotton, including special equipment to lay the film, plant seeds, and dispose of the mulch. See also short report of tests in *Industrie des Plastiques Modernes,* May 1961.

"Outlets for polyethylene film," *British Plastics,* February 1963. Discussion of packaging, construction, agriculture and other markets; six pages, 18 pictures, no quantitative data.

"Il film di polietilene nell'agricoltura" (Polyethylene film in agriculture), E. Ferraris, *Materie Plastiche,* November and December 1961 and February 1962. Thorough series with 91 illustrations presenting Italian usage. See also study of plasticized PVC film in agriculture, stressing weathering, by Pechiney personnel, in *Industrie des Plastiques Modernes,* May 1961.

Processing Technology: Polyethylene and General

"Shrinkable polyethylene film," K. C. Steffen (U.S.I.), *Modern Plastics,* September 1966. Description of process using normal resins in standard equipment. Other articles by Mr. Steffen on this topic appear in *Kunststoffe* and in *Plastica,* both February 1966.

"Trends in tubular film extrusion," R. E. Lowey (Gene Lowey, Inc.), *Paper, Film and Foil Converter,* May 1966. Practical discussion of principles and machinery now becoming more common.

"Low-density polythene tubular film," M. J. Hughes and M. J. Cawkwell (Bakelite Ltd.), *Plastics,* March 1966. Effects of extrusion conditions on optical properties and physical strength.

"Polythene tubular film production," J. N. C. Bourne (I.C.I.), *Plastics,* July 1965. Machinery and techniques are discussed, including comparison of air and water cooling.

"Improving tubular film extrusion," W. A. Haine (Union Carbide), *Plastics Technology,* August 1962. Tips on start-up and shut-down, cleanliness and thickness control; a trouble-shooting guide is included.

"What's the best blow-up ratio?" J. F. Pilaro *et al.* (U.S.I.), *Modern Plastics,* November 1961. Study of effect of blow-up ratio on properties; recommends range between 2 and 3.

"A comparison of blown and chill-cast polyethylene film," G. D. Murphy (Spencer), *SPE ANTEC* 1961. Shows effects of density of *film* on optical and mechanical properties, and effects of processing variables for each method. Other articles cover chill-roll casting (*Modern Plastics,* August and December 1959) and flat film extrusion into a water bath (*SPE Journal,* June 1958, March and December 1959).

"Polyethylene film extrusion," 50-page manual available from U.S.I. in English or Spanish, intended to serve as a practical guide to foremen and plant managers.

Processing Technology: Polypropylene

"Future courses of development in inflation processes," Jun Taga (consultant), *Japan Plastics Age,* July 1966. Description of Dow/Taga process, in which tubular polypropylene film is extruded downward into water.

"Crystallinity and orientation in tubular polypropylene film," G. A. Ermilova *et al., Plasticheskie Massi* (Soviet Plastics), December 1965. Control of crystal size by extrusion conditions yields better strength at all temperatures.

"Production, properties and development of polypropylene film," H. Komata, *Chemical Industry* (Japan), February 1964. Concerns method and conditions for chill-roll casting. In Japanese.

"Manufacture and properties of polypropylene film," R. A. Rose, *Transactions of the Plastics Institute,* April 1962. Discusses process and conditions, and compares with polyethylene film.

"Il Moplefan biorientato" (Biaxially-oriented Moplefan), A. Coen *et al.* (Montecatini), *Materie Plastiche,* December 1961. Thorough review of properties of Montecatini's product.

"Extruding polypropylene film," A. J. Gilden and W. J. G. McCulloch (Enjay), *Modern Plastics,* July 1960. Study of water-cooled and roll-cooled flat film extrusion. Other articles on roll-cooled polypropylene are in *Plastics Technology,* April 1961 (by Hercules), and *SPE ANTEC* 1960 (Spencer).

Processing Technology: PVC

"Anger film extrusion plant," *Plastics,* August 1966. Description of Anger machinery to make thin PVC films, down to 20 microns (0.8 mil), using powder blend in a twin-screw extruder. The line includes optional orientation equipment.

"Rigid vinyl film by . . . extrusion," R. P. Becker (Diamond Alkali), *SPE ANTEC* 1966. Describes machinery, conditions and their variation, and properties; uses vented cooled screw.

"La fabbricazione di film vinilici rigidi e semirigidi per estrusione" (The extrusion of rigid and semirigid vinyl films), G. Bonfiglio (Montecatini), *Materie Plastiche,* pp. 1051–56, 1965.

"PVC film by extrusion of blown tubing," P. Adamski, *Polymery* (Poland), July-August 1962. Covers equipment, operation, materials and uses.

"Extrusion of PVC blown film," *Plastics,* September 1961. Written with aid of ICI, and discusses formulations, equipment, conditions and properties. Covers both rigid and plasticized films. Additional articles on blown PVC film production are in *Plásticos* (Argentina), August 1961 and *British Plastics,* October 1959.

Processing Technology: Nylon

"Nylon film properties," R. L. Hughes and D. G. Simpson (Spencer), *Plastics Technology,* November 1959, and *SPE ANTEC* 1959. Extrusion variables and their effect on strength and permeability of flat film of nylon 6.

Die Design

"Flat film die design," J. R. A. Pearson (Cambridge University), *Transactions of the Plastics Institute,* June 1964. Calculations of dimensions, considering pressure and temperature variation, and based on non-Newtonian flow.

Descriptions of Complete Film Extrusion Systems

Reifenhäuser (cast film), *British Plastics,* December 1966.

Anger (PVC, flat, twin-screw), *Plastics,* August 1966.

Bielloni (polyethylene, tubular, for bags), *International Plastics Engineering,* February 1963. See also references under "Special Equipment" below.

Cooling

"Techniques de refroidissement . . . dans la fabrication des films de polyéthylène" (Cooling techniques in manufacture of polyethylene film), W. R. K.. Wilson, *Plastiques Modernes et Elastomères,* May 1966. Eleven-page review of cooling technology.

"Theory of bubble cooling in film blowing," W. A. Holmes-Walker and B. Martin (Metal Box Co.), *British Plastics,* April 1966. Development of mathematical model of free blowing, and application to practical operation.

"New blown-film cooling ring," W. A. Haine (Union Carbide), *Modern Plastics,* July 1963. Describes vacuum-chamber surrounding the rising bubble.

"Rotating cooling ring for blown film," F. Schwärzler, *International Plastics Engineering,* April 1963. Brief description of apparatus, which includes chain drive and integral blower.

"High-clarity blown polyethylene film," J. Pilaro and R. Kremer (U.S.I.), *Modern Plastics,* June 1960; reprinted in Italian in *Materie Plastiche,* March 1961. Describes the nature and advantages of the "annealing chamber" to get clearer blown film.

Measurement and Control of Thickness

"The measurement and control of blown film," F. W. Hannula and D. F. Kelleher (Tracerlab), *SPE ANTEC* 1967. Describes the "free air back-scatter" gauge, which measures thickness of bubble wall without contact. Further description in *Plastics World,* January 1967.

"Control of extruded blown film thickness," W. C. Hays *et al.* (Industrial Nucleonics), *SPE ANTEC* 1966. Nuclear-energy gauge, with feedback of average thickness. Can be used with rotating dies. Some discussion of economics.

"Film extrusion with beta gauges," G. R. Gerhard (Industrial Nucleonics Corp.), *Plastics Technology,* February 1964. Covers measurement of flat film (or flattened tubular film). Further description in *Modern Plastics,* May 1963, and in *SPE Journal,* November 1961.

"Automatic profile gauge control," H. E. Sponaugle (Black-Clawson), *SPE Journal,* June 1963, and *SPE ANTEC* 1963. Describes mechanical lip-adjustment device, actuated by operator who interprets results of standard nuclear-energy measurement and recording apparatus.

"New tool for control of transverse gage uniformity of blown film," F. A. Obstfeld (Kordite) and C. G. Heisig (Taylor Instruments), *Modern Plastics,* January 1962. A pneumatic sensing device is used to get continuous thickness profiles on cut strips of blown film.

Winding and Related Operations

"Guiding the plastic web," J. M. Deering (General Precision), *Plastics Design and Processing,* November 1966. Discussion of various systems.

"Flexible Plastic Film Laminations," S. Sacharow (Reynolds Tobacco), *Plastics Design and Processing,* September 1966. Describes techniques for laminating films to foil, paper and other films.

"Web alignment and wrinkle control of plastic films," staff written, *British Plastics,* August 1966. Discussion of sensing systems, mechanical devices, tension controls, etc.

"A guide to winding," S. F. Oakes and A. A. Arterton (Hobbs), *Modern Plastics,* August and September 1960. Reviews available machinery types, principles, advantages and applications of each. Revised, more recent editions are available from Hobbs.

"How to measure and control static charges on plastic webs," T. F. McLaughlin (du Pont), *Modern Plastics,* February 1960. Description of various devices.

"Avoiding wrinkles in polyethylene blown tubing," H. W. Woodham (U.S.I.), *SPE Journal,* April 1958. Causes of wrinkling and proposed solutions.

Printing and Treating of Films

"The printing of flexible plastic films," S. Sacharow (Reynolds Tobacco), *Plastics Design and Processing,* March 1966. Discusses treating, testing for adhesion inks and printing conditions, and compares different films.

"The technology of using flame for treating polyethylene surfaces," T. F. McLaughlin (du Pont), *SPE ANTEC* 1964. Tells what kind of flame used, conditions, thermodynamics and other theory. Lists 34 patents and some other references.

"Guide to corona film treatment," J. C. von der Heide and H. L. Wilson, *Modern Plastics,* May 1961. Principles and advantages of this system. Similar work in *Kunststoffe,* October 1961 by H. Frohlich.

"Oxidation in extrusion," R. C. Phelps and R. W. Kowal (U.S.I.), *Plastics Technology,* March 1960. Discussion of what causes oxidation, its advantages and disadvantages, and its control.

"Surface treatment of polyethylene film by electrical discharge," H. E. Wechsberg and J. B. Webber (Monsanto), *Modern Plastics,* July 1959. How to use this method; includes effects of different variables and slip agents, and effect of treating on heat sealability. Further information on sealability appears in some of the references in Chapter 6.

Orientation

"Conditions for orienting films of modified polyethylene terephthalate," G. V. Sagalaev *et al., Plasticheskie Massi* (Soviet Plastics), March 1967. Three-page summary of this process. In the same issue, Mr. Sagalaev discusses the variables of orientation in another article, in more general terms.

"Folienreckung—eine bedeutsame Entwicklung" (Film orientation—an important development), *Verpackungs-Rundschau,* December 1965. Comparison and discussion with illustrations of two-stage and simultaneous orientation.

"Zweiachsige Reckung und Schrumpfung von Polyolefinen" (Biaxial orientation and shrinking of polyolefins), D. Heinze, *Plastverarbeiter,* March 1965. Discusses properties and uses of oriented film, including shrink film, especially polypropylene. Mentions orientation conditions. Additional information on simultaneous orientation is in *Kunststoff und Gummi,* November 1963.

"Biaxial orientation of film," *International Plastics Engineering,* April 1963. Brückner, Kampf, Dalglish and NRM equipment discussed in detail. Brief notes on Dornier, Japan Steel Works, and Marshall-Williams. Earlier survey of orientation processes by J. Jack of BX Plastics appears in *British Plastics,* June and July 1961. This is a long article and emphasizes methods for polypropylene.

"Measurement of orientation in polystyrene film and sheet," C. P. Hathaway (Monsanto), *SPE ANTEC* 1961. Discusses methods used to measure orientation with preference for measurement of relaxation stress. Reprinted in *SPE Journal,* June 1961. Refractive index can also be used as a measuring method.

Special Equipment

"Operating results from high-speed screw extruders when processing blown film," *International Plastics Engineering,* January 1965. Review of Alpine high-speed machinery for polyethylene; brief mention of PVC. Further information on these machines in *Kunststoff und Gummi,* November 1964, in an article by E. Beck, the inventor of these machines.

"Extruder with rotating cylinder," G. Schenkel (Troester), *Plastics,* October 1964. Translation of article in *Kunststoffe,* December 1963, which discusses different machinery used to get level rolls of film, especially Troester's machine with a horizontal barrel that oscillates about its own axis. Further information in *Kunststoff und Gummi,* August 1964.

"The Rotatruder," *Kunststoff und Gummi,* November 1963, and *Kunststoffe,* August 1962. Review of Oerlikon's film making machine with a vertical barrel, that oscillates about its axis. Short review of this and other vertical machines in *SPE Journal,* February 1962.

"Multi-head extrusion unit," W. Eckhardt (Oerlikon), *Plastics,* December 1959, and *Kunststoffe,* December 1959. Illustrated descriptions of Oerlikon's extruder with four film heads extruding upwards. Each head has an individual pump.

Special Techniques

"A novel technique for the elimination of blocking in blown film," G. L. Dighton (Dow), *SPE Journal,* April 1967. Describes process in which an aliphatic acid is included in the polymer and reacted with gaseous ammonia inside the bubble, to prevent internal blocking. Printing of the outer surface is thus easier.

"The identification of flexible plastic films," S. Sacharow (Reynolds Tobacco), *Plastics Design and Processing,* December 1965. Techniques for quick identification of film material. Further information appears in *Modern Packaging Encyclopedia,* issued each year.

"A guide to radiation equipment," J. W. Olander (High Voltage Engineering Corporation), *Modern Plastics,* June 1961, Describes equipment used to irradiate polyethylene film and other plastic products.

"Polyethylene film scrap a problem?" G. Bamberger (Woloch), *Modern Plastics,* April 1960. Describes process for recovering scrap by shredding and compacting, making pellets suitable for extruder feed. Other machines now available to do this, not yet described in formal articles.

"Continuous blow forming," T. Kasahara (Polymer Industries, in Japan), *Modern Plastics,* August 1959. Describes "bulge extrusion"—upward extrusion of heavy polyethylene tubing with regularly varying pressure inside the bubble. Makes containers.

SIX

extrusion coating

PRINCIPLE

The extrusion-coating operation is shown in Figures 6.1, 6.2 and 6.3. The plastic is extruded through a slot-type die, downward between two rolls—one rubber, one metal. The substrate is fed between the molten plastic and the rubber roll, and plastic and substrate are joined by pressure between the rolls. This product is cooled by passing it around the metal roll, and is then trimmed and wound.

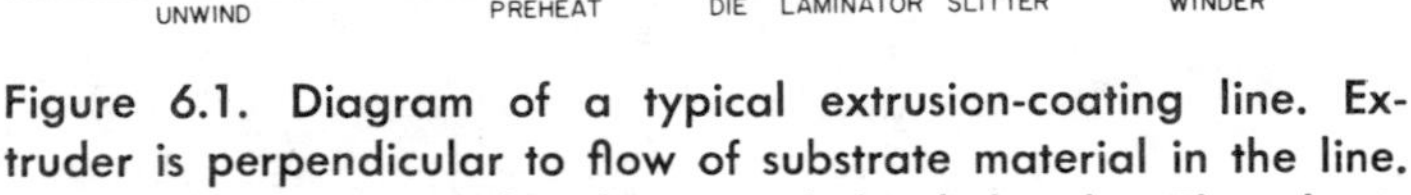

Figure 6.1. Diagram of a typical extrusion-coating line. Extruder is perpendicular to flow of substrate material in the line. It is in the center of this diagram, behind the die. The plastic passes from the die into the bite of a pair of rolls. The unwinding substrate (left) may be preheated on the very large roll, may be primed (not shown) to improve adhesion, and is then laminated to the extruding plastic in the bite just mentioned. The finished combination is cooled, slit and wound up at the right.

Figure 6.2. A complete extrusion-coating line. Substrate unwinds from the roll at the left, and finished product is wound at the right. The hopper of the extruder can be seen behind the die, in the center. The large hose in the foreground leads into the main cooling roll, which is one of the two rolls that laminate plastic and substrate together. (*Courtesy of F. W. Egan & Co.*)

Figure 6.3. Six-inch (150-mm) extruder for extrusion coating, showing center-fed die mounted parallel to extruder axis, as is customary. Entire machine is on rollers, to enable easy withdrawal from the laminating rolls, thus providing ease of die cleaning and accessibility to the rolls. (*Courtesy of F. W. Egan & Co.*)

MATERIALS

Polyethylene, especially low-density polyethylene, is by far the most important extrusion-coating resin, and milk containers are the most important market. Polyethylenes of rather high melt index are used (e.g., 4 to 12) as they flow more easily and adhere better, while their poorer mechanical strength is of no importance in a thin coating. Polypropylene, high-density polyethylene, and nylon have all been extensively studied as coating materials. The first two are not yet widely used as they have not been able to yield economically thin coatings, consistently, at high commercial speeds. Nylon is expensive and is only used where its unique properties are needed. Ionomers and ethylene/vinyl-acetate (EVA) polymers are newer materials, growing in importance.

The most common substrates (the materials to be coated) are paper, paperboard, cellophane and polyester films, and aluminum foil.

PRODUCTS

Coated widths from 24 to 72 in. (61 to 183 cm) are common, and at least one large firm has a 120-in. (3-meter) coater. Coating thicknesses range from less than 1/4 mil (6 μ) to 2 mils (50 μ), with 1/2 to 1 mil (12 to 25 μ) as the most common range. The thickness is usually measured in weight per ream of substrate (in the United States); such a ream of paper is 3000 sq ft. Thus, a 1-mil coating of 0.92-density polyethylene would be about 15 lb per ream.

Some production, notably for milk cartons, is simultaneously coated on both sides with two separate extruders operating in a single coordinated system. For other, smaller markets, two-sided coating is done with a single extruder in two successive passes. Sandwich laminates are also made, in which a single plastic coating layer is extruded between two substrate webs, bonding them together.

EXTRUDER SIZE AND OUTFIT

A speed of 600 ft/min (183 meters/min) is considered good operation, while 1200 ft/min (365 meters/min) is outstanding. A few commercial coaters have lately been built which can run over 2000 ft/min (610 meters/min) with certain materials, and even 2500 ft/min (764 meters/min) has been demonstrated in some laboratories. For the "good" (600 ft/min) and the "outstanding" (1200 ft/min) speeds, the outputs of 1-mil, 0.92-density polyethylene, 4 ft (1.22 meters) wide, are 750 and 1500 lb/hr (340 and 680 kg/hr). These are very high rates for extrusion, and therefore, very large machines—125 to 300 hp and sometimes even more (4½- to 8-in. barrel = 120 to 200 mm)—are customary. Long barrels of at least 24:1 L/D are common because much heat must be put into the plastic to reach the very high compound temperatures needed for good adhesion (e.g., 320°C). For the same reason, coating extruders must have adequate heating power (resistance or induction).

It is common to express output in terms of weight/hour per unit of die width. By current standards, 3 lb/hr per inch is low, 9 is very good, and 15 is excellent. (In kg/hr per cm, ½ is low, 1½ is good, and 2½ is excellent.)

EQUIPMENT

Dies. The majority of coating dies are center-fed (T-type), although small to medium dies, up to 48 in. (122 cm), are sometimes end-fed, giving a more streamlined flow path. Large dies, 54 in. (137 cm) and longer, are quite common, and are all center-fed so that the plastic in the die does not have to travel the entire distance from one end to the other. Typical die cross sections are illustrated in Figure 6.4. The die is mounted with the emerging web *parallel* to the extruder axis, allowing the substrate to move under the die perpendicular to the extruder axis. Therefore, with a center-fed die, one end must project back under the extruder head, and a heated adapter tube must be used to convey the plastic from the head to the feed inlet opening in the die center. Many large dies have extra support at the outer ends and sometimes at the center, either from above or below.

A conventional screen pack is placed in the extruder head behind the breaker plate, to filter out contamination. Some earlier installations used a cartridge-type (basket-head) cylindrical breaker-plate/screen-pack assembly which fitted into the adapter at the elbow, just over the entry to the die. This accomplished its aim of allowing easy and quick screen changes, but it was a source of resin stagnation and degradation, with consequent contamination of the product. A fully streamlined, unhindered flow through the elbow is necessary.

All these dies are usually heated with resistance heaters, although induction heaters are occasionally used. The lips form a sharp V in

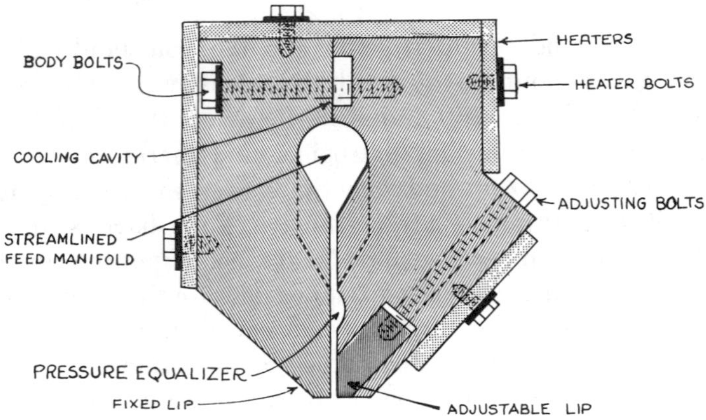

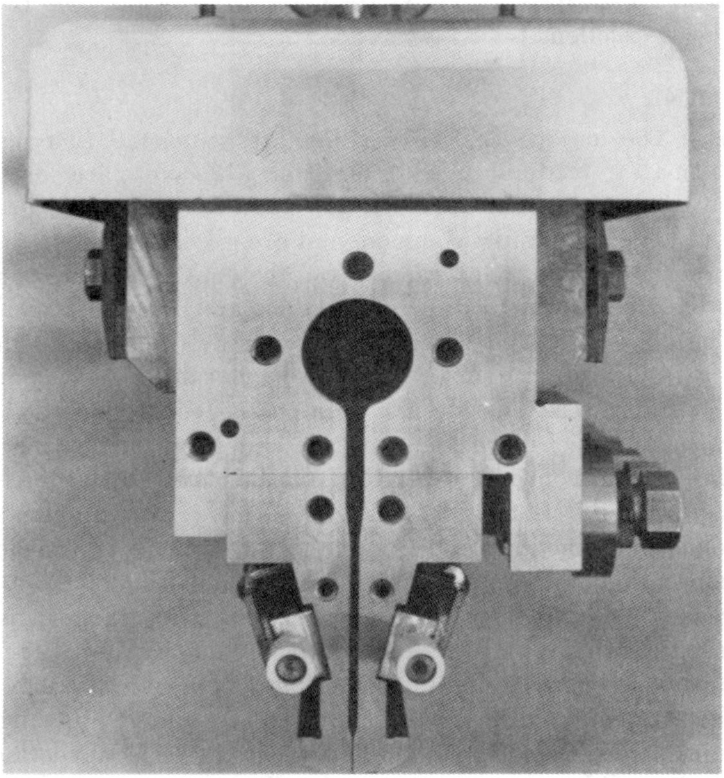

Figure 6.4. *Top:* Cross-section of an extrusion-coating die, showing many important features. (*Illustration by author) Bottom:* End view of another extrusion-coating die with end-plate removed. (*Courtesy of F. W. Egan & Co.*)

order to minimize the gap from die to roll. Figure 6.5 is a close-up photograph of a die in position. The lips are adjustable (by bolt settings) for regulation of the coating thickness, but there is no adjustable internal "choker bar" as on a sheet extrusion die. Instead, the interior is permanently constructed to help get the melt to the far end(s) of the die, with the most restriction to flow occurring between manifold and lips near the entrance to the manifold, and least restriction at the extreme end(s). See Figure 4.7. A rod or bar called a "deckle" is sometimes inserted between the lips or clamped over them to block off a variable length of flow, thus enabling many widths to be made from a single die.

Roll Assembly. On one side of the die are the payoff stations for the substrate rolls. There are two such stations, so that when one roll of substrate runs out another can be started with no interruption of

Figure 6.5. Close-up of extrusion-coating die in position. The die points downward into the bite of the rolls. The bolts to adjust the lip opening can be seen on the die. The roll on the left is the rubber roll; the other one is the metal roll. (*Courtesy of F. W. Egan & Co.*)

operation. This is done automatically on modern machinery and is called a "flying change." At the wind-up end, the machinery allows the product to continue winding on the same roll by making a "flying splice." When two substrates are used to make a sandwich laminate, two pairs of payoff rolls are needed.

Idler rolls guide the substrate into the nip of the rubber and metal rolls under the die lips. Often a substrate preheater is installed; this may be either a heated roll or a bank of radiant heaters.

The rubber pressure roll is made of neoprene or silicone rubber. Neoprene is more common and less expensive, it ages better, it is more heat-stable, and it can last longer if well cared for. However, silicone is often preferred as almost nothing will stick to it. With neoprene, there is always the danger that some hot, molten plastic will touch and stick, which usually means that the roll must be removed and reground to a new surface. With either material, the rubber is typically cured to a Durometer "A" hardness of 80 to 85, but different rolls are often used with different substrates, and their hardnesses may range from 50 to 90. Neoprene of Durometer 90 is typically used for a hard substrate such as cellophane. This rubber roll is often cooled both internally and externally. In some cases it can be eliminated entirely, as noted in "Special Techniques—Improvement of Adhesion" below.

The rubber roll is not driven, but rides in bearings and is pneumatically loaded to press against the metal roll. This pair of rolls is adjustable vertically and horizontally to change the distance between the die lips and the roll nip (or else the extruder itself is movable, to accomplish the same result). The metal roll is chromium-plated steel, and is about 30 per cent larger than the rubber roll. Typical diameters for metal and rubber, respectively, are 24 and 18 in (61 and 46 cm). The metal roll is driven with its speed continuously variable over a wide range. It is internally cooled with water, which flows at high speed in an annular space just under the surface. The chromium-plated surface is either smoothly polished or matte-finished. The water passage should be easily cleaned, as a fouled roll means reduced and uneven cooling capacity, hence reduced output and poorer product uniformity. From 10 to 30 gal/min of water are needed per 100 lb/hr of extrudate (5 to 15 cu meters/hr/100 kg/hr); because of this large requirement, the water must usually be recirculated through a heat exchanger.

From these rolls, the coated product is guided through adjustable trimming knives to the wind-up stations on the opposite side of the die from the payoff. (One machine is built with both payoff and wind-up on the same side of the die. This is safer to clean since the die and extruder are very accessible and can be moved up and away from the nip, thus reducing the possibility of damaging the roll surfaces.) There are two wind-up stations, often fitted with mechanical aids for switching to a new core when one is full. These stations are typically center-driven by constant-torque or constant-tension drives. Unwinding and slitting devices may be employed later, similar to their use for unsupported film.

OPERATION

Formulation. A high melt index is preferred for better adhesion. For colored coatings, maximum opacity is usually wanted, and a very high loading of black or white pigment is used. Also, antioxidants are desirable to prevent excessive degradation at the high extrusion temperatures. Too much antioxidant is harmful, however, as some surface oxidation is wanted to promote adhesion. Coating formulations are seldom mixed by the coaters, but are rather developed by the raw-material supplier and sold ready for use. Slip agents are sometimes added, as with film. One new improvement is the addition of an agent to the formulation which helps release of the web from the chill-roll at higher roll temperatures—which are often needed in humid areas to avoid condensation.

Start-up. With the roll assembly moved away, the plastic is started through the extruder at relatively low temperature (150 to 170°C) and a thin, unsupported film is run until thickness is evened out. During this time the temperature is gradually raised until thermal equilibrium at the final high temperatures (300°C or more) is reached. This way, no substrate is wasted and the first uneven plastic film is recoverable for regrinding and later use. Low speeds on the payoff and wind-up stations are recommended to simplify start-up, but the extruder itself should be brought up to full speed right away to help in reaching thermal equilibrium as soon as possible. If necessary, it may be momentarily slowed down while the rolls are brought into position.

Temperatures. Compound temperatures of 300 to 320°C are typical for polyethylene. The plastic should be extruded as hot as possible for best adhesion and minimum power requirements (highest capacity of machine), but temperatures above 320°C usually degrade polyethylene more than can be tolerated. Hence, this temperature is a practical maximum. Much heat is generated by the frictional working of the plastic in the screw, but some heat is also conducted in through the barrel walls. Adiabatic operation (more properly called "autothermal" operation), in which all heat comes from mechanical working in the screw, is thus rare, and the control instruments become very important as they are always in use. Sensitive, high-quality controllers are recommended, and the stepless types are preferred, especially for the die.

One coater has advertised products made at very low temperatures (e.g., 200°C). This technique uses a very thin highly oxidized film as an adhesive layer between the regular coating and the substrate. This permits very good thickness control of the coating, less edge curl, and no odor and discoloration problems, but the product is more expensive and cannot easily be used for very thin coatings.

The substrate is commonly preheated to 70 to 140°C, depending on its nature. Less heat is thus required from the extrudate to heat up the substrate, and any moisture adsorbed on the substrate surface is removed. This results in a hotter interface and better adhesion. Preheating is done either by heated rolls or radiant heaters.

The metal roll is held at around 50°C in order to cool the product. Its temperature becomes more important with the thicker coating weights, where the heat in the plastic is not dissipated as easily. Temperatures as low as 30°C may be needed for thick coatings, but this temperature cannot fall below the dew point of the surrounding air; otherwise, moisture would condense on the roll, transferring to the product and impairing good cooling.

Ordinarily, the product travels around this metal roll for 180° of its circumference. At the high speeds achieved by modern equipment, the contact time on this roll is extremely short (e.g., 0.2 sec at 1000 ft/min = 300 meters/min, with a 24-in. = 61-cm roll). A few units have been built with 270° of contact, but 180° is still the standard.

Die-to-roll Gap. The distance is critical. A gap that is too big allows more oxidation (better adhesion) but also cools down the film (poorer adhesion) and yields more "neck-in"—i.e., reduction of width and beading of the edges due to surface tension there. On the other hand, a gap that is too small will prevent the necessary surface oxidation, although neck-in and premature cooling are minimized. Optimum gaps are about 3 to 5 in. ($7\frac{1}{2}$ to $12\frac{1}{2}$ cm). The plastic should fall right into the nip and must touch the substrate first—just as it enters the nip. If it touched the metal roll first it would cool prematurely, impairing adhesion. The newest high-speed lines allow adjustment of gap (e.g., from 3 to 10 inches) during operation.

Roll Pressure. The rubber roll should press against the metal roll with about 100 lb force per linear inch (18 kg/cm). Greater force will yield satisfactory product but will shorten roll life, while substantially lower force will impair adhesion. The contact *area* depends on the contact length in the direction of product travel; this, in turn, depends on rubber hardness as well as the roll pressure, and can be visually observed from the ends of the rolls.

Thickness Control. The operator adjusts the coating thickness by means of bolts along the die, as with sheet or flat film. Since the product moves too fast for good measurements to be taken mechanically during production, it is difficult to know when thickness has varied and an adjustment is needed. Therefore, nuclear gauging has been used in recent installations: one gauge scans the moving substrate while another gauge scans the final coated product—the difference, which is the actual coating thickness, is automatically computed and recorded on a chart, which the operator uses to guide his adjustments.

Trimming. Most extrusion-coaters coat just under the substrate width and trim off the edges which bear thicker edge beads of coated material. These beads arise from neck-in (described above) and they represent a major economic loss to the coater. As yet, no economical way has been found to recover and reuse this plastic; it is now discarded as a total loss along with the substrate edges, which may even equal or exceed the lost plastic in cost.

One way of minimizing this loss is to "over-coat"—i.e., to coat a little *wider* than the substrate, so that the beads are *off* the substrate and can be trimmed off and recovered. However, this adds two additional requirements:

(1) The hot plastic edge beads must not stick to the rubber roll, which is now in direct contact with them. This is especially important with a neoprene roll. To avoid this sticking, the rubber roll may be kept cool by a small, cooled metal roll running in contact with it; or else, fluorocarbon-coated cloth strips are put around the rubber roll just at the points where the beads touch it.*

(2) A double trimming is necessary—one cut just to the *outside* of the substrate edge to recover the beads, and a second cut just *inside* the substrate edge to give a clean edge to the finished product. The intervening strip (perhaps only 0.2 inch = 5 mm) is lost, as with the regular "under-coating" method. However, this loss is much smaller, and the neck-in problems are eliminated.

Scrap. The only recoverable scrap generated in a coating process is that made at start-up and extruder changes, when substrate is stopped to await extruder leveling-out—and also that made by "over-coating" and trimming the edges. This scrap is so thin that it cannot be easily fed back to the extruder. It is best run through a compounding extruder (or other compounder) and repelletized. Any such reclaiming operation must be scrupulously clean to avoid contamination, if the scrap is to be reused in film or coating, because such contamination will cause pinholes and breaks in the surface. The scrap is much better worked off in less critical applications such as black agricultural film, some pipe, or injection molding.

Wind-up Speed. This speed is critical, and is usually the limiting factor to the output rate of thin coatings. The concept of "no-surge speed" is used, and refers to the maximum linear speed at a given output rate that can be run without unreasonable thickness variation in either machine or transverse direction. Therefore, exact control of wind-up speed is important.

Shutdown. A coating die must be cooled down as fast as possible —even faster than with polyethylene films because of the higher temperatures. This prevents decomposition and formation of hard

* Internal cooling of the rubber roll is inefficient, and water-bath cooling causes too much trouble with water carry-over; the small metal cooling roll is an acceptable alternative. However, the fluorocarbon-cloth technique is more commonly used, and the rubber roll is seldom cooled at all, except for cellophane coating. A cooler roll would last longer, but adhesion may be impaired by the low roll temperature.

particles that can catch in the lips and cause die lines. For shutdown, after all the desired good product is made, the machine is kept going slowly, as temperatures are lowered to around 150°C by cooling the barrel and the die, usually with air. The barrel may then be emptied by stopping the feed while keeping the screw turning; when the barrel is empty, the screw is stopped and the die is further cooled to room temperature as fast as possible. On start-up, likewise, plastic should be moved through the die as soon as possible at about 150 to 160°C—well before it reaches its operating temperature of 300 to 320°C.

Preferably the barrel itself is not emptied but is left full of plastic to avoid oxidation on the barrel surfaces. In such a case, the decomposed bits of plastic in the barrel are minimized, but care must be taken at start-up to prevent starting too cold and damaging the equipment.

PROBLEMS AND SOLUTIONS

The chart below suggests solutions to some common problems encountered in extrusion-coating.

Problem	*Solution*
Die lines	(a) Nicks in die. Refinish die lips, keep a spare set on hand. Dislodge particles from die lands with a brass shim, or change to clean lips, or disassemble and clean the die.
	(b) Particles in die. Review shutdown and start-up procedure to minimize particle formation.
Pinholes and/or breaks	(a) Coating too thin. Check with material supplier whether desired coating weight is unreasonable.
	(b) Contaminated plastic. Inspect holes for particles.
	(c) Decomposition. Reduce compound temperature.
	(d) Too cold (web tears). Raise compound temperature.
	(e) Too much drawdown (web tears). Reduce die lip opening or increase coating weight.
	(f) Moisture, with certain resins and additives. Use a hopper-dryer or vented extruder.

Problem	*Solution*
Poor adhesion	Raise compound temperature. Lengthen or shorten gap (see discussion above). Preheat substrate more. See that plastic contacts preheated substrate first. Coat the substrate with a primer—a substance that will promote adhesion.
Surging (varying dimensions)	Reduce extruder speed (and take-off speed). Increase back pressure in extruder. Check smoothness of operation of pull rolls and cooling roll. Reduce compound temperature.
Excessive neck-in	Reduce die-to-roll gap. Reduce material temperature, particularly at ends. Reduce die-lip opening, particularly at ends. Run faster, thus reducing residence time between die and nip. Use resin of lower melt index or lower density if product requirements allow. Change die lips to ones with shorter land length.

STANDARDS AND TESTS

Two groups have developed standard tests for determination of coating weight and moisture-vapor transmission; they are: The Polyethylene Extrusion Coaters Group of the Specialty Paper and Board Affiliates, and TAPPI (Technical Association of the Pulp & Paper Industry). There are also a number of other tests in use by coaters, described in the following paragraphs.

Adhesion. The "Scotch" Tape test (see Chapter 5) is generally *in*valid for coatings because the adhesion of the tape to polyethylene itself is poor. Practical tests, with results and judgments based on experience, include: scratching at the coating with a fingernail or a tool; repeated flexing; catching an edge and trying to peel off the coating; and making an "X" cut, flexing across the "X," and trying to separate coating from substrate at the corners thus exposed and stressed. Several commercial machines have been developed to test adhesion, such as the Southwick for porous substrates, and the Instron and the Suter for nonporous substrates.

Printability and Sealability. Tests are generally similar to those for film (see Chapter 5).

Permeability, Chemical Resistance, and Toxicity. Military specifications are used to test grease resistance (JAN–B–21) and oil resistance (MIL–B–131C). Other tests are similar to tests for film (see Chapter 5). Many of these are of major importance, since substrates are often coated to provide barrier properties.

SPECIAL TECHNIQUES

Improvement of Adhesion. Several techniques have been developed for this purpose:

(a) *Primers.* These are chemicals which modify the substrate surface and thus improve adhesion of the plastic to it. They are not actual adhesives. Titanium ester solutions have been quite common, and shellac wash coatings can also be used for aluminum foil. New two-component solvent-based primers are displacing some water-based systems, especially when very high speeds are desired, requiring maximum priming efficiency. Primers have become very popular, and on all modern machinery they are continuously applied by a specialized liquid-coating unit set between the substrate unwind and the coating head. They are particularly useful with nonporous substrates such as cellophane and aluminum foil, but even porous substrates are now primed—e.g., paper to be used in multiwall sacks. The improved bonding allows the plastic to be supplied at a lower temperature, which yields a more uniform product with less odor.

(b) *Inert-gas Blanket.* The heated inert gas (nitrogen) covers only the outer, non-contact side of the polyethylene web as it emerges from the die. Thus, there is less surface tackiness, less odor and discoloration development, and the unoxidized outer surface is more easily heat-sealed in later operations. This technique is in commercial use only where stringent high-quality specifications for sealability must be met (e.g., military uses).

(c) *Vacuum.* With porous substrates, adhesion can be improved by drawing a vacuum right through the substrate just after the polyethylene is applied. This pulls the polyethylene coating into tight contact with the substrate.

(d) *Air Knife and Electrical Field.* Either of these can be used to force the plastic into the substrate, even if it is a non-porous one. These systems, as well as the use of vacuum, eliminate the need for a rubber pressure roll.

(e) *Electrical Discharge Treatment of Substrate.* Recently, this has been used to improve adhesion, even in combination with a chemical primer. Both porous and nonporous substrates can be so treated. Flame treatment can also be used.

Converting. Much of the same technology that was described for film in Chapter 5 is also valid for coated products, which can usually be sealed, cut, folded and otherwise handled like film. Tetrahedral containers are a good market for coated paper, especially in Europe. In skin-packaging, the coated board can now be the base, while flexible, clear film, coated or not, can be the "skin." When the coated products are to be folded, it is important that the coating be able to resist this without cracking or separating at the fold.

Printing. It would seem that the most logical way to make printed coated paper is to print the paper rather than the plastic. Paper prints more easily, and the printing can be protected by the plastic coating. This is still done for small quantities, but it is much more economical to make unprinted coated stock at very high outputs. Individual purchasers of this product can then print it as needed. The polyethylene surface must be treated to achieve good ink adhesion, as with film. Similar treating techniques are used: flame treatment is common for milk-carton stock, while electrical discharge treatment is also widely applied.

Vertical Extrusion. One machinery manufacturer offers a complete coating line featuring a $4\frac{1}{2}$-in. (120-mm) extruder *suspended vertically* from the ceiling, and feeding a center-fed coating die (Figure 6.6). The roll assembly is located below the die and is conventional. The whole extruder, including motor, drive and hopper, hangs above the nip, and the payoff and wind-up are therefore very easily accessible. This allows easy and safe cleaning of the head and die by simply rolling the extruder away from the rolls. Other advantages include less load and more equal load on extruder thrust bearings, minimum floor space, and a more streamlined flow path from head to die (no elbow adapter is needed). This last factor might allow even higher compound temperatures for some materials, hence better adhesion.

Plastisol Coating. A technique has been developed and reported in the literature for extruding liquid PVC plastisols on conventional

Figure 6.6. In this 120-mm (4½-inch) coating line, the extruder is suspended on rails from the ceiling. Its barrel is vertical, and it extrudes downward through an elbowless die into a conventional roll nip, cooling and winding stations. (*Courtesy of Oerlikon Plastics, Ltd.*)

extruders. The liquid is pumped under pressure to the screw, and the heat of extrusion is enough to fuse the mixture. The PVC coating can be applied like polyethylene to textile substrates and is being used commercially to make tops for convertible automobiles. No postcure in an oven is needed, and high production rates are possible.

Curtain Coating. In both Europe and America, machines are offered which continuously extrude a "curtain" of molten polyethylene or polyethylene/wax mixture. The substrate is passed underneath the die and through this curtain while in a horizontal plane. Some machines do not use an extruder—the coating is melted in a pot and forced through a slit to form the curtain. Other systems do use an extruder to melt the polymer, which is then fed to the coating die. This technique is used where substrates are too bulky or too rigid to be conveniently handled as a coiled web. Also, it can be used as skin-packaging by passing the item to be packaged, placed on a mounting card, under the falling curtain.

REFERENCES

Markets

"Markets for extrusion coating," comments in The Plastiscope, *Modern Plastics,* July and August 1965. U.S.I. statistics, predictions, plus more comment on PE markets in general.

More information in the major industry surveys in the following category.

Industry Reviews

"New developments in extrusion coating," C. E. White and A. E. Irvine (Union Carbide), *Plastics Technology,* September 1966. Discusses variables, problems, new resins, power needs, required equipment, etc.

"Progress in extrusion coating," *Modern Plastics,* May and June 1963. Staff-written major survey, describes each market, how coatings are used; statistics, both for resins and substrates. June section covers equipment, sealing, and other technical items. Earlier such survey in March 1961 issue.

"Extrusion coating of paper and board," Hans Hambsch, *Kunststoffe,* May 1963. Thorough illustrated discussion of process, machinery, problems, adhesion aids, applications. Another article on coating foil and cellophane is in later issue of *Kunststoffe,* and a general discussion of extrusion coating (also in German) by H. Wigger is in *Kunststoff und Gummi,* October 1963.

"Plastic coatings on paper substrates," J. R. Young, *Transactions of the Plastics Institute,* April 1963. Six-page survey of processes and products.

Processing Technology—General and Polyethylene

"Low-density polythene extrusion coating," D. A. Hall and W. B. I. Small (I.C.I.), *Plastics,* March 1966. Effect of polymer variables on performance.

"Effect of screw and other variables on polyethylene melt uniformity of extrusion coating at high screw speeds," F. G. Grundman and J. Eckstein, *TAPPI,* June 1965.

"A study of curl in coated constructions," R. A. Bates and L. J. Marsella, *TAPPI,* July 1964. Hold maximum temperature difference between polyethylene and cooling roll.

"Extrusion coating with polyethylene," G. Pedrotti, *Poliplasti,* August 1964. Economic and technical aspects.

"The influence of extrusion variables on polyethylene extrusion-coating performance," H. J. Nalepa and C. C. Cooper (Union Carbide), *TAPPI,* May 1964. Studies of temperature, speed (to 480 m/min.), pressure and linear speed at die.

"Extrusion coating of polyolefins," L. R. Errat, Emballages (French), 1963, No. 208. Similar article in *British Plastics,* June 1962.

"Guide to polyethylene extrusion coating," comprehensive booklet issued in 1962 by U.S.I., the American resin supplier, including operating information, testing techniques, etc. May be revised periodically. Available also in Spanish. Other suppliers such as du Pont and Union Carbide also have similar processing guides.

"Laboratory processing studies of polyethylene, polypropylene and nylon-6 coatings," J. F. Morris (Spencer), *SPE Journal,* August 1961, adapted from *SPE ANTEC* 1961. Comparison of these three materials, studying adhesion, moisture transmission, slip, gloss, greaseproofness, tear strength, all as function of thickness. Extrusion conditions given.

"Coating of paper with polyethylene," R. Reiter, *Plaste und Kautschuk,* (German), 1960.

"Neck-in problem in polyethylene extrusion coating and film casting," D. Lewis and W. McDonald (U.S.I.), *Plastics Technology,* October 1958. Definition of neck-in and demonstration of effects of resin variables and processing variations. Short but significant article.

Processing Technology—Polypropylene

"New developments in extrusion coating with polypropylene," C. E. White and A. E. Irvine (Union Carbide), *Plastics Technology,* February 1967. Conditions, equipment needs, properties and applications, all on three pages.

"Extrusion coating with polypropylene," P. E. Waghorn and L. K. Sharples, *TAPPI,* February 1966. Discusses surging, minimum thickness, sealing problems, die construction, polymer properties.

"Polypropylene for coating," C. C. Cooper (Union Carbide), *TAPPI Conference,* October 9, 1962. Review of laboratory achievements, 300 ft/min. and 8 lb/ream weight.

"Extrusion coating of polypropylene," R. E. Christensen (Union Carbide), *SPE Journal,* July 1962. Conditions and applications. Similar article by H. A. Elliot (Hercules), *Modern Packaging,* June 1962.

Processing Technology—Other Materials

"Coating with thermoplastic urethane films," W. Dormon and A. Laidlaw (Goodrich), *Modern Plastics,* December 1965. Die design, screws, conditions, economics, properties.

"Recent improvements in the technique of coating with nylon 11," M. B. Lebrun, *Industrie des Plastiques Modernes,* July 1962.

"How to extrusion coat with nylon," J. C. Werner and G. D. Murphy (Spencer), *Modern Plastics,* July 1965. Trials on 3½-inch extruder, 24 inches wide, to 400 ft/min. Covers conditions, properties, problems, etc.

"The free expansion of PVC," E. J. Cram *et al.* (Whiffen and Sons), *British Plastics,* January 1961. Article deals mainly with cast vinyl foam, but also includes formulations and conditions for coating the foam onto textiles.

"Comparison of plastisol extrusion coating with conventional coating methods," C. H. Brower (Union Carbide) and R. A. Nunn (Waldron-Hartig), *SPE Journal*, June 1961, based on *SPE ANTEC* 1961. Describes extrusion-coating of PVC plastisols, showing advantages and disadvantages compared to knife- and roller-coating of the same materials. Costs, output and quality are compared.

Specific Equipment

"Advances in extrusion die design," R. B. Gregory (Egan), *Modern Plastics*, March 1967. Description of new dies, including costs and illustrations.

"Bone extrusion coating and laminating plant," *British Plastics*, July 1966.

"Effects of die geometry on extrusion-coating characteristics of polyolefins," R. Edwards, *TAPPI*, April 1966.

"Beloit high-speed extrusion-coating equipment," *International Plastics Engineering*, July 1965. Big and fast equipment, diagram, brief data.

"Les tendances modernes dans le materiel de couchage par extrusion" (Modern trends in extrusion coating equipment), R. E. Lowey (Black-Clawson, now private manufacturers' representative), *Revue Belge des Matières Plastiques*, February 1965. Discussion of equipment components, requirements, set-up. Similar article by Mr. Lowey in *Paper, Film and Foil Converter*, September and October 1961.

"Chill-roll heat transfer for extrusion coating," R. T. van Ness (du Pont), *TAPPI*, November 1964.

"Screw design for hi-speed coating," J. P. Goslin and R. T. van Ness (du Pont), *TAPPI*, April 1964, and *SPE Journal*, January 1963. Long-metering screw compared with more conventional PE types and recommended.

"Black-Clawson induction-heated extruder," *International Plastics Engineering*, September 1963. Staff report on this 4½-inch extruder with coating accessories.

"Gauging and automatic control of an extruder-laminator," R. W. Nelson and S. L. Eakman (Tracerlab and American Sisalkraft), *TAPPI Conference*, October 9, 1962; abstract in *SPE Journal*, January 1963. Paper describes beta-gauge system made by Tracerlab for American Sisalkraft, with two units—one on substrate and one on final product.

Adhesion, Priming and Treating

"Flame or electrical discharge of paper substrates in high-speed extrusion coating," R. E. Greene, *TAPPI*, September 1965. These alternate methods to chemicals are described. Flame temperature of 320°C is used.

"Priming paper for extrusion coating," A. R. Hurst and C. G. Morse, *TAPPI*, July 1964. Discussion of available materials, and the relationships between primer machinery, plastic and substrate.

"Oxidation and odor control of extrusion-coated polymers," P. E. Waghorn, *Paper, Film and Foil Converter*, February 1964.

"Improving adhesion of polyethylene extrusion coatings," G. R. deHoff and T. F. McLaughlin (du Pont), *Modern Plastics*, November 1963. Reflectors to improve air gap oxidation, and use of chemical primers.

"Effect of processing variables on adhesion strength of polyethylene-coated aluminum foil," V. G. Crolius *et al.* (du Pont), *TAPPI*, May 1962. Study of variables and process, includes use of titanate polar primers to promote adhesion.

"Solving polyethylene coating problems," D. A. Perino (Milprint), *Modern Packaging*, February 1962. Use of free film of oxidized polyethylene as adhesive layer between substrate and coating, which can then be applied at lower temperature.

"Surface treatment of polyethylene-coated paper and paper-board," S. Leeds (Union Carbide), *TAPPI*, April 1961. Different arrangements of equipment and effects of variables. Flame treatment is briefly mentioned. See also references after Chapter 5.

"Adhesion and sealability studies," H. A. Arbit (Union Carbide), *TAPPI*, March 1957. Basic work with polyethylene coatings on paper, cellophane and aluminum foil, including explanation of use of inert gas blanket to improve heat-sealability.

Special Techniques and Miscellaneous

"Extrusion-coating techniques for obtaining maximum polyethylene density and improved barrier properties," G. M. Violette, *TAPPI*, June 1966. Electrical charge thrown at plastic to densify it.

"Polyäthylen-beschichtetes Papier mit glatten Rändern" (Polyethylene-coated paper with flat edges), *Kunststoff-Berater*, December 1965. Short description of method in which the non-contact side of the substrate is treated with steam just before coating.

"Verbundfolienherstellung durch Kaschierverfahren oder durch Extrusionsbeschichtung?" (Lamination production by roll or extrusion coating?), J. Kumpf, *Neue Verpackung*, October 1965.

"Extrusion coating and polyethylene recovery," L. I. Naturman (SPE Staff), *SPE Journal*, March 1963. How to separate and recover both plastic and paper: what has been done, economics, trends.

"Improved polyethylene extrusion coating," C. A. Plaskett and K. A. Arnold (St-Regis), *Modern Packaging*, March 1963. Use of vacuum to draw polyethylene coating tighter onto substrate.

SEVEN

wire and cable covering

PRINCIPLES

The process of covering wire and cable by extrusion is shown in Figures 7.1–7.4. The molten plastic is extruded into a crosshead through which passes the wire to be coated. On emerging from the head, the coated wire is cooled in a water bath, continuously tested and usually wound on a reel.

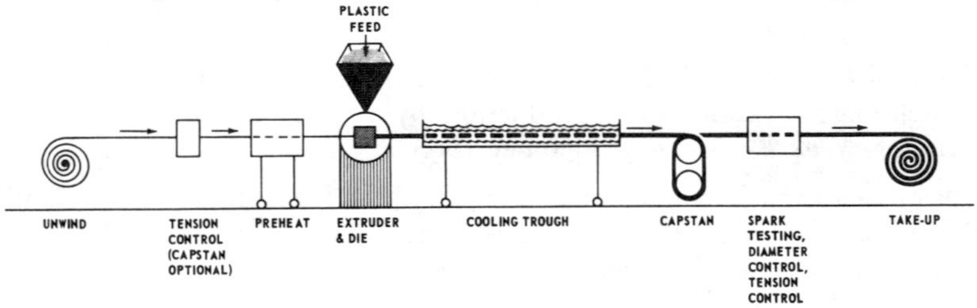

Figure 7.1. Diagram of a production line for the coating of wire or cable with plastic. The conductor to be covered unwinds at the left, is preheated, and passes into the crosshead die (center). The extruder is behind the die, and feeds it with molten plastic, which coats the conductor. The finished product is cooled, tested, and wound up at the right. (*Illustration by author*)

Figure 7.2. Photographs of extruders fitted with wire and cable covering crossheads. *Top:* Line for medium-sized wire covering, with 30° crosshead. (*Courtesy of Prodex Division, Koehring Co.*) *Bottom:* Line for large-diameter cable jacketing. (*Courtesy of General Engineering/Radcliffe Ltd.*)

Figure 7.3. Close-up of actual wire-covering line in operation. Preheated wire enters the die from the left, is coated with plastic, and is cooled in the water trough that begins just in front of the instrument panel. (*Courtesy of Waldron-Hartig Division, Midland-Ross Corp.*)

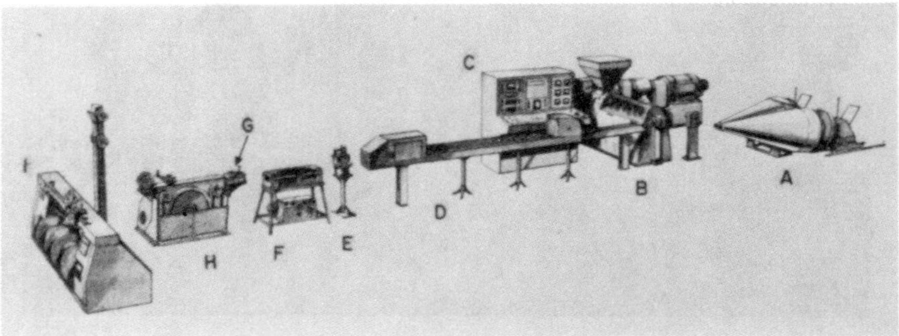

Figure 7.4. Layout of wire covering equipment. The components are as follows: (A) dual flyer payoff; (B) extruder; (C) control panel; (D) multipass cooling trough; (E) diameter control gauge; (F) spark tester; (G) length counter; (H) positive-pull belt capstan; (I) winding equipment with automatic reel changing. (*Courtesy of General Engineering/Radcliffe Ltd.*)

DEFINITIONS

Primary insulation means covering uninsulated wire or wires directly with plastic, to isolate them electrically.

Jacketing or *sheathing* means covering a wire or a group of wires with a plastic coating or "jacket" for mechanical or other nonelectrical protection. Jackets are often put on wires already having primary insulation.

MATERIALS

Polyethylene and vinyls make up the bulk of this business. Polyethylene—especially low- and medium-density polyethylene—offers excellent insulating ability, moisture and chemical resistance, flexibility, and light weight. Lately, the properties of polyethylene have been improved even more by cross-linking. Compared to the polyethylenes, the vinyls are heavier and are somewhat poorer insulators, but they are self-extinguishing and can be made very flexible. They therefore are frequently used as jacketing for other wires. Nylon is used because of its strength, heat resistance, and abrasion resistance, and some new plastics such as polysulfone are finding increased use in very specialized applications. Often two or more plastics are used to cover a wire, thus combining their advantages. Plastics are also used in combination with the other important covering materials—rubber, paper, and metal.

PRODUCT AND EQUIPMENT RANGES

Almost anything can be covered, from a tiny wire .008 in. (0.2 mm) in diameter on up to a cable 5 in. across (13 cm) and more. Consequently, extruders are found from 1½ to 10 in. in diameter (35 to 250 mm), putting out from 10 to over 1000 lb/hr (5 to over 450 kg/hr). Linear speed is usually the limit to output; the payoff and take-up speeds reach practical maximum speeds. Today, speeds of 4000 ft/min (1200 meters/min) are found in production for small telephone wires. Top speeds go down as diameter increases, and 20 to 100 ft/min (6 to 30 meters/min) are common for heavy cables. Here the rate of cooling becomes a limitation, especially for polyethylene.

EQUIPMENT

Payoff. For the smallest wires running at the fastest speeds, the payoff reel is stationary. The principle is much like a spinning reel used for fishing; the wire uncoils and passes over one flange of the reel. Since the reel does not turn, it can pay out wire at very high rates, unhindered by friction or the weight of the heavy reel of wire. Such stationary reels are only good for single conductors since they put a twist in the conductor for each turn coming off the reel. This twist is harmless to a single solid conductor but it would shorten a stranded conductor. For the latter, rotating reels must be used, which have a practical maximum of about 2000 ft/min (600 meters/min). Another advantage of a stationary reel is the very low tension possible; this allows the coating of very fine wire, and most usage of this type of payoff is with wire less than one millimeter (40 mils) in diameter. A typical design is seen in Figure 7.5.

The stationary reels are best set with their axis at an upward-pointing angle, as in the illustration. Angles between 20° and 45° are typical. Rotating reels are normally set with horizontal axes.

Both systems are installed in duplicate, with controlled transfer of payoff from one reel to the other when the first one runs out. For slow-moving cable lines this transfer can be manual, but at high

Figure 7.5. Dual flyer payoff with stationary reels used for covering thin wire at high speeds. (*Courtesy of General Engineering/Radcliffe Ltd.*)

speeds automatic switching is necessary if operation is not to be interrupted.

It is possible to run from a single payoff station if the conductor is stopped just before the reel runs out. The new reel is put in place, its end is attached to the cut end of the old reel and the equipment is started up again. It is best to leave the extruder running throughout this procedure to maintain steady conditions.

Rotating payoffs have internal braking systems to prevent the momentum of the heavy spinning reel from varying the tension in the conductor.

For medium size conductors (Nos. 12 and 14 AWG) it is possible to use a very large single rotating reel that holds enough wire for many hours of production. Larger wire requires lower speeds, so rotating reels are satisfactory, and the unit will work for either single or stranded conductors without frequent stoppage.

The largest wires and cables have rotating payoff reels from 40 to 72 in. in diameter (100 to 180 cm), or else are unwound directly from coils without any reel core. These heavy reels must be driven to insure constant tension between payoff and die, so control systems are used which sense the tension in the cable and thereby control the payoff drive.

Preheater. Preheating of the conductor prevents premature shrinkage of the hot plastic away from the metal surface. This shrinkage could otherwise cause stresses that make the plastic more susceptible to cracking when warmed. Also, preheated conductors yield better physical properties in the insulation, such as tensile strength and elongation. Finally, preheating affects adhesion; such control is important if the wire ends must be stripped in service. Small wires are best preheated by passing them over metal wheels which make the wire part of an electric circuit. The thin wire then is heated by its own resistance to the current and this is a clean, closely controlled method of preheating. For larger wires, resistance goes down but mass goes up, and this method becomes impractical. A gas-flame heater is used instead. Very large conductors are also preheated by high-frequency induction heaters.

Another benefit of preheating is the removal of any moisture or oil on the conductor surface. The flame is the better method for this. If preheating is insufficient to clean the wire, a mechanical wipe before the preheater may help.

Dies. The two basic types of wire-covering dies are "pressure" dies and "tubing" dies, illustrated in Figure 7.6. In both types, the conductor passes through the crosshead through a "guider tip." This tip is made of a very hard metal (e.g., "Carbaloy") to resist abrasion. Its position is often axially adjustable. The surface that contacts the hot plastic must be smooth and streamlined to avoid decomposition as well as surface roughness due to melt-fracture and similar phenomena.

In the pressure die, the plastic is still under some pressure *inside* the die when it contacts the conductor, which leaves the die already coated. The tubing die actually extrudes a plastic tube around the emerging conductor. This tube is drawn around the conductor just *after* the die face by a vacuum drawn through the same passage in which the conductor travels.

Pressure dies are best constructed with the opening equal to the outside diameter of the final product. The plastic swelling at the die will be compensated for by drawdown, and a coating with minimum strains will be produced. The clearance between the wire and its passage through the guider tip must be very small—e.g., 0.002 or 0.003 in. (50 to 75 μ)—to prevent the plastic from entering and impairing smooth and easy passage of the wire.

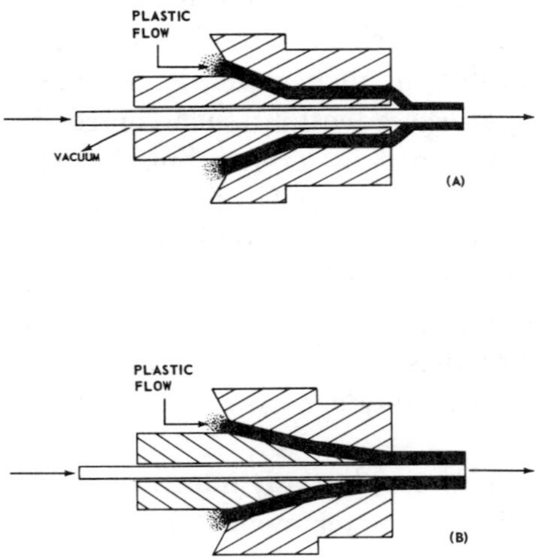

Figure 7.6. Two types of dies used to cover wire and cable: Type (A) at the top is a "tubing" die, which extrudes a thin-wall tube around the conductor. This tube is drawn by vacuum onto the conductor after it leaves the die. The vacuum is pulled through the clearance between the conductor and its passageway through the crosshead. Type (B) below is a "pressure" die, in which the molten plastic contacts the wire before it leaves the die while the melt is still under some pressure. (*Illustration by author*)

Tubing dies require a larger clearance between conductor and passage because vacuum must be drawn through that clearance. A good figure is 0.010 in. (1/4 mm) all around between the conductor and the passage. There is no danger of plastic backing up through it because the plastic is extruded through a separate annulus. The ratio of the area of this annulus to the final plastic cross-sectional area is called the "drawdown ratio." If this ratio is too high, a rough surface and/or internal stresses in the coating will result. Typical drawdown ratios are:

Nylon	4.0
Vinyls	1.5
Low-density polyethylene	1.5
High-density polyethylene	1.2

The annulus should be as small in inside diameter as possible, although this size is limited by the mechanical strength of the guider tip. The outside diameter is then set by the appropriate drawdown ratio.

Polysulfone is unique in its need for exceptionally high drawdown. Tubing dies are always used, and both diameter and wall thickness of the exit annulus are about 20 times the final outside diameter and thickness of the insulation.

Since each die is precisely built to required dimensions, it is seen that many products require many dies. This number can be reduced somewhat by using dies with adjustable guider tips. Also, parts of a die can often be interchanged or substituted to give a different size covering, and a tubing die can be made to give different thicknesses by varying drawdown. This latter method can only be done within narrow limits, however, to avoid roughness and internal stresses.

The inside of the die must be designed to produce equal flow all around the guider tip. Since the plastic must be bent through an angle in the crosshead, the flow would ordinarily be greater on the inside of the bend. The design of the flow passage must compensate for this by making flow easier on the far side of the bend.

Crossheads are set at either 90°, 45°, or 30° from the extruder. The 90° angle is far more common, and was universal until recently. The smaller angles (Figure 7.7) provide a more streamlined interior, and the corresponding machine location is better adapted to

Figure 7.7. Close-up of a pressure-type wire-covering die in a 30° crosshead. Note the valve to control pressure, set in the crosshead next to the extruder gate. (*Courtesy of Prodex Division, Koehring Co.*)

some plant layouts. It is sometimes preferred for vinyls because of the streamlining. At least one maker supplies a crosshead in which the angle can be varied.

The volumetric capacity of the crosshead is also important. If it is too large, degradation may occur, especially with vinyls. If it is too small, stresses (produced by turbulence in the head) will not have a chance to relax, and a rough extrudate surface (melt fracture) will result, especially at high speeds and small diameters.

For tubing dies, the lands are usually parallel, and their length is about equal to the final covered diameter. For pressure dies, there is no land, and the plastic goes through a converging, tapered annular passage, illustrated in Figure 7.6. Sometimes the passage is divided into two or three sections of successively decreasing angle of taper to reduce resistance to flow while still reaching the proper final size.

The vacuum drawn through a tubing die must be at least 20 mm mercury; if less, the "tubing" will not adhere strongly to the wire, and a loose coating may result.

The choice between pressure and tubing dies is a complex matter, determined by the material, the application, the available extrusion equipment, and many other factors. Pressure dies offer more resistance to flow since they have smaller diameters and cross-sectional areas than comparable tubing dies. Hence, a tubing die may be used to get maximum output with small wires and thin covering, especially with high-viscosity melts. A pressure die, however, is better able to fill in the spaces between conductor strands or separate wires. *Primary insulation* usually is run through pressure dies, which give the best adhesion and least strains. *Jacketing,* on the other hand, is usually run through tubing dies, especially where an irregular cross section, such as a group of wires, is to be covered. For such a group it would be very difficult to build a guider tip for a pressure die, with the small clearance needed to prevent material back-up. Vinyl, used for both jacketing and primary insulation, is run through both types of dies. Nylon, which flows very freely at extrusion temperatures, is best extruded through tubing dies for all applications. For very tiny wires, no matter what the material, tubing dies may be used to avoid the increased tension required to pull the wire through the short distance of molten plastic at the tip.

The whole head assembly must be built to take very high pressures. Because of the small size of some wires, the pressure in the extruder head may exceed 5000 psi (340 atm) and operation at this pressure is not uncommon.

Cooling Troughs. All covered wire and cable is cooled by passage through a water trough, except for some cross-linked polymers. The troughs are small in cross section—4 by 4 in. (10 by 10 cm) is typical, but larger troughs are needed, of course, for large cables. Troughs are very long because of the high linear speed, especially for polyethylene. This plastic not only needs much cooling because of its heat of crystallization, but also requires slow cooling to prevent the development of voids in the shrinking coating. Troughs for polyethylene are at least 30 ft (9 meters) long, and can go up to 100 ft (30 meters) or more. For vinyls they are about half the length required for an equivalent cross section of polyethylene. For large undersea cables the trough is often 275 ft (85 meters) long—almost as long as a football field—and it is actually doubled back on itself to obviate the need for a building with that much linear space! These lengths do not seem so great when it is noted that a wire traveling at 1000 feet per minute is in a 65-ft trough for only 4 sec, during which it must be cooled sufficiently to allow coiling without permanent distortion. (Metric equivalent = 300 meters/min in 20-meter trough for 4 sec).

It is common in Europe, and becoming more so in the United States, to run thin wire back and forth several times in the same trough, rather than use one very long trough. For example, a wire may make four passes through a trough 65 feet (20 meters) long.

There is an air gap between die and trough. It is customarily from 6 to 18 in., but is increased to 3 ft for the fastest-moving wires. Increasing this gap permits an annealing of the surface, which improves gloss and reduces stresses. If the gap is too long, however, cooling grows uneven and the entire line is made longer. Another way of getting better gloss on the covering is by directing a gas-air flame at the surface as it leaves the die.

Cooling of cellular polyolefins requires special conditions. The die-to-trough gap is longer (up to 6 ft = 2 meters) as the expansion must take place in this distance. Also, the troughs are very long, as cooling must overcome the insulation effect of the gas bubbles in the plastic. Troughs of 150 to 200 ft (45 to 60 meters) are common.

For polysulfone, it is very important to quench-cool the plastic only a few inches after it contacts the wire.

The wire enters and leaves the trough through or over soft barriers, such as cloth or urethane foam, which let the wire pass through but keep most of the water from flowing out. An air blast or mechanical wiper is frequently used to remove last traces of water before winding.

Thin-walled insulation can be cooled with tap water. However, for medium and large cross sections, the surface cannot set too quickly, or the plastic will shrink inside and separate from the conductor. Accordingly, the water temperature is controlled. For polyethylene, the coated wire must enter a bath at 60 to 80°C for best results. Often the trough is divided into several compartments of successively decreasing temperatures, ending with tap water (e.g., 80–60–40–20°C in four sections). No closed circulation is needed as the water is heated by the entering plastic (and heaters if necessary), and the temperature is controlled by the rate of overflow, which is replaced with cold tap water. These troughs are seldom insulated or covered, with the exception of undersea cable.

In-line Testing and Control. Wire and cable covering is tested in line more than any other extruded product. This is partly because it is rather inaccessible for many tests when on a reel, and partly because it is designed for long-term usage as opposed to disposable or short-term applications.

Spark testing is almost universal. The wire passes through a high-voltage field, and if there are any breaks, pin-holes or thin spots in the covering, a circuit is completed to the conductor and a signal buzzer or light is set off.

Diameter is measured mechanically, electrically, or photometrically. Mechanical devices must be very precisely constructed to measure the tiny differences in the wire. The electrical devices are more accurate, especially for the smaller wires. They usually measure capacitance of the covering at a given point in the wire, and the resulting electrical impulse is fed back to the extruder drive. In a photometric device, the coated wire passes through two intersecting beams of light at their point of intersection. The two shadows thus formed fall on sensing elements, which determine the diameter and feed it back to the drive. If the diameter is too small, the drive speeds up; if too large, it slows down. Frequently the diameter reading is fed to the take-up and payoff instead; they slow down or speed up the conductor travel while leaving the material flow constant from the extruder.

Measurement of eccentricity is as important as measurement of diameter, as the conductor must be centered in its insulation. This is checked with a cathode-ray source which projects the images of conductor and coating on a screen.

For the undersea cables, the diameter of the entering conductor is mechanically measured. The diameter of the finished cable is electrically measured (capacitance), and the difference is automatically computed and used to control the coating thickness.

Capacitance measurement is also used to determine degree of foaming in extruded cellular insulation. The capacitance is recorded on a strip chart, and the operator adjusts die temperature or die-to-trough gap to get more or less foaming, as he sees fit from the chart.

Take-up Equipment. The wire or cable is drawn through the line by a capstan puller, illustrated in Figure 7.8. These are either single- or double-drum and are grooved to hold the wires that are wrapped around them. Four or five wraps are customary. Diameters range from 12 to 20 in. (30 to 50 cm) for small wires through 32 in. (81 cm) for intermediates. Large cables use "tractor" pullers or "caterpillar" capstans—really pipe pullers adapted for cable hauling. Some drum capstans are not grooved, but have a smooth surface which joins with a driven belt to give a positive pull (as in Figure 7.4).

Many drum capstans are externally water-cooled. The several turns that the wire takes around the drums add to the effective cooling length and thus shorten the required length of the cooling trough.

Figure 7.8. Capstan puller for cable or heavy wire. The wire comes out of the water trough (upper left), and is wound around the two drums several times before passing to the winder (not visible). Water in the bottom of the capstan housing cools the wire as it passes through. (*Courtesy of Waldron-Hartig Division, Midland-Ross Corp.*)

For the high speeds currently encountered, these units must be both rugged and precisely built. At lower speeds, torque requirements increase, and below 50 ft/min (15 meters/min) the tractor type is preferred. These tractors must provide considerable tension, especially where the trough is very long. Without that tension the cable would drag on the bottom of the trough before completely cooling, and thus be marred. It is not unusual to have a motor of five horsepower or more driving the puller at the end of a cable line.

Tension controls are often used at the end of the line, and are similarly used between payoff and extruder. In fact, another capstan is frequently put between payoff and extruder to ensure constant conductor feed at constant tension. The tension control is commonly a mechanical device ("dancer") that senses sag in the cable; when it sags excessively, the device tells the take-up to pull harder, or perhaps increases braking in the payoff.

From the pulling capstan the wire is picked up on rotating take-up reels that are similar to the rotating payoff reels. Automatic devices are again used at high speeds to switch from a full reel to an empty one. Also, level-winding attachments are common and resemble those on a conventional fishing reel. For heavy cable, moving at slow speeds, a man can provide the level winding action. For the fastest production, the take-up reels are sometimes completely eliminated, and the wire is allowed to fall into an open fiber drum, coiling upon itself. Its fall into the drum is usually guided by a variation of the level winder, to ensure even distribution in the drum.

Other Equipment Requirements. Electrical speed variation of the extruder (AC–DC motor-generator system) is preferred for wire and cable because it can be controlled easily by the various electrical impulses produced on the line. Also, it is one of the steadiest types of drive. A pressure valve is seldom needed as pressures are very high, except for some large cross sections.

Just before winding, some cables are marked by continuous marking machines, similar to pipe marking. Smaller wires are also frequently marked at the same location, to enable their tracing in use (e.g., in switchboard wiring).

The controls and their synchronization for a large and complex line such as for undersea cable are very intricate and expensive equipment. In general, wire and cable covering is a high-quality operation, and this is reflected in costs and quality of equipment.

OPERATION

Formulation. For vinyls, many extruders have their own formulations, which they mix by dry-blending in ribbon-blenders or specialized intensive mixers. Such a blend, often preheated, is frequently fed directly to the final extruder, which is run with sufficiently high mixing to give a good, homogeneous product. Other times, the dry-blend is hot-compounded and formed into pellet feed for the extruder. Some prepared compounds are bought direct from the raw-material suppliers, especially those formulated to meet standard specifications.

Polyethylene and nylon, on the other hand, are almost always bought as finished compounds ready to extrude. Only the colorants are sometimes added as "master batch" just before extrusion. The manufacture of some polyethylene compounds in the suppliers' plants is very closely controlled; one major American extruder actually sets forth conditions of manufacture and sends inspectors to ensure their observance.

Where polypropylene is used on copper wire (because of its good temperature resistance) it must contain a stabilizer/inhibitor to prevent its reaction with the copper, which would otherwise hasten its degradation.

Stabilizers, antioxidants, carbon black and colorants, and stress-cracking preventers are the major additives put into wire and cable compounds. The *stabilizers* are necessary for all vinyls to prevent degradation under heat, and the *antioxidants* do the same for polyethylene. The antioxidants also enable the polyethylene to be used in air at prolonged high temperatures. *Carbon black* is the best known protector against ultraviolet light degradation and is a must for any outdoor use. Its type (channel black) and particle size (as small as possible) are important in getting the most protection for the least carbon black content (usually 2.6 per cent in polyethylene) as this additive impairs some electrical properties. *Colorants* are useful to mark wires for ease of tracing in complex circuits, and they also provide attractiveness in such consumer items as lamp cord and telephone handset cord. *Stress-cracking preventers* are added to polyethylene for those long-term uses in which the wire or cable will be subjected to unknown environments. This insures against the expense of locating and repairing a break in a long-distance cable which may be buried, submerged, or otherwise inaccessible. To gain this stress-cracking resistance, special resin combinations are used by the various raw-material suppliers; certain additives, such as butyl rubber, are also effective. Adding anything at all to a polyethylene resin is likely to impair its insulating ability, especially at high frequencies, and this must always be considered in product formulation.

Start-up. The conductor is threaded through the system first. Next the extruder is started and the proper conditions of speed and temperature achieved, while the plastic just squirts out of the die around the stationary wire. Only then is the wire started and brought up to full speed. If the covering is not evenly distributed around the wire, centering bolts on the die are adjusted, which displace the die with respect to the guider. They are omitted on "self-centering" dies, which are accurately machined to desired dimensions, and are most often used with very fine wires.

Temperatures. Typical compound temperatures are:

Low-density polyethylene	220°C
High-density polyethylene	260°C
Nylon	285°C
Plasticized vinyl	185°C
Polysulfone	365°C

The vinyl is not run hotter because of the danger of decomposition, and it cannot be run much cooler because the high pressures in a wire die produce appreciable frictional heat. The screws for the polyethylenes are even shallower than those for vinyls, and even more frictional heat is produced. Typical channel depths in the final (metering) zone are .125, .105, and .150 in. (3.1, 2.6, and 3.7 mm) for low-density polyethylene, nylon and vinyl, respectively ($3\frac{1}{2}$-in. = 90-mm extruder).

The rear of the extruder barrel is run somewhat hotter than usual in order to reduce the load on the extruder drive. This load is ordinarily quite high because of the high pressures in the head and die. Common practice is to run the rear of the barrel at the same temperature as the front, which should be at about the desired compound temperature, except in the case of vinyls, where the rear is run about 40°C cooler. With the other materials, if the feed bridges and becomes erratic, the rear of the barrel has to be cooled down. Cold water is circulated around the feed section of the barrel, in all cases.

Preheating of the Conductor. For most plastics, a preheat temperature of 150°C is desirable. High-density polyethylenes do better with a temperature of 175°C, while 120°C is sufficient for cellular polyethylenes, which have much poorer heat conductivity. If the wire is heated beyond these optimum temperatures, no further improvement in adhesion is gained, but more heat must be removed in the cooling trough. This, in turn, means slower rates or longer and costlier equipment, or both. On the other hand, just warming the wire is not sufficient; experiments have shown a steady improvement in properties and lowering of stresses with increasing conductor temperature, up to about 150°C.

Scrap. Very little scrap is generated in wire-covering operations as it is uneconomical to strip the coating from substandard wire. Furthermore, the little all-plastic scrap that is generated in start-up is usually unsuitable for re-use as the reprocessing and rehandling impairs the insulating ability. It can only be worked off in such noncritical applications as television lead-in wire. Therefore, for a wire covering plant to run economically, it must prepare for long runs on known first-grade material with a minimum of batch-to-batch variation.

STANDARDS

The wire and cable industry is called on to meet many formal standards set by both commercial and government groups. These standards not only apply to plastic insulation and jacketing but also are concerned with the entire construction of the wire or cable. They relate mainly to performance, and have been developed to ensure that the construction will perform satisfactorily in the proposed end-use. They test such properties as maximum safe operating temperature, chemical resistance, flame resistance, flexibility at low temperature, electrical properties, and resistance to deformation and abrasion.

Government and Societies' Standards. Some important American standards are listed below.

Aircraft Industries of America NAS 702	Vinyl hook-up and general-purpose wire, used from −55 to 105°C.
ASTM D 1047–49T	Vinyl jacketing, from 0 to 60°C.
D 1351–54T	Polyethylene insulated conductors, up to 75°C.
American Standards Association C–8.35–1957	Black weather-resistant polyethylene.
Civil Aeronautics Administration (FAA–1046B)	Polyethylene insulation and vinyl jacketing for telephone cable.
Civil Aeronautics Administration (893B)	600-volt control cable, polyethylene insulation, PVC jacketing.
Federal Specifications	
J–C–94b	Vinyl-insulated power cable, 600 V maximum.
J–C–129c	Flame-retardant (vinyl) building wire, 60°C and 600 V maxima.
Insulated Power Cable Engineers Association (IPCEA)	
S–19–81, 2nd Edition	General specifications for wire and cable with rubber and plastic insulations.
S–61–402, Interim Standard No. 2	Polyethylene insulation on power cable = NEMA specification WC–5–1961.
Military Specifications	
MIL–C–17C & 17D	Radio frequency cables, coaxial and twin. Polyethylene insulation, black vinyl jacket.
MIL–C–915A(SHIPS)	Wire and cable for shipboard use. Many types of insulations and sheaths covered; mostly vinyl, some polyethylene.
MIL–W–76B	Hook-up wire, polyethylene or vinyl insulation.
MIL–W–1904	Wire for submarine mines; must pass part of Federal J–C–129.
MIL–C–2681A(SHIPS)	Power cable for special use on merchant ships. Refers to MIL–C–915A. Vinyl.
MIL–W–3093(1)	Wires for telephone construction. Vinyl insulation.
MIL–W–5086A	Electrical aircraft wire, vinyl insulation.
MIL–I–3930B	Covers compounds for vinyl coverings.

MIL–W–6370B	Polyethylene-insulated wire for aircraft antennae.
MIL–C–10581B	Insulation and jacketing (polyethylene and vinyl) for telephone cables and equipment.
MIL–C–11311(SigC)(3)	Telephone station cable, vinyl insulation and jacket.
MIL–C–13294B	Polyethylene insulation, nylon-jacket telephone cable (called "infantry field wire").
MIL–C–13777D	Missile cable (polyethylene).
MIL–C–15325B	Vinyl jacket, polyethylene insulation, three-conductor tow cable.
MIL–C–15376(SHIPS)	Like MIL–C–15325B, six-conductor cable.
MIL–C–15479C(SHIPS)(1)	Submarine cable, polyethylene insulation, vinyl jacket.
MIL–W–16878C(NAVY)	High-temperature vinyl and polyethylene wire for instrument panels and meters.
MIL–W–19150(NOrd)	Brown polyethylene insulation, nylon jacket.
MIL–C–19547(SHIPS)(3)	Polyethylene jacket and insulation, shore communications cable.
MIL–W–21306(SHIPS)	Vinyl-covered colored switchboard wire.
MIL–C–27072	Flexible, nonportable missile cable.
MIL–C–55036(SigC)(6)	Telephone cable, polyethylene insulation and jacket, vinyl outer jacket.
MIL–LP–390	Polyethylene material specification.
Rural Electrification Administration	
PE–15	Distribution wire, black polyethylene insulation, vinyl jacket.
PE–17	Aerial distribution wire. Black polyethylene.
PE–19	Black polyethylene-insulated bridle wire.
PE–20	Vinyl insulation and jacket for telephone wires from outside to inside of homes.
PE–21	Black polyethylene. Line wire.
PE–22	Polyethylene insulation and jacket for telephone cables.
PE–23	Same as PE–22, for direct burial.
PE–24	Polyethylene insulation, vinyl jacket, telephone distribution wire for direct burial.

PE–27	Figure-8 distribution wire, high-density polyethylene as insulation, low-density jacket.
PE–28	Figure-8 distribution wire, same materials as PE–27.
PE–38	Figure-8 telephone cable, same materials as PE–27.
PE–44	Low-loss buried distribution wire, cellular polyethylene insulation, solid polyethylene jacket.
PE–45	Buried distribution wire, same materials as PE–27, plus a high-density polyethylene outer jacket.
PE–46	Figure-8 drop wire, high-density polyethylene insulation, semirigid PVC jacket.
PE–200	Basic raw-material specification for polyethylene insulation and jacketing.

Civilian Standards. Underwriters' Laboratories maintains a series of specifications to be passed in order to gain their seal of approval; these concern flammability and heat resistance. Some of these specifications for vinyls are listed in the table below. Western Union and IBM are among the major users of wire and cable who have set specifications for the products they use. And Western Electric, maker of most American telephones, has many specifications for its products.

Specifications of Underwriters' Laboratories for Vinyl-Insulated Products

Designation	*Description*
T and TW	General-purpose and moisture-resistant building wire, 60°C; also used for switchboards, machine tools and control circuits.
THW	Moisture-resistant, heat-resistant building wire, 75°C. Similar to TW wire in application.
NMC and UF	Nonmetallic sheath cable (corrosion-resistant) and underground-feeder cable. All-thermoplastic construction, for direct burial.
NM	Nonmetallic sheath cable.
SPT	Parallel cords, 60 and 75°C, for lamps and appliances.

Designation	*Description*
SVT	Flexible cord, light duty, for vacuum cleaners, mixers, etc.
SJT	Flexible cord, medium duty, for refrigerators, washing machines, etc.
ST	Flexible cord, heavy duty, for portable tools, motors, portable power supply.
SRT	Cables for electric cooking ranges.
TF	Wiring of electrical fixtures.
TFF	Flexible fixture wire.
TA	Switchboard wire.
XT	Two-conductor cord for Christmas-tree and similar light-duty uses.
CXT	For outdoor Christmas-tree wiring.
GTO	Gas tube, sign and oil-burner ignition cable. For neon signs.
Blanket Wire	Electric blankets
Appliance and Hookup Wire	Machine tools and radio control circuits.

TESTS

Thermal Aging. The plastic-coated wire is placed in an oven at a temperature that correlates with the particular end-use. After a standard period, e.g., seven days, the tensile strength and elongation are measured and compared with unaged samples. For jacketing, the jackets are carefully stripped from the wires and similarly aged and tested. The test for thermal embrittlement consists of wrapping the wire around itself as in Figure 7.9, and keeping it in an oven at or near its maximum service temperature, e.g., 65°C. The wire is peri-

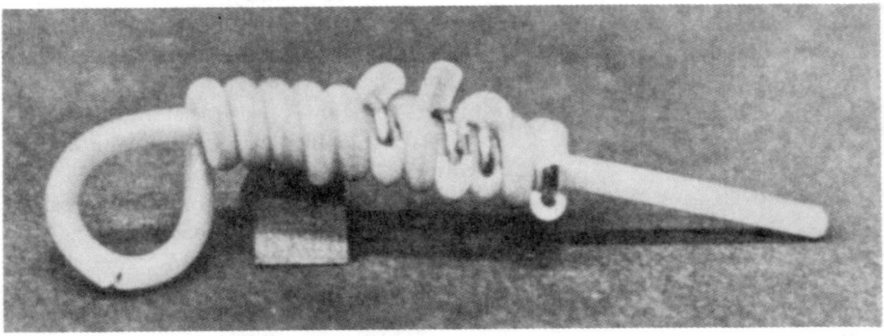

Figure 7.9. Thermal embrittlement test; this polyethylene-covered wire was aged 24 hours at 70°C. (*Courtesy of Union Carbide*)

odically inspected, and the time to develop cracks is determined. Testing for resistance to thermal cracking is included in specification PE–200 noted above.

Heat Shock. This test is used to determine stresses in the plastic insulation and is really a form of shrinkage test. The wire is wrapped around itself as in the thermal embrittlement test, and it is placed in an oven at a high temperature (e.g., 130°C) for 30 or 60 min. By this time, a highly stressed insulation will show cracks, while a good insulation will remain whole. The cracks only appear where the plastic is stressed further by the wrapping.

Physical Resistance to High Temperatures. This is measured by various mechanical tests at high temperatures. It is important in applications where the wire gets hot or remains in a hot environment, and is related also to resistance to oxidation. Cross-linked polyethylenes are valuable for such applications.

Resistance to Low Temperatures. As much wire and cable is used outdoors, it is important to maintain strength and flexibility at low temperatures. Tests on the wire itself are included in the IPCEA specification S–61–402 and REA specification PE–38, both noted above, as well as ASTM test D–746 for brittleness temperature of molded specimens.

Weather Resistance. This concerns resistance to degradation due to action of ultraviolet light. Carbon black is the usual protective agent, and its effectiveness is measured by testing its content (ASTM test D–1603, 2% minimum needed); average particle size (20 millimicrons or less); degree of dispersion (REA specification PE–200); and light absorptivity (ASA standard C–835).

Environmental Stress-cracking. This test—indeed, the whole topic—has been a subject of much discussion and study for many years. It is important for polyethylene pipe as well as insulation, because these both have long-term applications where physical properties must function without serious loss for years. The Bell Laboratories test is the most common; in this test a longitudinal slit is made in small test strips of the plastic and each strip is bent double with the slit surface in tension. The strips are submerged in tubes of a water solution of a known stress-cracking agent (a polyglycol such as "Igepal" or "Hostapal"), at a standardized solution concentration, and temperature (see Figure 7.10). The resistance to stress-cracking is measured by the time it takes to show cracks at the slit. Some samples fail in a few minutes, while others—the lowest melt indices, the polyethylene copolymers, and compounds containing butyl rubber—last a month or more. Other tests are also used, including service tests in which actual coated wire is immersed in the polyglycol. Details of stress-cracking tests and their results are in ASTM test D–1693 as well as in specifications PE–38 and PE–200 noted above.

Chemical Resistance. This is usually tested on compression-molded squares which are immersed in the reagents and later measured for dimensional change and weight loss. Where wires are to be used in industrial plants, or are to be buried in reactive soil, the expected reaction of the plastic must be known. Stability in contact with copper is measured by ASTM D–2342.

Abrasion Resistance. This is important where wires are to be strung unfixed either overhead or on the ground. Nylon and poly-

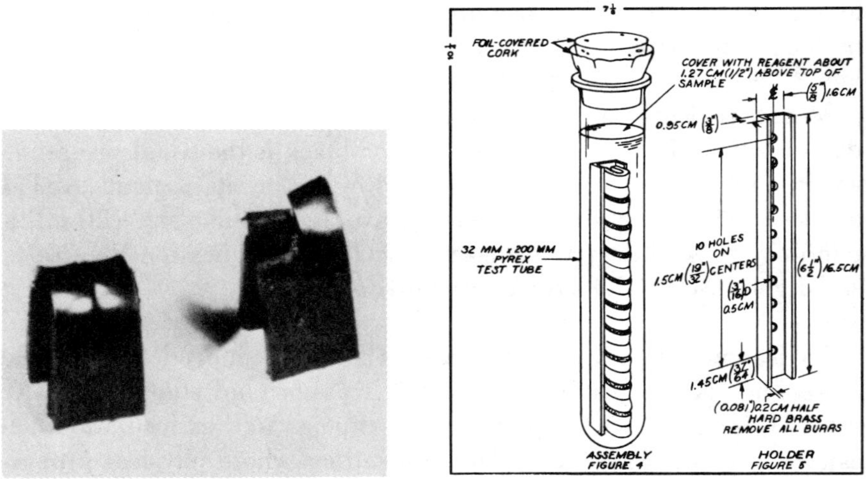

Figure 7.10. Left: Stress-cracking test specimens, before and after cracking. Right: Diagram of testing apparatus. (*Courtesy of Bell Telephone Laboratories and ASTM*).

sulfone are used where maximum resistance to abrasion and "cut-through" is needed, sometimes as an outer covering over another less expensive plastic. In extreme cases, steel braiding is placed around the entire insulated cable. Abrasion resistance is tested on molded bars or plaques which are rubbed at constant load against standard abrasive surfaces. The number of cycles necessary to wear the plastic down a given distance is measured. Practical testers have also proven useful; one tests the insulated wire by rubbing it back and forth over a metal abrading edge. When the insulation wears through, an electric circuit is completed through the now-exposed metal wire; a buzzer is set off, and the cycle-counter is stopped automatically at the exact number of "rubs" to failure. Maple wood is also used in abrasion tests, to simulate the rubbing of insulated wire in trees.

Electrical Properties. Plastics are able to insulate current-carrying wires with a minimum of distortion or reduction of the current. Their effectiveness in doing this is measured by a variety of electrical tests:

Power Factor. This is tested at various AC frequencies, such as 1 kc, 100 kc and 1 mc. It is a measure of how much energy is lost from the conductor into the plastic because of the constant switching of polarity in the plastic molecules as the AC voltage continually reverses direction. This energy appears as heat in the plastic. Non-polar plastics such as polyethylene are excellent in this respect, and they are therefore used for voice transmission (minimum distortion of signal) and long-distance power transmission (minimum loss of power).

Dielectric Constant. Where there are two metal lines separated by a plastic, as in a coaxial cable, the combination forms a capacitor, building up a voltage between the two lines over and above the transmitted signal. This voltage reverses itself with each cycle of alternating current. The current passing through the *wire* that is required to keep charging and discharging the capacitor represents an electrical loss; it is dependent on the *dielectric constant* of the plastics. This is measured in the same test as the power factor (ASTM test D–150, ASTM test D–1673 for cellular). The dielectric constant is most important where small signals are transmitted at high frequencies, e.g., telephone, undersea cable. With large currents at low frequencies, the corresponding power loss is negligible.

Dielectric Strength. This is a measure of how much voltage (AC or DC) is needed to break through or puncture a standard gap of the plastic (ASTM test D–149), making it a conductor; it is expressed in volts per mil.

Direct Current Resistance (DCR). This measures the amount of current that actually leaks through the plastic insulation and is lost. It is important for long-distance power lines and, once again, polyethylene is outstanding (ASTM test D–257).

The extreme concern that is often seen regarding the nature and dispersion of additives in wire insulating compounds, as well as contamination, arises from the fact that almost anything added to plastics impairs their electrical properties. Therefore, such additives must be efficiently utilized and thereby kept to a minimum.

These tests are ordinarily conducted on plaques of the plastic, but sometimes the final insulated product is tested. The outstanding example is the testing of huge reels of undersea cable by the immersion of the entire reel in water, in a specially built tank 12 ft in diameter and 5 ft deep (4 meters diameter x $1\frac{1}{2}$ meters deep). There is also an Underwriters' Laboratory test (UL–83) for electrical stability of insulated wire in water.

PROBLEMS AND SOLUTIONS

Common problems in wire and cable covering are discussed below.

Problems	*Causes and Solutions*
Rugosity (surface dullness or roughness)	(a) Melt fracture, with very small wires. Alter die to have longer lands or shallower approach angles, or run hotter, or run slower.
	(b) Poor mixing, especially with vinyl dry blends. Run at higher pressure, or preblend more thoroughly, or use shallower screw, or cool screw, or run slower.
	(c) Moisture. Dry material more thoroughly, or use vented extruder.
	(d) Porosity (trapped air). Use vented extruder or vacuum hopper. Increase melt pressure in extruder.
	(e) Cold compound temperature. Run hotter.

Problems	*Causes and Solutions*
	(f) Decomposition (especially with vinyls, accompanied by discoloration). Run colder, or use better-stabilized formula, or more streamlined head and die. (g) Too much drawdown from a tubing die.
Pimples and pocks	(a) Air bubbles on wire in trough during cooling. Shield water and wire entries to keep air out. (b) Steam bubbles on wire in trough. Move water faster to prevent boiling on plastic surface.
Lumps and bumps	(a) Water drops splashing on plastic between die and trough. Shield the area to prevent this. (b) Moisture or trapped air. See (c) and (d) in preceding category. (c) Bits of plastic drooling or sticking at die and then transferring to emerging hot surface. Possibly caused by moisture. Otherwise, apply flame to die face, or coat with silicone before starting, or change exit geometry.
Variable thickness (linearly)	(a) Surging in the extruder. Check uniformity of mechanical motions or causes of poor mixing (see rugosity, above). (b) Variation in take-off. Check steadiness of capstans, take-up reels, etc. (c) Controller lag too large. Check amount of variation necessary to actuate feedback. Check for malfunctions in the control system. (d) Varying temperature. Check for variation in temperatures of die and material, in pressure, and in motor load.
Off-center coating	(a) Die out of adjustment. Change bolt settings. (b) Hot or cold spots in die. Check with pyrometer or visually. Compensate with centering bolts if cannot change temperature.

Problems	*Causes and Solutions*
	(c) Internal die design not properly compensating for the bend in the crosshead. Change compound temperature or use a different head. (d) Plastic sags under its own weight. Run cooler, or change to higher-viscosity compound.
Grooves, parallel to wire	(a) Die lines. Grind off nicks or burrs on die lips. Clean out hardened decomposed bits of resin. (b) Abrasive point in take-off system near die, that is scraping against insulation while still warm. Example: heavy cables in long troughs will scrape against the trough bottoms unless very high tension is applied.
Electrical properties not normal	(a) Contaminated, oxidized, or reworked resin. Possible excess of some ingredient in the formulation. Test a plaque made from the feed pellets to make sure. (b) Test conditions not right. Check operation of instruments with a control sample of known quality. Check test sample for external dirt or moisture. Consider possible interference with instruments by radio transmitters, induction heaters, radio-frequency heat sealers, and the like. (c) Voids or porosity in insulation. Source could be moisture, decomposition, trapped air, too fast shrinkage during cooling. (d) Moisture in or on test sample. Dry sample before testing, or keep moisture out of extruder feed. (e) Off-center coating (see above).
Separation of covering from conductor	(a) Poor adhesion. Preheat wire more, or run compound hotter, or apply an adhesive. (b) Poor contact in tubing die. Use more vacuum (at least 20 mm mercury) or run hotter.

Problems	*Causes and Solutions*
	(c) Dirty or moist wire. Preheat more, or pass through cleaning device before entering die.
	(d) Decomposing plastic. Run colder, use stabler compound.
	(e) Cooling too fast. As plastic shrinks, it separates from wire. Run cooler, or use a temperature gradient in trough (hot end nearest die), or increase air-cooling gap between die and trough (too long a gap, however, will cause distortion of dimensions).
	(f) Air trapped along conductor. Apply vacuum to guider tip.

SPECIAL TECHNIQUES

Post-handling. After a simple wire is covered with plastic, many operations may still have to be performed. For example, telephone handset cord is twisted over special mandrels to give it a permanent curl. Wires may be twisted together along with paper or metal strip or other wires, or around a steel core, and this entire cable may be jacketed with more plastic. Frequently, the wire is printed with its rating and the manufacturer's name; this is done either right in line with the extruder or else afterwards, where the wire is uncoiled, passed through a continuous printer and re-reeled. Sometimes the entire color of a wire is thus applied, i.e., uncolored plastic is extruded over the wire and the finished wire is run through the coloring machine, which gives it a solid color, or stripes it or does both. This has the economic advantage of requiring only one natural-color material in stock, yet still enabling manufacture of many colors.

Undersea Cables. The manufacture of these huge cables (see Figure 2.8) has become quite a specialized operation. Western Electric has developed a process, illustrated in Figure 7.11, and they have licensed manufacturers all over the world. These installations, built around only a few large extruders [e.g., 100 hp, 4½ in. (120 mm) in diameter], are truly impressive. Before extrusion, the uncolored low-density polyethylene pellets are *visually* inspected in an antiseptically clean room to eliminate all chance of contamination and resultant electrical deficiency. An elaborate capstan system feeds the

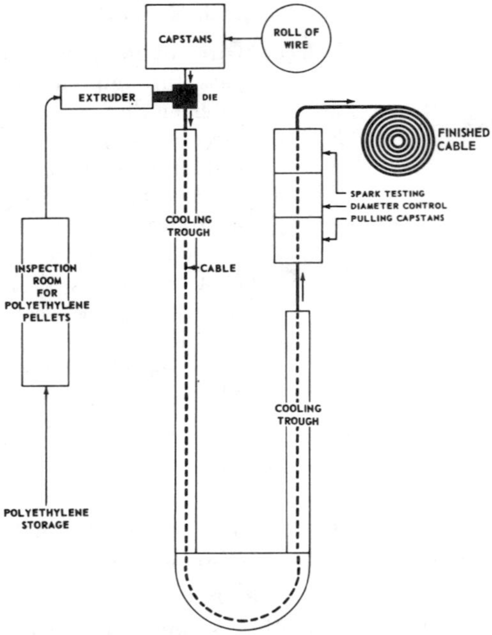

Figure 7.11. Plant for the manufacture of undersea telephone cable. This floor plan shows the exceptionally long cooling trough needed to cool the thick polyethylene insulation. (*Illustration by author*)

crosshead with 1/4- or 3/8-in. diameter (6 to 9 mm) solid copper wire over which about 1/4 to 1/2 in. (6 to 12 mm) of plastic is extruded. The wire passes at about 50 ft/min (15 meters/min) through over 275 ft (85 meters) of *enclosed* water trough, with carefully controlled and gradually decreasing temperature, starting at 80°C and ending at room temperature. The trough is so long that it is doubled back along itself around a 6-ft (2 meters) diameter to avoid the need for a building with more than 275 ft (85 meters) of free space in a line. The cooled cable is taken up through a capstan system as elaborate as the feed capstans, and the reels are subjected to numerous quality tests before final approval. Some extrusion lines for submarine cable differ slightly from the one just described, but they are all basically similar, and yield a highly-engineered product capable of years of service with minimum repair.

Cellular Insulation. Cellular polyethylene primary insulation is extruded by adding a chemical foaming agent to the compound before extrusion. Some extruders formulate their own compounds, but most buy the standard offerings of the major suppliers. The foaming agents are usually complex organic compounds which decompose at extrusion temperatures, liberating nitrogen. They are compounded into the plastic at low temperatures at which they do not decompose. Data are available on speed and temperature of gas evolution from the makers of these agents. Successful employment of these agents depends on careful control of material temperature to achieve "blowing" (gas evolution), just when the plastic is leaving the die. If extrusion temperature is too high, the agent blows too soon—too far back in the extruder—and the bubbles will coalesce before leaving the die. This yields a rough-surfaced, large-celled structure. If temperature is too low, the agent will not completely decompose by the time the wire leaves the die and starts to cool. Thus, not all the blowing potential will have been utilized, and the product will be too dense, as well as uneconomical. The die construction is critical. The land length (parallel or near-parallel section just before lips) should be as short as possible, so that once the agent blows, the plastic leaves the die as soon as possible.

Extrusion temperatures vary for the different foaming agents (over a dozen have been extensively studied) but most fall betwen 140 and 200°C. The take-up lines may have capacitance-measuring devices which measure the degree of blowing. Densities of 0.4 to 0.7 grams/cc are common for these products. Their advantages include light weight, which makes them easier to install and a lesser strain on their supports if suspended, low material cost per foot (meter) of wire, and a very low dielectric constant.

Cellular polypropylene created much interest from 1960–1962 because of its high abrasion resistance and service temperature. It requires special stabilization against aging and copper corrosion, and is extruded at higher processing temperatures (both preheat and material). It has found some special uses, but in general has not replaced cellular polyethylene, which is cheaper and easier to extrude.

Striping. Wires quite often show stripes or bands of color which aid in the immediate identification of the wire's capacity rating. These are commonly printed, but can also be extruded as a thin strip of colored plastic of the same kind as the insulation. This similarity allows quick permanent adhesion of stripe to base. Little extruders are specially designed for this purpose, and the entire unit may travel around the wire during operation.

Two-coat Simultaneous Extrusion. Another unusual technique is used to give a wire two coats in one process. For example, polyethylene might be used as a less costly primary insulation while nylon is employed as an abrasion-resistant outer coat. Two extruders do this by feeding into a common die specially designed to direct the two plastics over the wire in the right order. Some machine manufacturers offer a package unit—extruders and special die—just for this purpose.

Another technique uses one extruder with a screw reverse-flighted along its second half. There is a hopper at *each end* of the screw, and one end is driven while the other rides in a bearing. The two plastics are fed into the two ends and are extruded toward the middle, where they enter a die set in the barrel wall halfway down the screw. The original system uses a twin-screw machine, but a single-screw machine should work also. This method is simpler than the two-extruder method as it requires less equipment. However, it is less versatile because there is only one drive and the two extrusions therefore are not independent. Some independent control can be gotten by controlled feeding. It is interesting to note that in this machine the thrusts in the two halves of the screw tend to cancel each other out, leaving a reduced load on the thrust bearing. This is especially important with twin-screw machines, which are sometimes limited in their speeds (hence in their outputs) by the need for small thrust bearings.

Coating Multiple Wires. Double wires, such as television lead-in wire and lamp cord, are usually coated in a single operation. This is done through a die built to take and center the two wires. Pressure-type dies are preferred. Of course, cable jacketing through tubing dies is also an example of multiple wire coating.

Cable-In-Duct. Power cable for burial is often made with a polyethylene conduit continuously extruded over the cable. This eliminates the need for making conduit separately, and then threading the cable through it. The cable is fed from a reel through a crosshead die, which extrudes the conduit around the cable; the combination is then handled much like polyethylene pipe.

Dry-coloring. Dry blending of vinyls was mentioned above under Operation—Formulation. Often colorants are added to this blend as powders, and sometimes these powders are tumbled with precompounded pellets. Another very common technique is the use of a masterbatch, i.e., a concentrate of colorant in the plastic it is to be used with, sold in pellet or chip form. These coloring techniques allow the extruder to buy plastic in larger lots, all uncolored, which minimizes his obsolete materials and may reduce his material costs. They do give him the added responsibility of deciding what colorants to use and of operation in such a manner (good mixing) that a uniformly colored product results. Formulation of the colors is aided by the commercial availability of color formulas known to give many standard wire colors.

Cross-linking. Polyethylene can be cross-linked either with chemical agents or by irradiation. The latter method has been thoroughly studied but is too costly for most wire and cable uses (although it is in commercial use with polyethylene film). Chemical cross-linking, however, is less expensive, and is now being used in wire and cable coverings. Organic peroxides are the usual cross-linking agents. This cross-linked covering is more resistant to heat, solvents, and many other chemicals; in fact, the plastic can be so changed that the covering does not ever melt, but rather becomes rubbery at very high temperatures. Service temperatures for cross-linked polyethylene are, for example, 90°C continuous, with intermittent exposure to 130°C. Typical uses are as line wire, weatherproof jacketing, low-voltage power cable, and some building wire.

A large proportion of carbon black (25 to 60 per cent) is often used in compounds to be chemically cross-linked. This acts as a filler and increases rigidity, and its improvement in thermal conductivity aids in processing. It also improves weathering and abrasion resistance. The carbon black does not actually participate chemically in the cross-linking reaction, unlike the vulcanization of rubber. Cross-linked polyethylene compounds are often referred to as "vulcanizable" compounds, even though the chemical reactions are different. Some cross-linkable compounds do not contain carbon black at all; they are referred to as "unfilled" and thus have superior electrical properties.

The cross-linking agent is included in the extrusion compound, which is then extruded at a low temperature to prevent reaction in the extruder or the die. A typical temperature for unfilled material is 115 to 125°C. Temperature control is critical; if the extrudate is too cold, hard *uncured* particles occur in the insulation—and if too hot, the compound begins to cure prematurely in the extruder, also producing hard particles of *cured* material in the product. The nature of the particles, as well as the degree of normal cure, is measured by determining weight loss after boiling in a solvent such as decalin.

The temperatures of the head and screw are often controlled with steam or hot water. One supplier recommends a screw of the following dimensions for his unfilled compound:

Compression ratio: 3 to 1
Length/diameter ratio: 15 to 1
Melt zone: four flights of depth h, in accord with the equation:

$$h \text{ (mils)} = 60 + 18D \text{ (}D\text{ in inches)}$$
$$\text{h (mm)} = 1.5 + 0.018D \text{ (}D\text{ in mm)}$$

After leaving the extruder, the insulation is then cured as it passes through a long, heated chamber. Continuous vulcanizer tubes, as used for rubber, are usually satisfactory. Because of the low extrusion temperatures, it might be expected that the resistance and the corresponding internal melt pressure would be very high. However, this process is normally used for covering rather large cores, and the large cross-sectional area of the annular die opening yields a relatively low resistance even at low temperatures.

Chemical cross-linking can also be done by extruding a normal polyethylene on normal equipment, and then passing the covered wire through a bath containing the reactive agent. This method has been adequately studied, but it is somewhat limited and inefficient at high wire speeds, and consequently has not found much commercial use. This method may link the outer surface more than the interior of the plastic covering, which is sufficient for increased chemical resistance but not for improved heat resistance—where the plastic must be resistant all through its thickness.

Semi-conductive Insulation. Where some degree of electrical conductivity is needed, semi-conductive jacketing is applied. Large amounts of certain carbon blacks are added to the extrusion compound, somewhat similar to the carbon-filled cross-linkable compounds described above. The products are used in shielding of cables, replacing impregnated textiles or plastics. Extruders prefer these high-black compounds to be made by the material suppliers (and, occasionally, vice versa). Furnace blacks are used to get maximum conductivity, instead of the channel blacks, which are used in other extrusion formulation to give weather resistance. The percentage needed for semi-conductivity (30 to 50%) is so high that either type of black confers adequate weather resistance.

Extrusion is done on standard equipment, but care must be taken to avoid excessive shearing, which will separate the particles of carbon black and thus reduce conductivity. Screw channel depth is 25–50% greater than for conventional polyethylene, and a high *rear* barrel temperature is preferred. Typical melt temperature is 185°C, but the die and head, as well as the front barrel zone, are held at about 125°C. Extrudate temperatures much above 185°C should be avoided, because they yield a porous extrudate. Hopper-drying is recommended to remove moisture adsorbed by the carbon black.

Precoating of Wire. An adhesive is sometimes applied to the conductor or other core before extrusion. This precoating is more often done on metal conduits and even regular steel pipe, as noted in Chapter 3 on Pipe. The adhesive is continuously applied just before the conduit enters the crosshead die.

Covering of Things other than Wire or Cable. The technique of covering a cross section with plastic by passing it through a crosshead can be applied in many nonelectrical industries. For example, wooden rod and steel rod are covered both for protective and decorative purposes. Special feeding devices must be built for these inflexible cores, but the remainder of the process is like that for cable covering. Cotton rope and steel braid are coated to make colorful ropes resistant to water and other chemicals, but with the advantages of economy or strength of the base materials. Also, nylon, rayon, cotton, or virtually any thread or filament can be coated with plasticized PVC or polyethylene to combine the properties of the core and coating.

Simultaneous Conductor and Coating Production. In a recent development, an insulated cable is produced by extruding a polyethylene tube and simultaneously filling it with liquid sodium. This technique eliminates the whole payoff section of the equipment and simplifies the die. Furthermore, it allows the use of sodium as the conductor, giving advantages of flexibility, light weight, and high conductivity. Special equipment is needed to splice this product because exposed sodium is highly reactive.

REFERENCES

Applications

"Plastics in wire and cable," D. V. Rosato and staff, *Plastics World*, June 1967. Thorough review of markets for plastics in this field.

"Evolution of sheath construction for communications cables," R. C. Mildner *et al.* (Dow), *Wire and Wire Products*, June 1967. Nine pages, seventeen references, review past and present practices, particularly constructions with solid and foamed polyethylene insulation, and metal-plastic laminates (bonded jackets).

"Low-modulus materials for coaxial cable," R. M. Kell *et al.* (Battelle), *Wire and Wire Products*, April 1967. Study of materials—various polyolefin polymers.

Three articles on the use of plastics in telephone cables in Australia, France and the United States, presented at a conference of the *Plastics Institute*, April 11 and 12, 1967 (London).

"Service requirements and fault incidence in the field," A. F. G. Allan *et al.* (General Post Office), *Plastics Institute Conference*, April 11 and 12, 1967 (London).

"The progress made by polymers in the French electrical cable industry," H. Lemainque, *Plastiques Modernes et Elastomères*, March 1967.

"Polyethylene-insulated power cables in Europe," F. F. Humbel (Union Carbide Belgium), *Wire and Wire Products*, February 1967. Thorough, well-illustrated review of history, costs, design and experience in several countries.

Transactions of the Plastics Institute, special issue on cables, June 1961. Six full length technical articles on plastics and rubber in cables covering stabilization, cable construction, irradiation, etc. No extrusion methods discussed; only materials and applications.

"Engineering thermoplastics for ocean telephone cables," J. B. Howard (Bell Laboratories), *Plastics Technology*, May 1959. Design and properties of cable insulation, and description of tests.

"Thermoplastics in cables," *British Plastics*, February 1959. Construction and usage of electrical cables.

Polyolefins—Extrusion and Performance

"Developments in polyethylene cable material," D. A. Williamson (Shell), *Plastics Institute Conference,* April 11 and 12, 1967 (London).

"Low-loss polypropylene for electrical purposes," W. Reddish (I.C.I.), *Plastics Institute Conference,* April 11 and 12, 1967 (London).

"A review of polyolefin test requirements in wire and cable specifications," R. J. Martinovich and C. F. Leslie (Phillips), *Wire and Wire Products,* March 1967. Discussion of density, melt index, swell ratio and mechanical properties of raw materials, as they apply to electrical applications.

"Extruding propylene copolymers for wire and cable coating," W. M. McNally (Hercules), *Plastics Technology,* January 1967. Discusses wire preheating, extrusion and cooling temperatures, tests on wire, effect of extrusion variables on properties, comparison with other plastics. Earlier information in *Wire and Wire Products,* March 1963, G. P. Johnson (U.S. Steel).

"Tailoring of high-density polyethylene for high-speed wire coating," J. N. Scott *et al.* (Phillips), *Wire and Wire Products,* March 1966. To 5000 ft/min., with 0.95-density resin between 0.3 and 0.5 melt index. Properties, including rheology.

"New improved polyolefin wire and cable resins," R. J. Martinovich *et al.* (Phillips), *Wire and Wire Products,* March 1965, Review of low- and high-density polyethylene, high-molecular-weight polyethylene, cross-linked high-density. Physical properties and advantages of each. Earlier survey in *Wire and Wire Products,* November 1961.

"Evaluating the processing characteristics of wire and cable coating polyethylene resins," J. E. Hager and P. E. Jacobs (U.S.I.), *SPE Journal,* January 1965. Discussion of special extrusion tests which correlate with wire extrusion behavior.

"The continuous extrusions of flexible conduit over electrical cables," W. K. Atkins (Union Carbide), *Wire and Wire Products,* June 1963. Extrusion conditions, equipment needs, die design, cooling techniques, chart of problems and solutions. Polyethylene is the material.

"Cooling curves for extruded polyethylene wire," R. D. Biggs and R. P. Guenther (Bell Laboratories), *Modern Plastics,* May 1963. Thorough mathematical study of this topic with charts and drawings.

"Extruding hi-density polyethylene for quality and savings," E. P. Stiles (Grace), *Wire and Wire Products,* February 1963. Conditions of extrusion, advantages of high-density polyethylene, need for wire preheat, cooling, moisture, resin stability.

"Polymer requirements and insulation techniques . . . for high density polyethylene," R. J. Ettinger (Goodrich-Gulf), *Wire and Wire Products,* September 1961. Die design, material requirements and operating conditions, problems and how to avoid them. Similar information by Union Carbide personnel in *Wire and Wire Products,* July 1960.

"Properties and processing of ethylene copolymers for wire and cable," D. J. Vaughn and W. W. Spohn (du Pont), *Wire and Wire Products,* October 1960. Mainly properties, some operating conditions and die diagrams. Similar article by Phillips personnel in *Wire and Wire Products,* April 1960, with description of recommended equipment.

PVC—Extrusion and Performance

"The effect of the component parts of a PVC compound on high speed extrusion performance and on electrical properties," C. M. Thomas (British Geon), *Plastics Institute Conference,* April 11 and 12, 1967 (London).

"A vinyl chloride graft polymer resin for wire insulation and cable jacketing," H. M. Leeper *et al.* (Monsanto), *Wire and Wire Products,* May 1966. Properties of Opalon 740, typical formulations for solid and foam, rheology. No extrusion conditions. Needs less plasticizer, gives better electrical properties.

"The extrusion of wire insulation and cable jacketing from PVC dryblend," F. R. Nissel (Prodex), *Wire and Wire Products,* September 1965. Advantages, problems with dryblend; equipment recommended: vented Prodex, Henschel mixer, 30° crosshead, screen changer.

"Advances in dry blend extrusion," N. T. Flathers *et al.* (Escambia), *Modern Plastics,* May 1961. Similar material appears in *Wire and Wire Products,* June 1960 and June 1961. Article concerns vacuum hopper and fine screens to improve appearance and reduce porosity in dry blended vinyls. More information on dry blends in an article by E. J. George (Western Electric), *Wire and Wire Products,* March 1961, showing equipment and methods which Western Electric uses to mix their dry blend.

"Properties of various PVC materials for cable insulation," G. Buhmann and K. Grill (Standard Elektrik Lorenz), *Kunststoffe,* January 1961. Study of plasticizer absorption, particle size, etc., for plasticized dry-blended vinyl mixtures, with microphotographs and strength data.

"The manufacture of telephone cords," H. L. Wessel (Western Electric), *Wire and Wire Products,* October 1960. Complete description, from bare wire to finished coiled cord, of Western Electric's facilities.

Other Materials—Extrusion and Performance

"Nylon 6 as a wire jacketing material," C. W. Taylor (Spencer), *Wire and Wire Products,* March 1960. Conditions, material requirements and properties of nylon wire coating.

"Paste extrusion of filled TFE resin," R. W. Johnson (du Pont), *SPE Journal,* February 1961 and *SPE ANTEC* 1961. Ram extrusion of polytetrafluoroethylene. Information on melt extrusion of Teflon 100, by du Pont personnel, appears in *Plastics Technology,* August 1960.

Equipment, Design and Operation (except Dies)

"Developments in cable manufacturing equipment and in control and measuring equipment," B. K. Bigland (General Engineering—Radcliffe), *Plastics Institute Conference,* April 11 and 12, 1967 (London).

"High-speed wire coating: Its application in polyolefin product development," B. A. Eberhardt and R. G. Rodgers (du Pont), *Wire and Wire Products,* October 1962. Describes du Pont's 4000-ft/min. experimental line, $2\frac{1}{2}''$, 40 HP. Shows photos and diagrams of equipment, plus experimental data.

"The 'piggyback,' portable extruder for simultaneous multicolor application," J. M. du Pont (Thermoplastic Equipment Corporation), *Wire and Wire Products,* June 1962. Describes machine and its uses, with photos and some designs of the dual-entry dies.

"Wire coating with polyethylene resins," F. H. Skewis (du Pont), *Wire and Wire Products,* October 1961. Discusses instrumentation mainly on extruder but also on winding machinery.

"Simultaneous extrusion of polyethylene and PVC coatings on electrical conductors," G. E. Henning (Western Electric), *Wire and Wire Products,* 1961. Thorough analysis of equipment with some operating conditions. Two extruders feed one die.

"Synchronized extruder capstan drives for high speed insulating lines," A. W. Harrington and H. L. Kelley (Louis Allis), *Wire and Wire Products,* December 1960. Pictures and diagrams of various synchronized systems, with advantages of each.

"Wire speed calculations for extrusion machines," W. J. Owens (General Electric), *Wire and Wire Products,* September 1960. Four nomographs relating wire speed in feet per minute to extruder output and wire dimensions.

"Insulated wire manufacturing costs trimmed with diameter control gauge," W. F. Zimmerman (American Metal Molding Company), *Wire and Wire Products,* May 1960. Describes line for vinyl coated building wire including photoelectric gauge which feeds back to control the wire speed through the extruder.

"The control of eccentricity in electrical cables," F. E. Planer (Addison Electric Company), *Wire and Wire Products,* April 1960. Causes of eccentricity, how it is controlled (by capacitance measurements).

Die Design and Melt Fracture

"Analytical methods of wire coating die design," J. G. Williams (Imperial College), *Plastics Institute Conference,* April 11 and 12, 1967 (London).

"Flow instability of polymer melts," M. T. Dennison (Shell), and "The mechanism of melt distortion," F. N. Cogswell and P. Lamb (I.C.I.), *Plastics Institute Conference,* April 11 and 12, 1967 (London).

"Absolute non-eccentric crosshead of screw extruder," N. Okazaki (Consultant), *Wire and Wire Products,* April 1966. Design and calculations for head with spirally-grooved core, and special die design. Original work in *Wire and Wire Products,* June 1964. Applies to film and pipe too.

"Precentered pressure coating die," N. C. Wyeth (du Pont), *SPE Journal,* October 1965, and *SPE ANTEC* 1965. New wire coating die combining advantages of pressure and tubing systems.

"Extrusion die design to reduce melt fracture," A. G. Ferrari (Western Electric), *Wire and Wire Products,* July 1964. Uses capillary rheometer to study shear/surface relation, entrance angle effects; discusses multiple-taper dies.

"The effect of die land length on the surface appearance of low-melt-index polyethylene extrudates," J. S. King and A. J. Durno (Canadian Industries Limited), *Wire and Wire Products,* April 1964. $3\frac{1}{2}$-inch extruder, average output only 105 lb/hr, black material. Land varied up to 4 inches, results justify long-land dies to get best appearance.

"The design of dies for high speed wire coating," L. R. Hammond (Eastman), *Wire and Wire Products,* June 1960. Application of theory to die design, using tapered internal entrance angles to avoid melt fracture.

"Causes of melt fracture and its relation to extrusion behavior," R. M. Schulken and E. M. Boy (Eastman), *SPE ANTEC,* 1960. Analysis of melt fracture with corresponding experiments. Many references, including reference to basic work by J. P. Tordella (du Pont), *SPE Journal,* August 1957 and February 1956.

Aging, Weathering and Stress Cracking

"Aging of PVC compounds," D. M. Pugh (Distillers) and "Aging of polyolefins," S. L. Fitton (Shell), *Plastics Institute Conference,* April 11 and 12, 1967 (London).

"Stress cracking phenomena," J. B. Howard (Bell Laboratories) and "Factors affecting the stress crack resistance of polyethylene cable jacketing," G. W. Daues and F. P. Evans (Monsanto), *Plastics Institute Conference,* April 11 and 12, 1967 (London).

"Effect of extrusion processing on environmental stress cracking resistance of polyethylene sheathing compounds," J. R. Apen (Bell Laboratories), *SPE Journal,* July and August 1965. High extrusion temperature, low output, and deep metering zone on screw, all yield less cracking.

"Heat aging characteristics of cross-linked polyethylene," P. R. Tully (Cabot), *Wire and Wire Products,* July 1965. Contains 37% carbon black . . . compared to regular polyethylene, five pages, much data. Further information, also covering unfilled compounds, is in *Wire and Wire Products,* April 1961.

"Spannungsrissbildung bei Polyäthylen" (Stress cracking of polyethylene), I. L. Hopkins and O. W. Baker (Bell Laboratories), *Kunststoffe,* November

1959. Discussion and differentiation of thermal and environmental stress cracking, with causes and conditions of each. More information in *Kunststoffe,* September 1959, by E. Gaube (Hoechst). Long and comprehensive article, "A review of stress cracking in polyethylene" by J. B. Howard (Bell Laboratories), was presented at *SPE ANTEC* 1959. It described major tests, appearance and causes of failures, and included a list of almost three hundred environments with their effects on polyethylene.

"Wetterbeständigkeit von weichgemachten PVC-Masses für die Kabel- und Leitungstechnik" (Weather resistance of plasticized PVC for the cable industry), W. Birnthaler and G. Falk, *Kunststoffe,* September 1959. Study of changes in mechanical and electrical properties.

"Embrittlement of polyethylenes," C. S. Imig (Spencer), *Plastics Technology,* May 1959. Description of the nature of embrittlement and which polyethylenes resist it best.

Testing and Controls

"Some special electrical test methods for telecommunication cables," T. Hazen (Union Carbide), *Plastics Institute Conference,* April 11 and 12, 1967 (London). For additional testing information, see the ASTM Book of Standards, Part 29.

"Controlling polyethylene insulated wire capacitance," A. E. Hartman (Western Electric), *Wire and Wire Products,* October 1966. Describes installations at Western Electric's Omaha plant for telephone transmission cable. Electrical details; long (five-page) article.

"Specifying polyethylene insulation for wire and cable applications," W. H. Joyce and H. F. Tomfohrde III (Union Carbide), *Wire and Wire Products,* August 1966. Classifies material types, shows which are used for what applications, discusses test methods and data.

"Impulse dielectric testing applied to insulated wire," D. C. Alexander (International Telephone and Telegraph Co.—ITT), *Wire and Wire Products,* June 1966. Tests on reels, various plastics. Information on dielectric testing of polyethylene at high frequencies, by Union Carbide (Canada) personnel, is in *Wire and Wire Products,* October 1963.

"Controlling scrap at the extruder," Robert Sterns (Logic Systems), *Wire and Wire Products,* February 1966. Use of automatic outside-diameter gauge, keep oversize down, save material and save money. Shows how to calculate savings.

Cellular Insulation

"New coating process for foamed polyethylene insulated telephone cable," M. Sugi *et al.* (Sumitomo Electric), *Wire and Wire Products,* August 1967. No extrusion used, wire run through 30% polyethylene solution in xylene, novelty is in water-dip step which raises speed from 160 to 800 ft/min. See also *Wire and Wire Products,* August 1962.

"Foamed polyolefins as primary wire insulation," R. H. Hansen (Bell Laboratories), *Plastics Institute Conference,* April 11 and 12, 1967 (London).

"The use of chemical blowing agents in direct extrusion expansion," E. R. Dilley (Whiffen and Sons), *Transactions of the Plastics Institute,* February 1966. Concerns polyolefins and nitrogen-producing agent.

"The influence of extruder geometry in . . . cellular insulation," E. O. Bauer and P. E. Fox (Bell Laboratories), *Wire and Wire Products,* March 1962. Study of die design and flow problems with foamed polypropylene.

"Foamed polystyrene insulation coaxial cable," M. Yoshimura and A. Nago (Fujikura Cable Works), *Wire and Wire Products,* October 1963. Use foamed polystyrene *tape,* plus polyethylene film wrap; heat to weld.

"Extruded cellular vinyl," A. D. Varenelli (U. S. Rubber), *Wire and Wire Products,* July 1961. Techniques and results, especially for wire covering.

"Drying polyethylene sheathing compound prior to extrusion," F. H. Kruger (Western Electric), *Plastics Technology,* February 1961. Discusses moisture content and drying machinery for black polyethylene.

"Cellular insulation for communications cable," B. B. Pusey (Union Carbide), *Signal Corps Symposium,* 1961. Comparison of cellular polyethylene DGDA-2580 with other solid and cellular products; includes economics and extrusion conditions.

"Cellular polyethylene," W. T. Higgins (Union Carbide), *SPE Journal,* January 1955, and *Modern Plastics,* March 1954. Basic information on this material including processing and properties.

Semi-Conductive and Cross-Linked Polyethylenes

"Preparation and evaluation of conductive polyethylenes," B. Wargotz (Bell Laboratories), *Plastics Institute Conference,* April 11 and 12, 1967 (London).

"Vulcanizable polyethylene and hi-voltage cable insulation," J. R. Flesher (Union Carbide), *Wire and Wire Products,* April 1967. Five-page article on unfilled vulcanizable (cross-linkable) polyethylenes, with properties, test data, and recommendations for use. A thorough, illustrated application/market survey by M. M. Suba (Union Carbide) is in *Wire and Wire Products,* April 1965. Earlier, more technical reports on these materials are in *Wire and Wire Products,* April and May 1963 (French author); *Electrical World,* September 1963 (Union Carbide); *Rubber Age,* March 1962 (Union Carbide); and *Wire and Wire Products,* August 1962 (F. A. deMelio of Union Carbide). The latter article recommends machinery and extrusion conditions.

"Performance characteristics of semi-conductive polyethylene," H. J. Pazinski (Union Carbide), *Wire and Wire Products,* March 1966. Composition, properties, data, 2½ pages. See also article by A. E. Tarbox (Union Carbide), *Wire and Wire Products,* October 1961.

Other Special Materials

"The pigmentation of polyolefins and PVC for cables," E. J. G. Balley (I.C.I), *Plastics Institute Conference*, April 11 and 12, 1967 (London).

"Sodium as an electrical conductor," E. E. Griesser (Nacon Corporation, associated with Union Carbide), *Wire and Wire Products*, December 1966. Illustration of process, cable construction, test data, joint seals, properties. Liquid sodium is fed to crosshead, and passes into tube.

"Connectors and insulated sodium conductors," E. M. Scoran (Burndy Corporation), *Wire and Wire Products*, December 1966. A six-page description of a newly evolved device, for Union Carbide's new process.

"New insulation and jacketing materials," W. J. Abrams and R. Clark (Polymer Corporation Limited), *Wire and Wire Products*, July 1963. Butyl rubber and polyethylene blends (60:40), nitrile rubber and PVC blends (percentages not given).

"New flame-retardant polyethylene compound," T. Yoshida (Furukawa Electric), *Wire and Wire Products*, July 1962. Discussion of formulation and properties—100 parts (0.3 melt index) low-density polyethylene
50 parts chlorinated polyethylene
5 parts chlorosulfonated polyethylene.

"Extrusion of polysulfones for wire and cable covering," J. J. Helbling (Union Carbide), *Signal Corps Symposium*, December 1965.

EIGHT

monofilaments

PRINCIPLE

To make monofilaments, plastic is extruded through a plate containing many small holes. The strands that emerge from these holes are taken down through a quenching bath which cools them. Then they are reheated in line and are oriented in air or liquid of controlled temperature; in this step they are drawn down to a much finer size, and take on great strength. Finally, the filaments are wound on spools. The process is illustrated in Figures 8.1, 8.2, and 8.3.

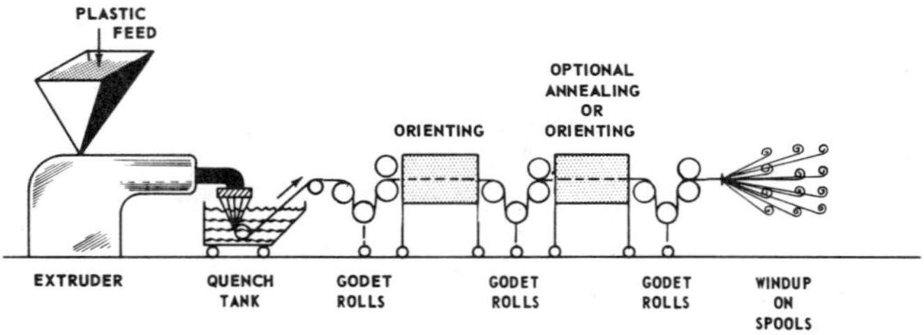

Figure 8.1. Diagram of a monofilament extrusion line. Strands are extruded downward into a quenching tank and then oriented. Sets of rolls hold the filaments firmly on both sides of each orientation station. The finished filaments are usually wound on individual spools. (*Illustration by author*)

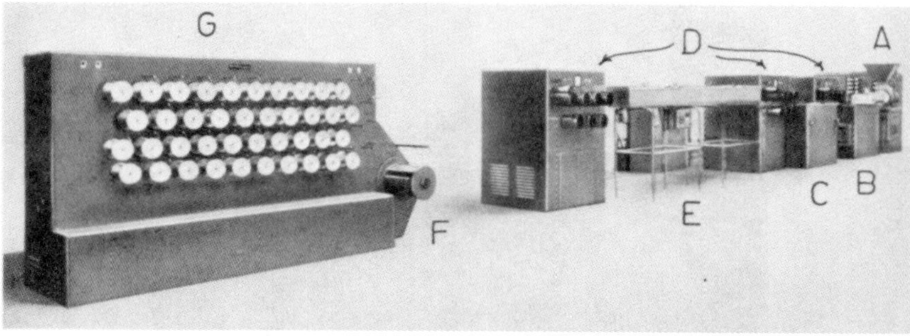

Figure 8.2. Complete commercial filament line, including (A) extruder, (B) quench tank, (C) liquid bath orienter, (D) godet rolls, (E) air-oven annealer—optional, (F) start-up and scrap roll, and (G) individual winders. (*Courtesy of Barmer Maschinenfabrik AG*)

MATERIALS

Polypropylene, polyethylene, polystyrene, PVC, nylon, and saran are all extruded into monofilaments. The highest molecular weights of these polymers are preferred in order to gain maximum strength. Nylon and polystyrene have rather exclusive and continuing markets —nylon because of its great strength, abrasion resistance, and flame-retardance, and polystyrene because of its rigidity and low cost. The others are more or less competitive with each other, and the industry leans toward polypropylene as the leader because of its light weight, ease of extrusion, high strength, and low cost. It is even replacing some polystyrene now. All materials, of course, compete with natural fibers and hair.

PRODUCT AND EQUIPMENT SIZES

Monofilaments are distinguished from multifilaments at about .005 in. (0.12 mm) in diameter. Strands below this size are extruded, but are seldom handled individually in the process, and therefore modified equipment is usually required. Monofilaments range from .005 to .060 in. (0.12 to 1.5 mm) in diameter. This represents about a 150-fold range of cross-sectional area, and begins to show why a monofilament line must be capable of a tremendous range of speeds and output rates in order to be truly versatile. Current equipment runs at 300 to 600 ft (90 to 180 meters) per minute, with from 20 to 100 filaments at a time. A recent trend is to many more strands per line—

Figure 8.3. Close-up of filament line shown in Figure 8.2. Above, filaments are extruded downward into quench tank and pass up around the first set of rolls and into the liquid orientation tank at the left. Below, the filaments pass out of an air-annealing (or orienting) stage, around the final set of rolls, and on to the winders. (*Courtesy of Barmer Maschinenfabrik AG*)

e.g., 192—to increase output, especially where the filaments are wound all together, not on individual spools. This is done for most brush bristle manufacture, twine, and some other products as well. Even with individual winders, 192-strand lines are now sold. These high-output lines (e.g., $3\frac{1}{2}$ in. = 90 mm diameter; 200 lb = 90 kg/hr) are still the exception, rather than the rule, and the smaller, more common machines produce from 5 to 100 lb/hr (2 to 45 kg/hr), depending on material and diameter. These are 2 in. (45 mm) and $2\frac{1}{2}$ in. (60 mm) extruders, with drives of 15 to 25 hp. The operating limit to output is usually the linear speed with which the filaments can be drawn, stretched and wound.

Monofilament dies and handling equipment are offered by many leading extrusion equipment makers. However, many units have been built by the monofilament producers themselves, who incorporate features which fit their particular situations and markets. Winders are usually made by a few specialized textile machinery firms.

EQUIPMENT

Extruder. Conventional extruders are used for all polymers except saran, which requires special materials of construction and equipment design. The drive should run efficiently and powerfully for all materials at all speeds, since it is likely that a wide range of speeds will be needed. Therefore, mechanical or electrical speed variation, or else geared noncontinuous variation, are preferable to the magnetic-clutch type of drive. Several speed *ranges* are recommended; these are achieved with either changeable gears or belts, or a gearbox with a clutch. A good versatile machine should run at from 5 to 125 rpm or higher. For a known and limited market such versatility may not be needed, and its expense may be avoided.

Die. Most monofilaments are extruded downward into a liquid bath. This means that the head assembly must contain an elbow to turn the plastic stream 90°. The problems are similar to those in a tubular film die, and the plastic is best led around through a gooseneck to enter the die section from the top, rather than from one side. This center-fed die allows equal plastic flow to all holes, which are set in a circle. Filaments of PVC, which decomposes easily, are sometimes run horizontally, straight out over a cooling roll, to avoid the elbow in the die. This is not necessary, however, and PVC is today being run successfully through streamlined center-fed elbow dies.

The dies that are used where the filaments are individually spooled are circular in shape, with the holes equidistant from the center. This ensures equal flow through each hole, hence equal filament size. For large dies with many holes, two or more circular rows of holes are used. Also, some flat (linear) dies are used for bristles. For ribbons used in twine or rug backing, a flat film die may be used, with the film slit into ribbons right after the first water quench and prior to stretching; one system uses a *tubular* film die, with air cooling.

The die holes themselves are from 3 to 4 times the diameter of the final filaments in order to allow for drawdown during orientation, as well as to reduce head pressure and restriction to flow. For PVC, the drawdown ratio is often no more than two, because at the normally low extrusion temperatures its drawing ability is poor. New extrusion techniques for PVC allow higher extrusion temperatures, and holes can be correspondingly larger for equivalent product diameter.

The die plates are made of hardened steel, or other metal if needed for corrosion resistance, and sometimes each of the holes is countersunk on the inside. The included angle of countersinking should be as small as possible to prevent melt fracture; 20° is ideal. Land length of these plates is relatively short, i.e., about two to five times the hole diameter. It cannot be much greater because excessive pressure would be caused in the head from the greater restriction to flow. Such plates are interchangeable in the die assembly, and many are ordinarily kept in stock by a monofilament producer.

Behind the holes, inside the die, a guider plate is sometimes used. It also has holes in it that correspond with the holes in the die. However, these guider holes are larger and also are countersunk on the upstream side. The function of the plate is to guide equal streams of plastic to each of the extrusion holes. Sometimes die plate and guider are one piece of metal.

An independently-driven positive-displacement pump, such as a gear pump, is sometimes included in the die assembly to ensure a constant flow of plastic to the die plate. This pump is used particularly for nylon and polypropylene which have low melt viscosity at extrusion temperatures and would otherwise flow unevenly. The pump eliminates the effect of surging in the extruder. A pressure control valve can do the same thing with less cost, but also with less precision. Constant flow is very important because any thinning of a filament might mean a break in the take-off farther on.

Quenching. The quenching bath is usually warm water. Formerly, ethylene glycol or glycerine were used to obtain temperatures over 100°C, which are more desirable (but not necessary) for polystyrene. However, part of these liquids is always carried out on the filaments, and the consequent expense has effectively curtailed their use. Quenching is infrequently eliminated for polystyrene and PVC, and the filaments are run directly out into the orienting stage.

The purposes of quenching are to cool the filaments quickly to preserve a tougher molecular structure of smaller crystallites, and also to cool them enough to allow them to wrap firmly without distortion around the rolls preceding the orientation section.

The quench bath temperature is controlled by electric immersion heaters (or steam coils) responding to a thermostat. In simpler systems, a thermometer is merely hung in the bath and cold water is continuously bled in to keep the bath from becoming too hot. With such systems, the water level is kept constant by an overflow tube or weir in the bath. Adjusting the height of this weir is a convenient means of raising or lowering water level, thus controlling the draw distance from die to water. The bath temperature is ordinarily kept between 40 and 90°C. A little soap is sometimes dissolved in the water to reduce surface tension, thus inhibiting carry-over of water out on the filaments. At the bottom of the bath, which is from 1 to 2 ft (30 to 60 cm) deep, the filaments pass around a large idling roller. They then come up through the bath, around another roller, and thence to orientation. The bottom roller usually can be raised or lowered, both to change the residence time in the bath and for ease in "thread up" at the start.

The quenching bath is clearly visible in Figure 8.3 which shows a typical monofilament line, from the die to the entrance to the orienting stage.

Orientation. Various orientation systems are shown in Figure 8.4. In the orientation stage, the filaments are put under high tension while in a semiplastic state, which stretches them and makes them much thinner. The molecules straighten out and line up parallel to the filaments, thereby giving them great strength. Oriented tensile strengths of filaments are sometimes more than ten times the strength of the unoriented polymer. The temperature of the plastic during the orientation is very critical, and several methods have been devised to maintain this temperature:

Liquid Baths. The filaments pass through one or more baths similar to the quenching tank and held at the orientation temperature. This is the most uniform method, but it requires the most equipment and

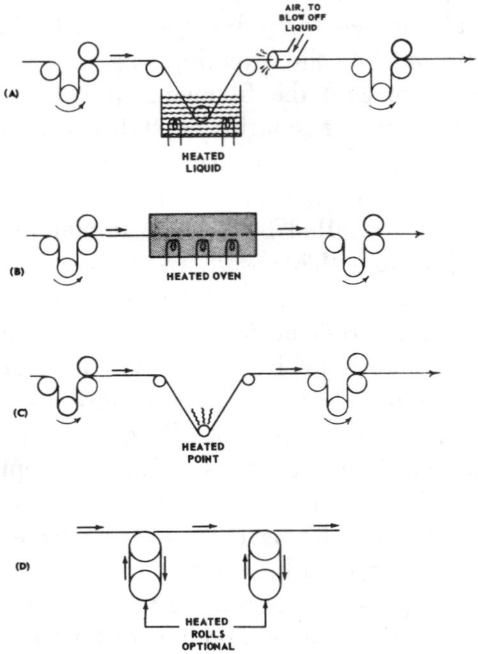

Figure 8.4. Principles of orientation. A = liquid-bath method; B = air-oven method; C = heated-point method; D = cold drawing. In all cases, the set of rolls at the right are run faster than the set at the left. (*Illustration by author*)

adds the problem of removal of bath liquid carried out on the filaments. It is still used on older U.S. equipment, and was used extensively on European lines, but is being replaced by radiant-heat ovens, which can better handle high-temperature plastics such as polypropylene and nylon.

Heated Oven. All the filaments pass between two heaters or heated surfaces which quickly bring the plastic up to orientation temperature. The filaments do not touch the heaters, which form a sort of oven between them, and there is no liquid removal problem. This is one of the simplest common methods, and is increasing in popularity. Radiant heaters are most often used, while circulating hot air was formerly employed in this method. In a few installations, the entire orientation area is enclosed and the resulting space is heated by steam or electricity; such units are currently used in polystyrene bristle lines. In their design, it must be remembered that convection will produce different temperatures at the top and bottom of the oven.

Heated Point. In this lesser known method, the filaments pass over a small heated roller; they are heated by contact with the roller and oriented at and just beyond that point. This simple method is limited to smaller filaments, especially flat filaments, because the heat transfer is by conduction from one side and the rate of this conduction may limit extrusion speed.

Heated Rolls. In this method, all the filaments pass around steam-heated rolls before being oriented right in the open. The temperature of orientation is controlled by the roll temperature, linear speed, and number of wraps around the rolls. This method is used for polystyrene, especially heavy bristles; and it has also been used for polyethylene. It avoids the cost and drying problems of a liquid bath and allows the rigid polystyrene filaments to travel around the rolls without breaking or taking a permanent curl. Saran is also run out in the open like this and, in this case, heating the rolls is not even necessary.

For most systems there are two sets of large rolls, one on either end of the orientation stage. The stretching is done by running the second set of rolls much faster than the first. The rolls in each set must run at the same surface speed, and the filaments must not slip on them when the high orientation tension is applied. These rolls are called "godet" (pronounced "go-day") or "snubber" rolls, and all of their drives are synchronized so that either over-all speed or speed ratio between the two sets can be changed at will. This ratio, which determines the degree of orientation, varies from 2:1 up to perhaps 12:1. The rolls are cantilever-supported in a frame, and must be held very strongly to withstand the high tensile stresses. Typical size is 10 in. long by 10 in. in diameter. These rolls are made of steel or aluminum and are matte-finished enough to discourage slipping, but not enough to actually abrade or cut the filaments. One roll in each set may be made of rubber, forming a nip on the filaments with one of the metal rolls.

Most units built now have three or four rolls in each set, mounted in permanent positions on the frame, with their axes perpendicular to the vertical plane of the frame. The filaments go around each roll only once, and often pass through a nip, as noted above, to keep them from slipping. This has replaced the older system of *pairs* of rolls, both metal, where the filaments wrap around these rolls several times. One roll in this older system is skewed with its axis displaced from 6° to 12° from the axis of the other roll. This allows the several wraps to lie separated from one another as they pass around the rolls. Many of these roll-pair systems are still in operation, although very few are built this way today.

Little combs are often mounted in various places throughout the system to keep the filaments separate. If they bunched together as they travel around the rolls, they would be more likely to distort or even break.

Annealing. In this stage, the filaments are warmed while held firm and unable to shrink. In this way, the molecules that were uncoiled and stretched out in the orientation stage now lose the inclination to return to their original coiled, unoriented state. The strength remains, and the shrinkage under heat is greatly reduced. Some monofilament lines today omit the annealing step. It is employed only where the end-use requires the reduced shrinkage.

Annealing can be done right on the line, after orientation, by passing the filaments through an oven or a liquid bath held at the proper annealing temperature for the material; as an example, the proper temperature for polystyrene would be about 135°C. There is another set of rolls after the annealing bath, running at the *same speed* as the set just before it. Thus orientation is neither increased nor lost. Another method of annealing, sometimes used for polystyrene, is to wind the filaments around huge frames and place them in an oven for perhaps ten minutes at 135°C. Disadvantages of this method include possible oxidation and degradation of filaments, need for the huge oven and the additional labor costs. The advantages are: no liquid-removal problem and no need for an annealing bath or oven.

Winding. A large roll called a "scrap" roll is used to wind up all filaments at once—during start-up or during operation when a strand temporarily cannot be wound on its spool. A similar large roll is used for applications where the filaments will be used in groups, such as twine, and also for applications such as bristles, which are sold as cut lengths. Such bristles are so stiff that a large reel, 3 ft or more in circumference, is needed to prevent them from developing a permanent curl as a result of the winding. Some of the thickest polystyrene bristles develop a curl on any reasonable diameter of reel. For such bristles, the reel consists of four bars (Figure 8.5) around

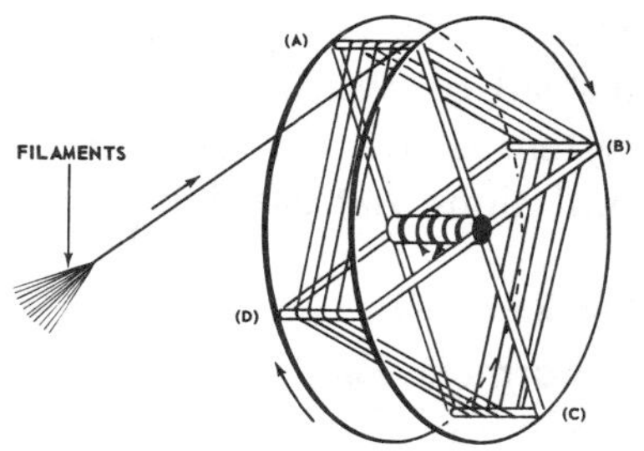

Figure 8.5. Device used to wind heavy filaments for bristles. A big wheel, 6 feet (about 2 meters) in diameter, winds the filaments around the four posts A, B, C and D. When the wheel is full, the filaments are cut off at the posts, thus making straight lengths. A short length at each post is bent and is discarded. (*Illustration by author*)

which the filaments are wound. The lengths of bristles between the bars thus remain straight, while the sharp bends at the corners are cut off and reworked or discarded.

In most monofilament extrusion, each individual filament is wound on its own spool. Each spool is independently driven, and it varies in speed with the buildup on the spool. Most drives are constant-torque devices, which give a fairly uniform pull on the filament throughout its winding. The most precise winding is achieved with controlled constant-tension winders; the tension is mechanically sensed and fed back to the motor on the spool. The tension or torque settings of these drive systems are adjustable for different filament diameters, speeds, and compound strengths.

One might ask why all the spools do not run at the same speed. This is impractical, as the spools are all started at different times and the buildup on each spool is different. Different buildups mean different rotational speeds of the spools because the linear speed of all the filaments is the same. Thus each spool must have its own speed, which can compensate independently for the buildup and wind the spool tightly. Some units are built with one motor, and suitable mechanical linkages to allow various torques among the spools. These will work, but generally the individual-drive systems are more reliable and easier to maintain, and are accordingly preferred despite their extra cost.

Small level-winds are fitted to each spool and traverse it. The level-winds are sometimes driven by a single unit operated mechanically (by cams) or hydraulically. In some high-quality equipment, the level-winds are independent of each other and each is driven by the motor of its own spool.

When the level-winds move back and forth across the spool quickly, the filaments can remain firmly on the spool without flanges.

This makes possible the use of cheap, disposable cardboard tubes instead of returnable wooden or metal spools. This disposable-tube system is still not universal, but is slowly replacing many of the installations that use flanged spools.

OPERATION

Formulation. Monofilament compounds are usually purchased already formulated but uncolored. Colorants are added as master-batch and are mixed in right at the extruder. It is quite important to select pigments that have good heat and light stability as monofilaments are extruded at high temperatures and are often used outdoors. A rather high loading of pigment is necessary if high opacity is wanted in the filaments.

Additives are often included by the material suppliers to improve the heat and light stability. This is necessary for polypropylene, the leading monofilament material. Filament grades of polystyrene are not specially stabilized, but are the highest molecular weights available. Plasticized grades are never used; rather, the grade sold as "high-heat" is preferred, as it yields the best physical properties.

Flame-retardant polyolefin filaments have been made with antimony oxide and/or chlorinated paraffin additives and sold for Christmas trees and wreaths, and similar household decoration. However, the product is dull and hard to extrude, and the price is still too high. PVC filaments are much more common in this market; they are inherently nonflammable and attractive, and need not be oriented. Die design and extrusion conditions must take into account the ease of decomposition of PVC. Even some saran filaments have been used for Christmas trees, but this is because of a temporarily depressed price situation. They make a good product, but saran is normally too expensive and difficult to extrude.

Start-up. Threading the line is a necessary operator's skill. The line is started slowly and the filaments are carried in a bunch through the quenching bath. The roll in the bath may be raised to assist in threading. It is advisable to preheat the quench water rather than lose time by waiting for it to heat up from room temperature merely from the heat of the plastic. The filaments are then carried through the system and wrapped around the scrap rolls. They are separated by hand in the "combs" along the line and one by one are taken off the scrap rolls and wound around their individual spools.

Temperatures. The plastic temperature at the die must be closely controlled to guarantee a uniform melt, which is necessary to avoid diameter variation and resultant breakage. Material temperatures are kept rather high as the plastic flows and draws better, the surface is glossier and smoother, and less head pressure is produced (which means more pounds per horsepower). Handling problems are minimized because the filaments are extruded downward and the melt never supports its own weight. Also, cooling capacity is not normally a limit to output because the quench bath can cool filaments at speeds faster than are practical to run in the rest of the equipment. The possible disadvantage to very high material temperatures are: slight decrease in tenacity, and—if the temperatures are too hot—increased chances of degradation and/or breakage. The loss in tenacity is attributed to the beginnings of polymer degradation. Typical material temperatures are 200 to 230°C for polystyrene, 230 to 260°C for high-density polyethylene and polypropylene, and 280 to 310°C for nylon.

The quench bath must cool the filaments down to a temperature at which they can be handled. It cannot be too warm, or the filaments will not be sufficiently cooled and will flatten out as they proceed through the system. However, the bath cannot be too cool as the filaments become weaker with decreasing bath temperature. A temperature of about 50°C has been found best for both polypropylene and high-density polyethylene. Polystyrene requires high temperatures (at least 85°C), and liquids other than water have been used, running between 110 and 130°C.

The effect of orientation bath (or oven) temperature depends upon the method of orientation. With a liquid bath orientation, the filaments become only as hot as the bath, but are very uniformly heated. Near-boiling water is used for polyethylene and polypropylene. Polystyrene requires higher temperatures; it can be run in water, but has also been run at 120 to 135°C in glycerine or ethylene glycol. Often it is run in air over heated rolls (as described earlier in this chapter) or else in an oven. Nylon and polypropylene require higher temperatures than are available with common liquids so they too are oriented in an oven. The temperature of an oven can be considerably higher than that of filaments—depending on the construction of the oven, the place of measurement of oven temperature and the speed of the filaments through the oven. One investigator reports 180°C as best for polyethylene and polypropylene in his equipment. The temperature of nylon would be a little higher, and is quite critical because of nylon's low melt viscosity. If it is heated even a little too much, the filaments soften and break. The other materials have a little more processing latitude, but the same thing will happen to all of them if they are overheated during orientation. If the orientation temperature is too cold the filaments will not stretch as easily, and the increased tension thus required may be enough to break some of the filaments, especially the smallest sizes.

Drawdown. The filaments are drawn down at the die before entering the quench bath. This drawdown is ordinarily kept as low as possible; in other words, the quenched filaments have about the same diameter as the die openings, for medium and heavy filaments—e.g., 40 mils (1 mm) and over. This condition of filament diameter = hole diameter allows for some drawdown to compensate for swelling as the plastic leaves the die, and it allows for the normal shrinkage on cooling. For thinner filaments, more drawdown is usually employed, because if a die were made with small holes only as big as the filaments, the holes would clog easily, breaks would be more frequent, and the extruder pressure may become too high. High drawdown leaves less stretching to be done during orientation and correspondingly poorer strength. Also, a thinner filament must be handled between die and godet rolls, which may result in frequent breaks. Drawdown is kept low (assuming suitable die design) by matching the take-off speed to the extruder speed, and by bringing the level of the quench bath right up to the die face. A gap of an inch (25 mm) or less is common. Another advantage of this small gap is the lessened chance of surface oxidation, especially at high material temperatures. An oxidized surface will discolor the filament and, moreover, will cause breaks in handling.

Removal of Bath Liquids. Filaments cannot be wound wet. If they were, the liquid would be trapped in the spool and would add to its weight, detract from its appearance and possibly affect the subsequent converting process; therefore, wherever baths are used, the liquid that carries over on the filaments must be removed. This is less important after the quenching bath especially if there is liquid orientation. However, if the orientation is in air, droplets of the quenching liquid on the filaments will cause uneven heating, hence uneven orientation. Of course, after liquid orientation, the filaments must be dried completely before winding. One technique is to blow air, preferably hot air, back at the filaments as they leave the bath. Such air can be blown through tubes through which the filaments are passing. Running the filaments over a perforated vacuum or blower roll is another method. Where all the filaments are wound at once, they are sometimes simply cut in bunches and hung up to dry. This problem of liquid pick-up becomes even more important when the liquid is expensive (e.g., glycerine).

Removal of Static Electric Charge. The filaments can acquire a large static electric charge as they go over all the rollers—especially if the entire process is dry. When this becomes troublesome because of the dust pickup, shock hazard, or converting problems, the filaments can be run through a dilute detergent solution in water just before winding. This, of course, introduces the water removal problem noted above. An alternate method uses no water but is less permanent; it drains off the static charge through grounded metal wires or chains laid loosely over the spools during winding. Still another method is to blow ionized air at the filaments at or near the winders. Finally there are plastic compounds sold containing static-dissipating additives which give more or less permanent protection.

Scrap Reclamation. Scrap monofilament, like other thermoplastic scrap, must be reclaimed to make the process truly economical. It is seldom possible to use the scrap again in filament extrusion unless it is kept scrupulously clean, or unless large filaments are being run. This is because the dust and similar contamination picked up in the reclaiming operations will cause breaks in the filaments during re-extrusion. Thus, the scrap is usually sold to plastics reclaiming firms or, if practical, ground up and worked off in a less critical application. If fine filaments must be ground, the grinder must be of a type built to handle them. It is often easier, if the equipment is available, to melt the filaments on a roll mill and grind the cooled milled sheet in a conventional granulator—or else to use one of the various compounding and scrap-handling extruders sold for such purposes.

PROBLEMS AND SOLUTIONS

The following chart gives solutions to some common problems encountered in monofilament extrusion. As with any complex take-off system, it is important to locate as closely as possible the exact place in the take-off where the defect originates.

Problems	*Causes and Solutions*
Filament breakage	(a) Surging in extruder. Run at higher pressure, or with better-mixing screw, or with cooled screw, or with gear pump, or more slowly. Try to determine period of the surging, and relate to drive, controllers, etc. (b) One or more holes partially blocked. The same filament(s) will break every time. (c) Uneven temperature or distribution of material in head and die. The same group of filaments will break every time. Look at blueprint of die, to find asymmetrical construction, if any, which could cause the uneven patterns. (d) Melt too cold. Not enough ductility to draw down. Raise temperature. (e) Melt too hot. Too fluid to hold together under drawing tension. Reduce temperature. (f) Drawdown too great (use smaller holes in die) or drawdown too fast (slow down whole system). (g) Moisture in material. Use a hopper-dryer. (h) Contamination in material. Inspect feed, stop using reground or reworked plastic. (i) Oxidation and weakening of filament surface. Run at lower temperature, or

Problems	*Causes and Solutions*
	reduce the gap between die and quench bath. (j) Decomposition in extruder, head, or die. Bits of decomposed material clog holes, or come through them but weaken the filaments. Clean die, run at lower temperature; examine die design for possible stagnation spots, change die or head if necessary, use better-stabilized plastic. (k) Too much orientation. Ratio of roll speeds may be too high causing excessive tensile stress in orientation stage. (l) Too-cold orientation. If plastic is not hot enough, too much tensile stress will be produced in the filaments. (m) Too-hot orientation (air oven). If plastic becomes too hot, it loses ductility and will break under the normal orientation stresses. (n) Nicks or abrasive areas on any of the rolls. Grind them smooth, or wrap them with tape. (o) Erratic drive. Check smoothness of operation of all moving parts.
All filaments varying in diameter	(a) Surging in extruder, drawdown too much or too fast, or erratic drive. See (a), (f), and (o) above.
Some filaments different from others, in size and/or strength	(a) Temperature gradients in die. Unequal flow in die. Check for drafts, burned-out heaters, poor die design. See (c) above. (b) Take-off not exactly aligned with die and extruder.
Poor surface appearance	(a) Moisture in material. Possible with polystyrene, nylon, or any pigmented or filled plastic. Dry the feed. (b) Melt fracture. Reduce linear speed, or use larger holes (more drawdown), or reduce entrance angle into die holes, or use lower viscosity compound, or run hotter. (c) Plastic too cold. Run hotter.

Problems	*Causes and Solutions*
	(d) Compound inherently dull. Change compound, or run hotter, or flame-polish.
Low tenacity	(a) Not enough orientation. Increase ratio of roll speeds. Use proper orientation temperature. Note that changes of orientation may also change the filament size, requiring compensating adjustments elsewhere.
	(b) Degraded plastic. Use lower material temperature, or change to better-stabilized formula.
	(c) Plastic unsuitable for filaments. Check with its manufacturer.
	(d) Nicks or cuts in filaments. Examine breaks and examine unbroken filaments. Look for a repetitive pattern and examine take-off for the source.
Heavy filaments do not lie straight	(a) Filaments wound around a roll while still warm. Adjust filament or roll cooling.
Oval cross section	(a) Filaments too hot while passing over rolls, especially in quenching bath. The rolls themselves could be too hot, also.
	(b) Die holes not truly round.
	(c) Temperature gradients in die affect roundness. Most likely in T-type die, with all holes in a line.
	(d) Too much tension in wrapping around any rolls.

STANDARDS AND TESTS

The monofilament market in the United States is relatively small compared to other plastics extrusion markets, such as those for pipe and film; and few industrial uses have been found for the filaments as yet. Therefore, no national standards have been set for the filaments themselves. Standards do exist for the rope and textiles made from them.

Important tests for monofilaments include the following: knot strength, tenacity, shrinkage, creep, chemical and ultraviolet-light resistance, and abrasion resistance. *Knot strength* is a tensile strength test on a filament that contains a simple knot. It is usually measured in pounds per square inch (kg/cm^2) based on original filament cross section and is important in certain textile constructions. *Tenacity* is the tensile strength itself. It is usually reported as grams per denier. The denier is a measure of the fineness of the product and is the weight in grams of 9000 meters of filament. Thus, a higher denier means a heavier filament, and a higher tenacity means a stronger filament. Tenacities of 2 to 5 g/denier are commonly encountered with plastics, and values of 6 to 10 are outstanding.

Shrinkage is really a measure of relief of orientation strains, and is found by suspending the filaments in an oven where they can shrink without restraint. The temperature and time depend on the end-use and the material. Results are reported as percentage reduction of original length. To measure orientation, the test is run at a temperature higher than that of the filaments during orientation, and the results are reported as a percentage: decrease in length $\div$ shrunk length $\times$ 100. If the filaments have been annealed, the shrinkage is much lower. Then orientation cannot be measured in this way, but must be found from differences in cross section or from tensile strength increase.

Creep is a measure of permanent filament elongation under load, as in instrument strings, tennis racquets, and many rope applications. A practical test consists of suspending bags of lead shot (calculated to give a known stress) from the filament. The increase in filament length is measured at regular intervals. Ordinarily, the smaller the change, the better is the filament.

Chemical resistance tests are most important for brush bristles and for special uses of rope and textiles. Tests on the raw material usually suffice, but sometimes tests are run right on the bristles. Attack produces softening, swelling, or actual solution, or else shrinking and deorientation. *Resistance to ultraviolet light* is measured by means of accelerated exposure tests, measuring loss in tensile strength (tenacity) and elongation. Such resistance is very important in outdoor furniture and rope, for which specially stabilized compounds are used.

Abrasion tests are most important for bristles and rope. Commercial abraders are sold and it is also easy (and recommended) to set up an abrasion comparison test on one's own end product. It must be remembered that the heat of abrasion can be just as damaging as the physical loss of material. This heat can soften and fuse the filaments, or else shrink and deorient them.

SPECIAL TECHNIQUES

Ribbons and Other Flat Filaments. A large market has recently developed for polypropylene monofilament ribbon, e.g., 0.004 by 0.100 in. (0.1 mm by 2.5 mm), in markets such as twine, sacks, and carpet backing. These ribbons mainly replace jute and other natural fibers largely unpredictable in price and availability. The plastic ribbons are also superior to natural fibers in most physical properties, and they can be heat-sealed. Two techniques are used to make them:

(a) A flat film is extruded from a conventional film die—e.g., 24 inches (60 cm) wide. The film is slit into strands after quenching with water, but before orientation in a radiant-heat oven. If desired, flat film may be first made and rolled up, and oriented later.

(b) A tubular film is first extruded, as shown in Figure 8.6. In this illustration, a tube of 8-inch (20 cm) diameter is extruded upward, cooled in air, and flattened as in normal film production. It passes through two sets of nip rolls, between which it is slit into ribbons. The second nips take the place of the first set of godet rolls, and the ribbons pass from there directly into an orientation oven. At the other end, each pair of filaments has its little "godet" pull roll, which feeds two individual winders. The number of filaments and winders can be built up as needed. Other tubular film lines for ribbon are available which employ conventional godets and orienting/annealing ovens.

In all of these methods, ribbons are stretched on standard orienting equipment and either wound on a single roll or on indiviual spools. Spools are needed where the product is to be subsequently woven (carpet backing, sacks); for twine, the ribbons are wound up in bunches, often interconnected because of deliberate incomplete separation during slitting.

Other flat filaments (e.g., 0.004 in. x 0.020 in. = 0.1 x 0.5 mm) are also extruded for textiles such as automobile seat covers. They make a thinner fabric and are less likely to twist. The die holes themselves are often made to conform to the desired shape. Alternatively, round filaments are extruded and squeezed to a flat cross section with rollers right after the die. The latter method saves building new dies if round-hole dies are already on hand.

Figure 8.6. Production line for oriented polyolefin tapes for sacks, carpet backing, etc. At upper left, a tubular film is extruded upwards, and is then slit, stretched and wound on the winders in the foreground. There are no godet rolls; rather, the small black rollers in the top row of the winder provide the force to pull the tapes through the orienting stage. (*Courtesy of Plasticisers, Ltd.*)

Foamed Filaments. A foaming agent is added to the extrusion compound, and the resulting filament is lighter, softer and bulkier. It is used in twine, for example. If the quenching bath is kept cool enough, the surfaces of the filaments will be quickly cooled, and will stay solid and unfoamed, but the inside will foam and expand. This yields a stronger, more abrasion-resistant product.

Unusual Cross Sections. Various sections such as triangles, crescents and stars have been made, to give greater rigidity for such uses as brush bristles and Christmas-tree needles. These are done by using specially cut holes in the die plate. The design of these holes and the passages behind them is critical, as they must be streamlined despite their odd shapes, and the hole shape must be properly different from the final shape (since drawdown and differential cooling will alter it).

Unoriented Filaments. For certain uses (some rope, decorative items, Christmas tree needles) orientation is unnecessary. This greatly reduces the cost of the equipment and the difficulty in operation, as the filaments can be run right out of the quench tank into a chopper or winding assembly.

Crimping. Bristles are often crimped into a wavy shape (Figure 8.7) by passing them through simple heated gears with rounded teeth, just after orienting and before cutting. This gives the finished brush more bulk and "body" with the same weight and cost of filament.

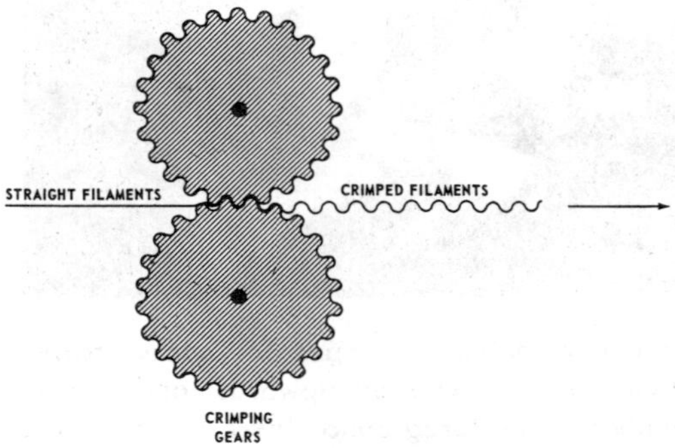

Figure 8.7. Simple heated gears such as these are used to put a permanent crimp into filaments for use in brushes. (*Illustration by author*)

Coating of Filaments. Detergent coatings for antistatic protection have been mentioned above, but other coatings can also be applied from a bath after orientation. These might be compounds to decrease slipperiness in rope or they might be coloring matter.

Tapered Filaments. These duplicate the taper inherent in most natural bristles and are preferred for paint brushes. They are made from nylon by varying the take-off speed in phase with the cutting of the bristles. The bristles get alternately thicker and thinner and are cut accordingly.

Flagging. The ends of some cut filaments are deliberately and mechanically made rough and fuzzy. These "flagged" ends are in demand for some brushes, where they yield a softer, less irritating brush end.

Netting. Filaments are extruded through a special die with moving parts, which weld the strands together at specific points. The welding can take place either just inside or just outside the die. By varying the location and frequency of the welds, as well as the diameter of the filaments, different nonwoven netting structures can be made, as illustrated in Figure 8.8. Four such processes have been developed (in England, Germany, France, and the United States) and have been licensed by different manufacturers in many countries. In the United States, netting has been on sale for over five years, but markets only lately are starting to develop.

Applications include packaging of onions, candy, fruit, and baked goods; chicken coops (no rusting); carpet backing instead of jute (no rot); and structures on which to build ladies' hats and other garments. The netting, especially when foamed, can be used for bottle protection, eliminating the need for corrugated board separators in cases. Orienting the filament components of the netting improves strength but must be done carefully—perhaps with mechanical aids—to avoid distortion of the pattern. Polypropylene and high-density polyethylene have been the materials most used for the netting, and cellular low-density polyethylene and polystyrene have been used for

Figure 8.8. This large roll of non-woven netting and the smaller piece of netting in the foreground are typical products of the extruded-netting methods. Note the fountain-pen included as a comparison of size. (*Courtesy of Union Carbide*)

bottle protection in order to reduce cost and weight while increasing resilience.

Net-like structures can also be made by slitting or perforating film and sheet, and then heating and stretching to widen the holes or slits.

Saran Monofilaments. Saran (polyvinylidene chloride) was for some time the leading monofilament material, especially for outdoor furniture and automobile seat covers. Its advantages include excellent weather resistance and flame retardance. However, polyolefins (polyethylene and polypropylene) have replaced saran because of:

(1) lower cost,

(2) superior strength—5 to 7 grams/denier compared to 1 to 2 grams/denier for saran,

(3) lighter weight—50 per cent,

(4) ease of processing—the polyolefins can be extruded on conventional equipment, while saran requires special materials and design to avoid decomposition.

The good weather resistance is obtained in the polyolefins by addition of ultraviolet light absorbers.

Some saran is still extruded into filament today. It is done in a short-barrel extruder (12:1 L/D ratio) with a sharply pointed screw. The screw is constant in pitch, with the channel depth decreasing to a quite shallow value (0.075-in. for a 2-in. screw) at the end. There are flights over the entire length of the screw except in a short and smooth final torpėdo section. All metal surfaces that come in contact with saran must be made from alloys of magnesium, nickel or cobalt because most other metals (notably iron) catalyze the decomposition of this polymer. This means that expensive alloys like "Hastelloy," "Stellite," "Stoody," or "Xaloy" 306 must be used. Not all of these can be easily worked or machined into the required extruder and die parts. The filaments are extruded at 175 to 190°C downward into a quench bath of *cold* (10°C) water, and oriented only 300 per cent (polyolefins receive from 500 to 1000 per cent orientation). The orientation is done in ambient air with no heating, cooling or enclosure. The spooling is handled as usual.

Continuous Multifilament Yarn. The *mono*filaments discussed in this chapter are large enough to be handled and used as single-strand filaments (although for certain uses they may actually be handled in bunches). Different techniques are used to make *multi*filament yarn; that is, yarn composed of many tiny, continuous threads. These threads are too thin to be handled by most normal monofilament equipment. The material is forced, as a very fluid hot melt or as a solution, through spinnerets containing many fine holes. The familiar nylon, "Dacron," rayon and glass fibers commonly are made this way. It is interesting to note that in the U.S. nylon rope is almost exclusively made from *multi*filament, partly because the price of the multifilament is as low as the price of the resin needed for monofilament extrusion.

Mixtures. Many textiles and ropes are made by mixing or blending two or more different filaments. Sometimes the difference is merely one of color such as is common for rope or for textiles of different designs. Some rope is made with an uncolored center and a colored exterior to reduce cost. Different plastics may also be blended; one rope manufacturer offers a blend of polyethylene and polypropylene to obtain properties intermediate between the two. A manufacturer of carpet backing now offers a woven product with polypropylene monofilament ribbon running one way, and polypropylene multifilament yarn running the other way. Three-dimensional fabrics are made by weaving in as part of the structure a filament that shrinks more than the other filaments in the fabric. When the completed fabric is put in an oven, the high-shrinking filament puckers and draws the fabric into a three-dimensional configuration. Finally, plastic filaments are mixed with natural bristle and plant fiber in order to get an infinite variety of property combinations for brushes and brooms.

REFERENCES

Applications

"Progress in monofilaments," *British Plastics*, July 1966. Summarizes established applications for all thermoplastics. Similar article, with properties, in *Rubber and Plastics Age*, February 1966. Further information in articles dealing with specific plastics, as noted below.

Equipment

"Monofilament production plant," F. Hensen and W. Utermann (Barmer), *International Plastics Engineering*, November 1965. Long article with references, covering components of production line, testing of product, and costs. Originally appeared in German in *Kunststoff & Gummi,* December 1964.

Monofilament line with 192 strands made by M.P.M. is briefly described with much data, *Modern Plastics,* August 1963.

"Samafor: Ensemble pour l'extrusion des monofilaments," *Industrie des Plastiques Modernes,* January 1962. Illustrated description of Samafor machinery, which uses liquid-bath orientation.

Processing—Polyolefins

"The production of monofilaments from high-density polyethylene," D. F. Stewart (Phillips), *SPE ANTEC* 1967. Property data, conditions, and definitions. Brief description and illustration of equipment.

"Fabrication of polypropylene monofilaments," A. Calza, *Poliplasti,* September 1966. Further information on polypropylene filaments is in *Kunststoffe,* December 1962 (short article on choice of melt index); *Plastics Technology,* July 1961 (Eastman personnel describe effect of extrusion variables); *Modern Plastics,* June 1961 (discussion of uses of polypropylene filaments); *Materie Plastiche,* February 1961 (chemical, electrical, weathering and basic properties of polypropylene and its filaments); and *SPE ANTEC* 1960 (equipment for monofilament production, including statistical analysis of variables).

"High-density polythene monofilaments," E. Overgage and G. John (British Resin Products), *British Plastics,* January 1961. Illustrated description and study of effects of variables. Brief notes on application.

Processing—Other Materials

"The manufacture of wires and bristles from polyamides," K. Riggert (Zimmer Plastic AG). Survey of methods of producing monofilaments from polyamides (nylon).

"PVC coated fibers take off," *Modern Plastics,* May 1963. Describes Pasquetti process, with some statistics and many illustrations.

"Extrusion and properties of saran filaments," J. Jack and R. A. Horsley (BX Plastics), *Journal of Applied Chemistry,* April 1954. Discusses materials, equipment, techniques, and properties of filaments. Basic work in this field.

Die Design

"Phenomenon of draw resonance in polymer melts," A. Bergonzoni and A. J. DiCresce (Enjay). Two-part paper, describes tests in ribbon die. Twenty references plus study of conditions.

"Designing dies for monofilaments," D. Poller and O. L. Reedy (Enjay), *Modern Plastics,* March 1964, and *SPE ANTEC* 1964. Study of land length and entry angle for rheometer and filament die.

Monoaxially Oriented Ribbons, Tapes and the Like

"Reifenhäuser mono-oriented tape line," *British Plastics,* February 1967. Discussion and illustration of one-stage and two-stage processes.

"Monoaxial verstreckte Folienbänder" (Monoaxially oriented film ribbons), K. Burggraf, *Kunststoffe,* February 1967. Discussion of method, machinery and applications of polyolefin ribbons.

"Oriented tapes of polypropylene and high-density polyethylene: process and equipment," T. Okada (Toshiba Machine Company), *Japan Plastics Age,* September 1966. Includes information on weaving. Further plant description, this time by Kasamatsu Plastic Laboratories, is found in *Japan Plastics Age,* May 1966. Kasamatsu equipment makes strapping from polyolefins, and an article on their process is in *Japan Plastics Age,* March 1966.

"Uniaxially stretched polyolefin films," D. J. Barrett (Bone Brothers), *Plastics,* March 1966. Bone's extrusion line is described for manufacture of stretched ribbons for rope, strapping and sack making.

"Polyethylene stretched tape and woven fabric," P. E. Schmidt and T. Hirayama (Amoco Plastics Research Incorporated), *SPE ANTEC* 1966. Blow film, slit it, stretch it and wind it all in the same line.

Non-Woven Netting

"Extruded netting set to snare big markets," *Modern Plastics,* November 1963. General description of du Pont (Plastic Textile Accessories) and Nalle processes, with much application information.

"Plastics mesh structures," *Plastics,* May 1963. Illustrated description of the Plastic Textile Accessories process, and a list of all licensees. Spicers, the sales agent for Plastic Textile Accessories, issues a 36-page illustrated brochure entitled "Netlon Integrally-Extruded Mesh," describing properties and uses of their netting, also with a list of licensees and patents.

"Production of net from thermoplastics," *Kunststoffe,* August 1962. Well-illustrated review of various die designs.

"Plastic netting," L. J. Zukor, *Plastics Technology,* July 1959. Short description of Nalle system.

Miscellaneous

"Strengthening monofilaments by drawing through die," T. G. Mitranova *et al., Chimiya Volokna,* No. 4, 1966. Comparison of die-drawing with stretching.

"Stabilization of linear polyethylene against ultra-violet light," G. A. Clark and C. B. Havens (Dow), *Plastics Technology,* December 1959. Weathering test results on high-density polyethylene monofilaments, using various light-stabilizers.

NINE

miscellaneous contours

PRINCIPLE

This chapter covers the extrusion of rods, tubes, tapes and the hundreds of specialized irregular cross sections that are used as edging, gasketing and window channeling, and in many other decorative and industrial applications. Most of these are extruded horizontally through dies of a corresponding cross section and are cooled and sized with air jets, water troughs, cooling sleeves, mechanical aids and any other devices the extruder can invent to fit the irregularities of his products. Typical products are shown in Figures 2.9 and 2.10.

MATERIALS

Vinyls, both rigid and plasticized, are the most popular materials largely because they can be custom-formulated for many different uses. Other contours are made from high-density polyethylene, polypropylene and polystyrene (both clear and high-impact), and some uses have been found for all thermoplastics in contour form.

PRODUCT AND EQUIPMENT SIZE

The product size is as limitless as the number of applications. Successful extrusions have ranged from tiny polyethylene tapes $\frac{1}{8}$-in.

(3 mm) wide, and capillary tubing of fluorocarbon plastic with $\frac{1}{16}$-in. (1.5 mm) *outside* diameter, to semicircular polystyrene lighting fixtures 1 ft (30 cm) across and complex vinyl barriers for concrete dam sections, weighing over one pound per linear foot (1.5 kg/meter).

Contour shops are generally specialized small companies with rather small, versatile equipment. Their personnel have acquired the experience and skill needed to run the great variety of cross sections encountered. They often have their own machine shop and formulating equipment, and frequently design both the die and the formula to fit the end use. Extruders are usually small because of the small size of most orders and the small cross section of most jobs. Machines of 10 to 25 hp ($1\frac{1}{2}$ to $2\frac{1}{2}$ in. [35 to 45 mm] in barrel diameter) are common, with a few larger machines (40 hp, $3\frac{1}{2}$-in. [90 mm] diameter) in use for the largest sections. Output for the smallest cross sections is often limited by the maximum take-off speed and cooling capacity. Another limit is also possible; i.e., at very high screw or take-off speeds it may become impossible to hold accurate part dimensions. Rarely is power the limit to a contour extruder's output.

EQUIPMENT

Dies. Contours fall into two classes: standard and custom. For the standard contours—tubing, tape, belt and rod—die construction is fairly well established, and a stock of dies is kept on hand to handle orders as they are received. For a few special and/or large-volume products such as strapping and drinking-straws, a line may be set up to make one product all the time, from a single die. At the other extreme many jobs require a unique, non-standard cross section; a special die must then be built, the cost of which is reflected in the quoted price of the finished contour. These dies may become the property of the end-user or else remain with the extruder in order to keep future business in his shop. The extruder applies his experience to the design of the custom die and, more often than not, several trials and changes are needed before an acceptable product is made.

For tubing, miniature pipe dies are used which are mostly the straight-through type. In these, and for any other hollow section, there are the usual problems of supporting a core by a spider-type device in the die, as discussed more fully in Chapter 3. For all dies, the internal passageways are usually streamlined, frequently by hand. The land length (length of constant cross section just before the opening), the angle of entry into the final land section, and the percentage oversize, are all critical die dimensions.

A land is needed to provide back pressure in the head and barrel. This pressure, in turn, promotes better mixing, which makes a more homogeneous product. Pressure also helps to seal weld lines, if any, resulting from the separation of flow streams by the spider arms. The lands are often about ten times the width of the lip opening; e.g., in a tube where the die and core form an annulus $\frac{1}{16}$-in. thick, the land would be $10 \times \frac{1}{16}$ or $\frac{5}{8}$ in. long. This ratio is reduced for very viscous polymers and lengthened for very fluid ones. The resistance to flow, hence the back pressure produced, varies as the *cube* of the changing thickness dimension, so small openings (below $\frac{1}{8}$ in. = 3 mm) are very effective in producing back pressure. For large cross sections, such as $\frac{1}{4}$- and $\frac{1}{2}$-in. rod (6 to 12 mm) the calculated land length may require a very long die. The greater cost of such a die often leads to land-to-opening ratios of less than 10 to 1, and output may be limited by the rough surface produced at high rates. This is not necessarily bad; it is an economic balance of the costs of production, and the correct compromise varies for different jobs.

The internal entrance to the land is an important variable. The entrance angle, in particular, is as small as practical. However, small angles mean longer dies, which cost more. Again an economic balance must be made. Entrance angles of 60 and 90° are common, as is hand finishing such that the included angle is large at the start (90 to 120°) and smoothly tapers down to a small angle (20 to 60°) at the entrance to the lands (see Figure 9.1). For small openings, where melt fracture is a possibility, effort should be made to keep this angle close to 20° at the land entrance. Melt fracture is the rough surface found on melts extruding at very high exit speeds (and very high shear rates) through small cross sections. It has been found that if the entrance angle is kept small there is less sharp change of direction of flow, and there is much less chance of melt fracture.

Designing the die oversize is necessary because a certain amount of tension is needed to pull the plastic away from the die, resulting in a smaller cross section. Also, the melt shrinks considerably as it cools—as much as 20 per cent for some plastics. Counteracting this is the tendency of the plastic to swell as it leaves the die in which it was under pressure. Both the swelling and the shrinkage depend on

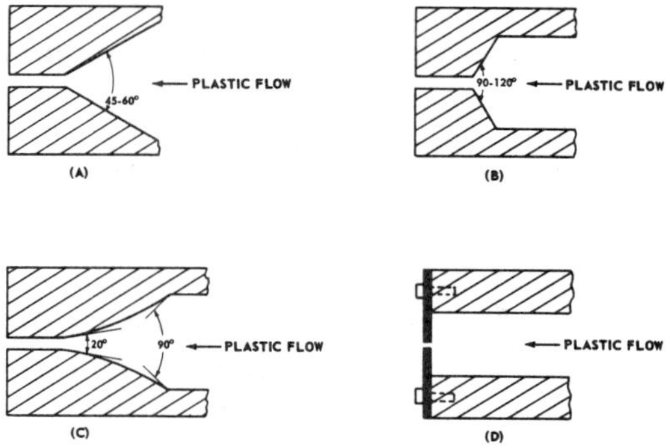

Figure 9.1. Cross-sections showing various internal entrance angles to contour dies. Die (A) is common and is simple to make. Die (B) is shorter and even easier to make, but the large entrance angle may cause material stagnation and/or melt fracture. Die (C) is the best, but it is the most expensive. Its streamlined approach inhibits stagnation, and its small angle (20°) at the beginning of the parallel land section allows exits at high speed without melt fraction. Die (D) is an inexpensive construction used mainly for trial or short-run work. (*Illustration by author*)

the nature of the plastic (especially its crystallinity), on the internal geometry of the die (short lands and large entrance angles yield more swell), and on the temperatures and pressures involved both before and after the die. Several publications treat this theory quite well, but only a few have applied it practically to contour die construction. This is partly due to the newness of our understanding of these phenomena and partly because few contour shops are big enough to bear the cost of such work. Much practical experience exists, rather closely kept by its appliers. A general guide for optimum oversize design is:

	(%)
Low-density polyethylene	15
High-density polyethylene	20
Polystyrenes, all	10
Nylon	20
Cellulose acetate/butyrate	10
Unplasticized PVC	10
Plasticized PVC	15

One might ask: Why not make the die highly oversize, not worrying about the exact amount of oversize, and draw down to desired dimensions? In fact, this is commonly done, but it is not always desirable, because (1) the drawdown introduces stresses which may weaken the product and (2) the drawing would proceed unequally in all but circular cross sections, leading to further dimensional distortion which must be accounted for in die design.

The unequal cooling of complex cross sections must be considered in designing their dies. All sharp projections and edges will cool faster (more surface area per unit volume) and these will therefore shrink first. This tends to make them smaller than the rest of the section, which can be compensated for by enlarging the die at these points. In practice, this means distorting the die profile, as illustrated in Figure 9.2, where curved sides are needed to get a square product.

It can now be seen that in the same die some parts must be designed differently from others, even where the section thickness is the same. This concept of a cross section as made up of many smaller elements is very helpful in design.

Another concern in the internal die design is stagnation and burning. Flow is laminar, and any corners or pockets will collect non-

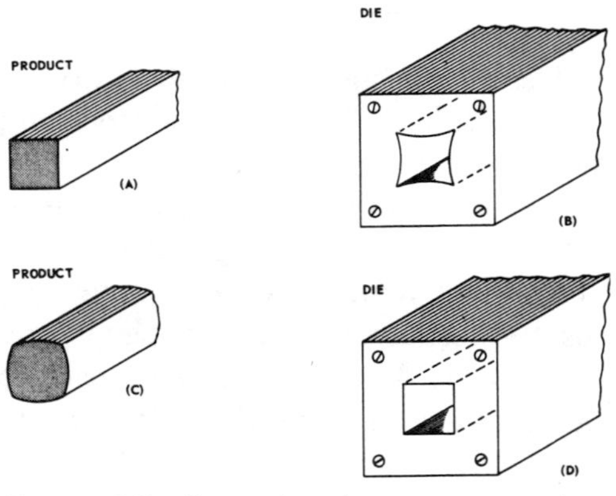

Figure 9.2. To make the square section shown in (A), a distorted die (B) must be used because the corners shrink more than the rest of the section. If the die were made rectangular, as in (D), the product would have bulging sides, as in (C). (*Illustration by author*)

moving plastic, which will decompose there. This is critical with vinyls, especially unplasticized vinyls, as any decomposition will spread quickly through the extrusion, spoiling it. Smooth finishes, streamlined design, and small entrance angles (90° or less) are all important.

Most contour dies are built to extrude horizontally. However, for very fluid melts such as nylon, it is hard to hold proper shape with a horizontal die. In such a case, as well as for many other flexible contours that are difficult to cool in air, the melt is extruded downward into a water bath. The die must then contain a streamlined internal bend, usually 90°, unless a vertical extruder is used; and its construction is similar to that of monofilament dies (Chapter 8). Downward extrusion is seldom used for rigid vinyls because: (1) the internal bend creates a potential place for stagnation and decomposition and (2) the product is often too rigid to be bent back to a horizontal travel path after cooling.

Sometimes all these principles of streamlining are completely discarded, as in die (D), Figure 9.1. Here the die is made simply by cutting out the section from a thick sheet of metal and bolting it to a cylindrical adapter, corresponding to an entrance angle of 180°. This technique is used for short or noncritical runs, on non-vinyl plastics, where economy is all-important. Even vinyls can be run on such dies where only a small amount of product is needed.

Virtually all dies are made in more than one piece. The decision of how to "break up" a complex cross section can seriously affect its cost as it affects ease (and cost) of machining. In Figure 9.3, for example, the die is made with one outer piece and two small central pieces held in place with external bolts.

Steel is the common material of die construction. Internal surfaces are chrome-plated only for long runs of vinyls, and even this practice is of questionable value. Fluoroplastic coating of inner surfaces prevents corrosion and improves gloss, but heat transfer is impaired, and much care must be taken in cleaning to avoid scratching. Lips are sometimes hardened; this practice is recommended for stock or long-run dies, after their final dimensions have been set, to prevent nicking. Brass and beryllium-copper dies are also made. They are easier to work with but are less durable. They are most useful for custom jobs where many trials and changes are expected. The stock dies, expected to be used many times, should be made of steel.

Figure 9.3. A die used to make a complex contour—a slat for roll-up window shutters. Note the small center pieces held in place with small bolts. (*Courtesy of Technical Machinery Corp. and Amut SpA*)

Heaters. Die heaters are usually band-type resistance heaters in cylinders or strips, or custom-made to fit the contour, as in Figure 9.3. Such heaters should be designed and mounted to ensure good contact with the die surface; a heavy steel bar is often fastened over the heaters for this purpose, preventing them from pulling away from the die. Cartridge heaters are cleaner and easier to use than bands, but require a bigger die and are a possible source of nonuniform heating. For many dies, cast aluminum block heaters are usable, and work quite well.

Take-offs. Each contour passes through a shaping device, sometimes built specially for it, and then into a water trough or over a conveyor belt to cool in air. For rod and tubing, the shaping system may be a cooling sleeve or a set of brass or steel sizing plates, such as are used for pipe (Chapter 3). These devices can be horizontally split for ease in start-up; the top halves are left off until the contour is threaded through the take-off. They are then placed in position to obtain accurate sizing. Plates and sleeves are also used for many other contours, especially hollow ones. A vacuum cooling sleeve is most popular in Europe. The most complex shapes are handled by passing them through a great number of adjustable brass fingers (sometimes fluoroplastic-coated) so placed as to make the final cross section as desired. Air jets are sometimes directed at critical points on the emerging extrusion to both cool and shape it.

With all these devices, the cooling is completed in a water trough, under a water cascade, or on an open conveyor belt (sometimes under air fans). Final sizing may still be taking place during this cooling. Air is often adequate for vinyls and polystyrene but water is preferred, and water is needed for almost all contours of polyethylene and nylon, which require much cooling. The simplest products, like belt and tape, occasionally can be carried off right from the die by the conveyor belt itself without any sizing or other treatment. Most troughs and conveyors are from 10 to 25 ft (3 to 8 meters) in length and are mounted on casters. Portability and interchangeability of these take-offs is important as the production coming out of a given extruder is constantly changing, and the take-off must often be changed to fit the cross section. Not the least part of learning how to run a complex job is deciding how to size and cool it.

From the cooling line, the contour is either rolled up on a reel, coiled in a drum, or cut in lengths. Various cutters are used to suit the flexibility, size, and linear speed of the product. They include traveling saws, guillotines, high-speed rotating blades, and people with scissors. When the product is reeled, the windup is the kind normally used for flexible pipe—a vertical reel driven by a constant-torque motor. Both winders and cutters are fed by pulling units, mostly the caterpillar type, with two treads. A pair of guide posts prevent the contour from drifting to one side or the other in the puller—a possible problem with some irregular contours. Simple rubber pull rolls are also used.

At the end of the take-off, it is sometimes convenient to use bent lengths of ordinary metal or plastic pipe to guide the moving contour, for example, into a drum.

Finished contours must frequently undergo many hand operations after extrusion, and special equipment is sometimes necessary. These operations include cutting, sawing, bending over heated forms, gluing, welding, polishing, stapling to cards or to itself, etc.

OPERATION

Formulation. The formulation is usually set by the *product* requirements but occasionally can be altered to aid in *processing*. For example, inorganic fillers can reduce swell at the die and aid in sizing, while also reducing costs. However, their compounds require drying, are opaque, and are less glossy. In other examples of formulation as a processing aid, small amounts of plasticizers can reduce power needs; stabilizers can inhibit decomposition and thus allow higher processing temperatures; and selection of the plastic base resin can affect colorability, ease of extrusion, resistance to decomposition and many other properties. *Caution:* with rigid PVC, small amounts of plasticizer may actually *increase* stiffness and brittleness of the product, even though melt viscosity is reduced.

Vinyl compounds are often made to order for special jobs. The formula must be adequately premixed—ribbon blenders and paddle mixers are still used, but high-speed intensive mixers are the preferred way to mix PVC powder blends. Also, the compound can be hot-processed in any of numerous ways to make pellet or granular feed for the extruder. The decision to hot-process or not depends on available equipment and product requirements as well as on the cost of the extra operation. Not all extruders can produce a well-mixed product from dry-blend at acceptable output rates.

All materials are often custom-colored. A small shop can hardly afford to keep dozens of colors of many plastics in inventory; nor can it afford to wait for a supplier to make up a batch in his color, especially when all he needs is a half-ton or less. Dry pigments can be used directly, particularly with powdered plastic resins, if they are tumbled or otherwise blended into the feed prior to extrusion. High pressures may be necessary in the extruder to eliminate streaking. A masterbatch of the color in the base resin has less chance of streaking but is more expensive. Wetting agents or mineral oil can be used to help even distribution of powdered pigments, preferably by predispersing the pigment in the agent or oil.

Sizing and Take-off. Product dimensions are often critical, as the contours may fit into or over other established parts. Skilled operators are essential for this work. They must know how to adjust the many brass fingers, the distance between sizing plates, the unsupported distance between die and cooling/sizing elements, the temperatures of plastic and coolant, extruder and take-off speeds, etc. It is mostly trial and error except for simple shapes like rods and tubes; for the others, the operators try to estimate how the contour should run, and start it that way. If something is not right, then they must figure out the appropriate adjustment.

Further operating information is included under "Special Techniques," at the end of this chapter.

PROBLEMS AND SOLUTIONS

The following chart shows suggested solutions for some common problems encountered in contour extrusion.

Problems	*Causes and Solutions*
Wrong shape	
(a) Proper cross section, but too small or too large	(a) Too much or too little pull on the contour. Change draw distance (from die to take-off)—longer distance = thinner section. Change take-off speed or material temperature. Also, swelling can increase cross section, and is greater with shorter lands, lower temperatures.
(b) Distorted cross section	(b) Unequal die temperatures may make some parts too small, others too big. Incorrect die construction can do the same; a few trials and alterations may be necessary. Sometimes, changing the shape or location of sizing plates or fingers will help.
(c) Proper section, but warped	(c) Improper cooling and take-off. Inspect line and locate point of warping. Try to give uniform cooling, and keep section supported until it is cool enough to hold its own shape. Slower linear speed and/or less intense cooling may be necessary.
Low gloss	
(a) In strips	(a) Cold or rough spot in die or on sizing device. Visually locate source before taking action.
(b) All over	(b) Extrudate too cold, die too cold, or compound inherently dull. Higher temperatures and/or slower cooling help gloss, but may reduce rate. Flame-

Problems	*Causes and Solutions*
	polishing and buffing done right on the take-off line can make major improvements.
Rough surface	Contaminated material. Die too cold. Extrudate too cold. Sizing device or take-off is scratching surface. Moisture in material. Melt fracture—exit speed at the die is too great (raise temperature, reduce rate, put job on smaller machine, use die with more streamlined entrance, larger opening or longer lands).
Surging (cross section linearly varying)	Take-off speed not uniform—check for broken gear teeth and worn belts. Periodic impact of cutter may be transmitted to material at die. Extruder may not be putting out uniform melt at uniform speed—run at higher pressure, lower rate, check for broken controller, use better-mixing screw or pre-mix dry-blend better. Try to locate the origin and period of the surging by testing all shaft speeds for uniformity, or watching for visible variation in temperature, pressure, screw speed or motor load. Weigh two-second lengths cut at die and plot weight vs. time to analyze surging. Finally, insufficient pressure in the die may be the cause—if so, run colder or slower, or use a die with longer lands.
Die lines (continuous lines in machine direction)	Clean out hard particles hung up in die. Also, a nick or burr in take-off system may be responsible. Try to locate origin visually.
Pits on surface	
(a) On bottom of flat surface	(a) Extrudate laid on conveyor belt while still too hot. Use different manner of cooling, or cool underside of section with air before it touches the belt, or blow air up through holes in the belt itself.
(b) All over	(b) Contaminated material, moisture in material, droplets of water striking section right after leaving die, water bath too hot and causing localized boiling.

STANDARDS AND TESTS

Many of the finished extruded products must meet certain standards related to their end uses. Specialized tests accordingly have been devised to run on plaques, or on the finished extrusion. For example:

(1) For plasticized vinyl refrigerator gaskets, the vinyl material is held in contact with high-impact polystyrene for a specified time to see if the plasticizer in the vinyl attacks or mars the polystyrene. The polystyrene, as a breaker (edging) strip, would be contacting the vinyl door gasket in actual service in a refrigerator.

(2) Lighting fixtures are exposed to ultraviolet light to test their suitability for prolonged use with fluorescent or ordinary incandescent light. Yellowing and embrittlement are observed.

(3) Many PVC contours are used because of their flame resistance. Burning tests often are run, therefore, either on the finished product or on test bars. ASTM or Underwriters' Laboratories test procedures are used in the United States.

(4) Vinyl tubing used for blood transfusions or other medical purposes must conform to strict chemical requirements—notably the complete absence of lead. Consequently, chemical analytical tests are run.

SPECIAL TECHNIQUES

Special Formulations. These are common as many contour jobs are formulated specifically for their end uses. Examples are as follows:

(1) A powdered magnetic iron compound is added to plasticized vinyl to make the entire compound magnetic. It is used for refrigerator gaskets, eliminating the need for a lock and making a "child-safe" refrigerator.

(2) Pearlescent (nacreous) pigments are added to give a unique iridescent pearl-like effect.

(3) Bits and chips of some non-thermoplastic are added to give a speckled effect. Screens must either be omitted or else be coarse enough to permit passage of the particles. The particles must not be abrasive to the extrusion machinery, and they must be mixed with the plastic uniformly both before and during extrusion.

(4) Numerous fillers are used in vinyls, mainly to lower cost. These include clays, diatomaceous earth, asbestos, and calcium car-

bonate. They increase rigidity and opacity, and decrease shrinkage, but also are likely to increase moisture absorption, density, and brittleness.

(5) Stabilizing additives include ultraviolet absorbers, which prevent yellowing and embrittlement in the sun or other ultraviolet-containing light source such as fluorescent fixtures. They are always used in polystyrene extrusions for lighting applications. They are expensive ($3.00/lb or $6.60/kg, and up) but are very efficient, and only a small amount need be used. Antioxidants are also added for many applications, especially in polyethylene, and stabilizers are always needed in vinyl formulations to prevent their degradation and consequent formation of hydrogen chloride.

(6) Colorants include carbon black, which is an excellent light-stabilizer, and fluorescent pigments that are unusually bright, as well as the range of conventional pigments and dyes for plastics. Not all colorants are stable under all processing and end-use conditions, and this must be considered when selecting them.

(7) Fiber-reinforced contours. Extruded profiles and tubing have been made from polycarbonate, nylon, and acrylic plastics filled with glass fibers or asbestos. These compounds are more often injection-molded, but they can also be extruded.

Flame Polishing. A very high gloss can be applied to the surface of a plastic contour by directing a gas-air flame (like a Bunsen burner) at the extruding surface just as it leaves the die. This technique not only works for vinyls, which are self-extinguishing, but also for flammable plastics like polystyrene and polyethylene. In the latter cases, the burners must be carefully installed to prevent accidental fires, and the contour should be quenched in water right after polishing. For added safety, the gas feed can be controlled to stop if the plastic contour stops moving for any reason.

Buffing. Another way to get high gloss, especially on high-impact polystyrene, is to buff the surface mechanically and continuously. This is done at a point where the contour is fairly cool and rigid, and rather soft buffs are used with mildly abrasive buffing compounds. No commercial equipment is currently offered for precisely this purpose, but available machinery can be adapted easily. More than one buffing station may be needed for small, fast-moving contours. The wheel(s) may have to be specially assembled with buffs of various diameters in order to apply equal pressure to irregular contours. This technique avoids the degradation and discoloration possible with flame polishing and is entirely safe, but it is generally more expensive to use.

Cross-head Techniques. Some contours include a core or imbedment which must be introduced through a cross-head; this is similar to a wire-covering operation. An example is the welting used in automobile upholstery, where a metallized strip or braid is covered with clear vinyl, and an integral strip for sewing is included as part of the contour.

Two-color Extrusion. Several techniques exist for this:

(1) A small extruder is mounted over the head of a larger one, and the die is designed to weld the two streams together just before they emerge. This is common in wire coating, and commercial machinery is available. Two machines of similar size can also be used, feeding into a common die; this is done for lighting fixtures, as well as film and other extrusions.

(2) A single extruder is built with screw(s) such that material is conveyed from both ends towards the center. One color is fed in at each end, and the extrudate leaves in the middle through a hole in the barrel wall and a die similar to (1) above. This particular extruder utilizes a twin-screw system.

(3) With two extruders, each extrudes a set of small tubes or rods of a different color. The dies are set next to each other, and the tubes are welded together just as they leave the dies, in any preset pattern. They need only their own heat to effect a good bond. The striped or patterned sheet that is thus formed can even be thermoformed. Alternatively, the tubes can be formed into pails or jars by using the proper auxiliary equipment at the die (e.g., mandrels).

Any of these techniques can be used to coat a solid or hollow plastic section with another plastic or another color of the same plastic, all in one operation.

Reinforced Tubing. In one method, a plasticized vinyl hose is reinforced with a spiral of rigid, unplasticized vinyl. Two concentric vinyl tubes are extruded through a single die, and a knife travels around the combination, cutting through the outer layer only. Because of drawdown, this outer layer then separates into a helix, reinforcing the continuous flexible internal tube. In another method, two extruders feed a single die with moving parts, and a helix of rigid PVC is continuously extruded and embedded in a soft PVC matrix. Plastic tubing and pipe can also be reinforced with fabric mesh or metal helices, fed into a cross-head die similar to wire coating. A machine is available which coils a continuous helix from a spool of wire. The conduit industry has generated several types of "reinforced" tubing. The best known is a corrugated tubing (noted at the end of Chapter 3) made by extruding a tube into a moving chain of split molds. Other methods involve extrusion of a thin sleeve over (or inside) a metal tube, often corrugated, and winding of a narrow extruded plastic strip into a helix form. In the latter method, the same strip can be used for different diameters, but cannot make very small tubes such as are needed for most conduit unless thin and/or semi-flexible plastics are used.

Pseudo-metallized Contours. These are made by extruding a transparent plastic (vinyl, cellulosics) into a cross section that contains a slot to hold a strip of paper coated with aluminum foil. If the plastic is uncolored, the product has a silvery metallized appearance. Adding dyes to the plastic can produce a gold appearance or any other tint.

Post-forming. Sometimes a complex contour is most easily made by extruding a simple shape (e.g., a tube or a sheet) and continuously post-forming it after the die (see Figure 9.4). Postforming is done

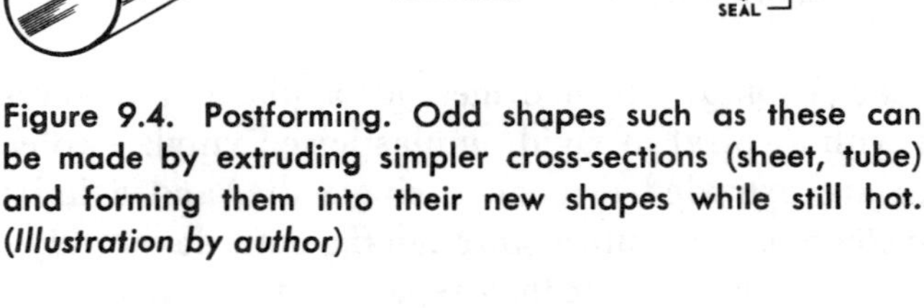

Figure 9.4. Postforming. Odd shapes such as these can be made by extruding simpler cross-sections (sheet, tube) and forming them into their new shapes while still hot. (*Illustration by author*)

with brass fingers, or with a special sizing plate or sleeve or custom-built forming shoe. It has the advantages of versatility and simpler extrusion dies, but dimensional control may require more care, and design of post-forming equipment may use up the savings gained by the use of a cheaper die. Every case must be calculated individually in order to decide whether to extrude directly or to postform.

REFERENCES

Commercial Aspects

"Purchasing plastics: extruded profiles," *British Plastics,* March 1967. Illustrated article discussing ownership of dies, length of runs, pricing, and how to specify product.

Technology

"Influence of cross-section of entry of extrusion heads on extrusion conditions," S. I. Klaz and I. F. Kanavets, *Plasticheskie Massi,* January 1967. Abstracted in *Modern Plastics,* May 1967. Discusses shaping of entry zones to match cross-section of profile.

"In-line annealing: a new approach to extrusion of rigid vinyl products," R. P. Socha (B. F. Goodrich). Adaptation of paper in *SPE ANTEC* 1965. Annealing reduces linear shrinkage potential to near zero. Experiment made on a window channel.

"Standard for extruded PVC soft profiles," *Kunststoffe,* January 1963. Review of standard DIN 16941, covering dimensions of extruded profiles.

"Extrusion of bi-colored profiles with twin-screw extruder," H. Marhenkel (Schloemann), *Kunststoffe,* May 1962. Short review of Pasquetti machine for this application.

"A continuous process for producing polyamide rod," J. L. Voigt, *Kunststoffe,* August 1961. Special equipment described which keeps the rod under pressure as it cools.

"Extrusion of high-density polyethylene tubing and profiles," D. J. Schmidt (Grace), *SPE Journal,* May 1960, and *SPE ANTEC* 1960. Includes die design comments, illustrations, extrusion conditions and take-off techniques.

"Section extrusion of rigid PVC," G. H. Burke and A. E. Parker (British Geon), *British Plastics,* March 1959. Economics, equipment and operation know-how given for rigid PVC contours, such as hand rails, window channels, and electrical conduit.

"Nylon-6 tubing," J. L. O'Toole *et al.* (Allied Chemical), *SPE Journal,* October 1958, and *SPE ANTEC* 1958. Description of equipment with photographs. Analysis of properties and effect of extrusion conditions. See also article by du Pont in *SPE Journal,* July 1957, which covers nylon profile extrusion in general.

TEN

extruder selection and operation

HOW TO SPECIFY AN EXTRUDER

This section is written especially for the new buyers of extruders who have had to base decisions on advertising literature and fragments of knowledge of what established competitors are using. It is strongly recommended that a prospective purchaser know what goes into an extruder, what is costs, and what it will do. Thus he can intelligently fit his machine to his needs. He need not become an expert, and he need not always specify the best and costliest. With the basic knowledge, however, he should be able to insist on and buy only enough "machine" to fit his needs. Nobody else will be as considerate of these needs as the purchaser himself.

Early in the process of buying extrusion machinery, it is important to decide on the quality level that will be acceptable. This is not easy to do: everyone wants the best quality from the cheapest equipment. However, if one is aware of the practices in his field regarding dimensional tolerances, surface appearance, and the like, and if he decides at what level he wants to produce, this will help in specifying the proper equipment for the job.

A "decision checklist" is presented on page 278. Comments intended to help make these decisions are on the following pages. Further description and illustration of the various extruder components are found in Chapter 1.

Capacity. The purchaser *must* decide the quantity, annual or hourly, that he wants to be able to produce. Ultimately, the best buy is that with the least cost per pound or kilogram per hour of *salable product,* but with only as much capacity as wanted. This sounds like a simple premise, but it is often violated.

Horsepower. Once the capacity is decided, and assuming the materials to be extruded are known, the expected efficiency can be applied to calculate the necessary horsepower (hp). *Efficiency* here means output per horsepower. It is one of the most important concepts of extrusion, and it is understood and applied in any good and successful extrusion operation. Typical efficiencies for a single-screw extruder are:

	Per hp	
	lb/hr	kg/hr
Rigid PVC	7–10	3–4½
Plasticized PVC	10–13	4½–6
Impact polystyrene	8–12	3½–5½
ABS polymers	5–9	2¼–4
Low-density polyethylene	7–10	3–4½
High-density polyethylene	4–8	2–3½
Polypropylene	5–10	2¼–4½
Nylon	8–12	3½–5½

These figures assume normal operation at medium temperatures, with 80 to 100 per cent of the heat in the extrudate supplied by the drive. The lower numbers are for extrusion at relatively high pressure, while the higher numbers are for low pressure. For unusual operations such as polyethylene extrusion-coating, or with twin-screw machines, where much heat is conducted in through the barrel walls, even higher efficiencies are possible.

To get the required horsepower, divide the desired hourly production by the expected minimum efficiency. Add a safety factor—and *this* result is the figure with which to ask for bids. For example: "I want a machine to produce 300 lb/hr of good XYZ-plastic pipe. I expect a minimum efficiency of 6 lb/hr per hp; therefore, I need 50 hp. Just to make sure, I want a 60-hp drive. Give me your machine description and price to do this job." In this manner, *performance* is specified—not merely machine dimensions, which may have little to do with the actual abilities of the extruder.

CHECKLIST

1. Projected Capacity = lb(kg)/year __________ hr/year __________ lb(kg)/hr __________.
2. Required hp = Efficiency, lb(kg)/hr/hp ________, Total hp ________.
3. Extruder Size = __________ machines at __________ hp each. Barrel Diameter each __________.
4. Single Screw ______________ or Double Screw ______________.
5. Screw Design ____.
6. AC–DC Drive ____________, Magnetic-clutch Drive ____________, Mechanical Drive ____________, other ____________.
7a. Belts = Yes __________ No __________.
7b. Gears = Herringbone __________, Helical __________, other __________.
7c. Lubrication = Forced ______________ Splash ______________.
8. Thrust Bearing = B-10 Life __________ hr at __________ psi (kg/cm^2) and __________ rpm.
9. Barrel Length, L/D ratio: __________.
10. Barrel Material = "Xaloy" ____________, "Xaloy 306" ____________, Nitrided ____________, other ____________.
11. Screw-flight Material = "Stellite" __________, Nitrided Steel __________, Flame Hardened ____________.
12. Heating System = Induction __________, Cast Heater Blocks __________, Bands __________, other ____________.
13. Cooling System = Air/Induction Heating ________, Water Fog ________, Liquid Evaporation ________, Outside Fins ________, Inside Channels ________, Air Over Heaters ________.
14. Controllers = Regular Time Proportioning ________ or Stepless ________.
15. Pressure = Safety Bolts' Setting ____________. Maximum Safe Operating Pressure ____________.
16a. Valve Type = Adjustable Screw ______________, Plug ______________.
16b. Gauge Type = Grease ______________ Transducer ______________.
17a. Screw Pushing = Yes ____________ No ____________.
17b. Screw Cooling = Yes ____________ No ____________.
18. Screen Changer = Intermittent ____________, Continuous ____________.
19. Installation Service ____.

Extruder Size. Once total horsepower is known, extruder sizes and barrel diameters can be set. Machinery manufacturers will have preferences, and should specify the diameter to fit the horsepower they are asked to use. If diameter is too small the machine is less expensive, but screw speeds may have to be unrealistically high, yielding higher material temperature, with possible cooling and/or degradation problems. A diameter that is too big will usually work well, but is unnecessarily costly. Remember that more than one machine may be needed, depending on product and take-off speed requirements.

Single Screw or Double Screw? Double screws may be used for vinyls, but it is not an inevitable choice. Double (twin) screws are used widely in Europe for vinyls, both for compounding from dry resin mix and making finished products from pellets or dry mix. Seldom are they used for other plastics—even in Europe. In the United States, for all materials, single screws are most common for finished-product extrusion, despite a few imported twin-screw installations for PVC pipe and sheet. This is due, in part, to the fact that no twin-screw machines suitable for finished-product extrusion are made in the United States. There have been recent rumors that one or another manufacturer in the United States would add a twin-screw extruder to his line, most likely licensed from Europe, but at the present time this has not yet happened.

Some twin-screw machines are used in the United States for compounding in the processors' plants, but this is not a common operation, and finished compounds are usually bought from the suppliers (except for some PVC dry-blends mixed by the processors). The suppliers themselves use both single- and double-screw machines in their compounding plants. The arguments in favor of twin screws for vinyls are that they mix better at lower pressures and temperatures, and the reduced frictional heat helps in moving the vinyl through with less chance of decomposition. Sometimes this means money saved as less stabilizer is needed. However, single-screw advocates reply that the twin-screw machines require small thrust bearings which impose a load limit (hence, output limit), cost more per unit output, and may give maintenance problems. Makers of twin-screw machines may then say that their systems develop less pressure and need not turn as fast, so that their thrust bearings are satisfactory. Sidewise thrust remains a problem. The controversy is still active at this time, and although twin-screw units remain neglected in the United States, they are very much alive in Europe. This is emphasized by the recent introduction of twin-screw models by two leading single-screw manufacturers.

It is impossible to recommend one or the other, except that single screws are preferred where the materials are not particularly heat-sensitive. For PVC, both types have been running successfully for years. The situation is much like the driving of a car: most of the success depends on the driver rather than the machine. The choice is best based on one's past experience, availability of service, confidence in the manufacturer, and such commercial considerations as price (*always* per unit of output) and delivery time.

Screw Design. This should be decided by the purchaser and extruder manufacturer—perhaps with the help of the suppliers of the plastics to be extruded (see pp. 318–325). Many firms can now test designs with a computer, saving much time and money. Make certain that channel depths are specified, not just compression ratio. If several designs are possible, the shallowest channels should be made, and the screw should not be plated or nitrided. Then it can be tried, and the channels can be deepened if necessary. Often, more than one screw must be bought to enable proper extrusion of different plastics.

Drive Type—Speed Variation. Most extruders are equipped with a motor and transmission that can drive the screw at a continuously variable speed. After choosing the number of horsepower needed, another choice must be made: the type of drive system. A *mechanical* system with pulleys with changeable diameter is inexpensive and is usually adequate, except where a single speed will be always used. The pulley sheaves tend to wear in specific spots and the unit loses its precision. This system, as well as other mechanical drive systems, runs at constant available horsepower; that is, it is capable of transmitting its full power over its entire speed range. This may be an advantage, especially with very viscous materials, but it must be used with caution, as the application of full horsepower at low speeds may create too much torque, and the screw might break. If such drives are to be used often with new materials or at new conditions, a safety device is recommended which will stall the machine at excess torque or otherwise protect the screw.

For big units—50 hp and above—the *magnetic-clutch* drive is virtually as cheap as a mechanical system and is a more precise unit. However, it may not give enough horsepower at low speeds or with stiff materials, since it is a constant-torque system; to counter this, several sets of gears or pulleys are often provided to give several overlapping speed ranges. This is entirely satisfactory where the line usually runs at a constant speed with only minor changes; where frequent major speed changes are required, quick-change gears or a different drive system are recommended. The magnetic-clutch drive requires cooling, especially at low speeds. This may be inefficient, and in some areas quite costly—either because a closed-circuit exchange is required or because much tap-water is used.

The third common type is a *generator or rectifier* operating on AC and driving a *DC motor*. Rectifiers have now replaced the generators in most new systems of this kind, and their lack of moving parts makes them easy to maintain. These drives are generally considered the most precise, and are also the most expensive. In some horsepower ranges, however, the price is now close to the magnetic-clutch system. There are several types: some have constant available torque throughout the entire speed range, and others have a combination of constant torque over the upper portion of the range and constant horsepower over the lower portion. The speed range is very broad and the system operates very efficiently over most of this range. It is therefore often used in experimental extruders, or for operations such as profiles or monofilaments where the machine is called upon to run at many different speeds. It is also often used where a sensing system is measuring the extrudate and sending back an electrical signal to control screw speed.

These three systems comprise virtually all the variable-speed drives used in the United States. The AC-DC system with solid-state rectifiers is gaining in popularity, and some sources predict that these drives will take over most of the market. They have made their gains mainly at the expense of the magnetic-clutch drives; the mechanical ones are still the least expensive, and hold on to their low-cost end of the market—about 25%.

A few machines are designed to avoid a variable drive altogether, and have several *fixed speeds* instead, similar to a standard-shift automobile. This is often adequate and may save some money, but a good and precisely built nonvariable gear system can often cost as much as a variable-speed unit. Stepless speed variation itself is not as essential on the extruder as it is on the take-off, particularly if there is a pressure valve to gain further control. A few of these nonvariable systems do not have clutches, which means that the whole machine must be stopped (and thermal equilibrium thereby disturbed) in order to change speeds.

One other speed-variation system, a shifting-brush commutator system, is also prevalent in Europe, but it has never caught on in the United States and is slowly losing ground to the other methods even in Europe. It has a rather narrow speed range (3:1) and is said to be more expensive to maintain than the mechanical system. However, it is more precise and more adapted to one-speed operation than the mechanical drives with which it competes favorably in price in Europe. In the United States, it is apparently priced too high to compete.

Hydraulic drives are used on a few extruders, particularly those used in injection-molding and blow-molding machines, where the screw is also used as a reciprocating piston. The hydraulic system is well-suited to start-stop (intermittent) operation and to power the reciprocating motion as well. It is a constant-torque system, and thus may have the problem of insufficient power at low speeds. Its advantages include ease of maintenance, small space needed at the extruder, and a wide speed range.

In choosing a drive, keep in mind that the motor and drive are sometimes located right under the extruder barrel. This saves space, and makes it possible to have both sides of the extruder equally accessible. In such cases, there should be ample air circulation to prevent overheating.

Transmission and Lubrication—Speed Reduction. An electric motor runs far too fast to be directly driving the extruder screw, so some speed reduction is always needed (except for a few special high-speed machines). Also, there is a choice of the means by which the driving unit (motor) is coupled to the driven element (screw). This coupling can be through a belt, a chain, a flexible non-reducing coupling, or directly through gears. Many extruders, including some of the best ones, use belts. The belt system offers mechanical insulation between drive and screw, and it facilitates changes of speed range, as sheaves and belts can be easily changed. The direct gear linkage gives a more positive, precise drive than ordinary belts, but a "timing-belt" or "timing-chain," which cannot slip, is a useful compromise. Some extrusion machines can achieve mechanical insulation with a flexible coupling between drive and screw. Advantages of mechanical insulation, no matter how achieved, include less need for exact alignment, as well as protection of one component from mechanical shocks given to the other. Belts and chains provide some speed reduction, but more is almost always needed, and this is done with gears. These are preferably herringbone or helical; ordinary spur gears are more liable to breakage and less efficient (more energy lost). Helical gears produce some thrust, but this can actually be used to advantage if designed to oppose the backward thrust of the screw. In addition, they are easier to change and are cheaper than the herringbones, which is important when different sets of gears are required to give a number of speed ranges.

A "splash" lubrication system (like the one used in an automobile) that is well designed to reach all parts of the gears is usually adequate. A forced (pumped) system is sometimes used as it permits filtration and can be designed for very uniform oil distribution. For high-powered extruders where full power is transmitted most of the time, many manufacturers recommend the forced lubrication, and some include a heat exchanger to cool the lubricating oil, whether or not the lubrication is forced.

Thrust Bearings. Manufacturers should be asked to provide their expected thrust bearing life. The thrust bearing takes the backward thrust of the screw as it pushes the material out the front end, and tremendous forces often are encountered. An extruder buyer should ask for his bearing rating in hours of life at a given pressure (e.g., 5000 psi/340 atm) and screw speed (e.g., 100 rpm). The rating is given as "B-10" life, which means that there are 9 chances in 10 that the bearing will last that long. The average life (5 chances in 10) is about 5 times the B-10 life. The B-10 life is inversely proportional to the speed and also to the cube of the back-pressure; thus, if high-pressure operation is anticipated, an oversize bearing is advisable.

Example: If B-10 life is 63,000 hr (7 years) at 2500 psi (170 atm) and 50 rpm,
it would be 31,500 hr at 2500 psi (170 atm) and 100 rpm, but
only 8000 hr at 5000 psi (340 atm) and 50 rpm, and
only 4000 hr at 5000 psi (340 atm) and 100 rpm.

This is a rather light bearing. If either the higher speed or especially the higher pressure is anticipated, an oversize bearing is needed; in this case it might be a B-10 of 250,000 hr at 100 rpm/5000 psi (340 atm). Such bearings are not unusual on extruders. Bearing life is somewhat like automobile liability insurance; a large increase in bearing life can be gotten for relatively little additional cost, and insures against failure.

Where powders are being extruded, care should be taken to keep all powder out of the bearing region. Although a few plastics are actually used as oil additives, many others will not help lubrication and may even decompose in the bearing region, promoting corrosion and/or early failure.

Special attention should be paid to bearing design and rated life in twin-screw machines. Bearing life may not be too long because bearings necessarily must be smaller (except for a few designs with nonparallel, converging screw axes). Tandem bearings, i.e., several small bearings on the same shaft, sharing the load, are often used in twin-screw machines. There is still some controversy as to whether they actually work as planned, sharing the load equally among the individual bearings. However—and this is important—in a multiple-screw extruder the operating pressures can be appreciably lower, which could compensate for the smaller bearings. Since twin-screw designs vary greatly, their manufacturers should be asked to provide the expected operating pressure as well as the bearing life.

Barrel Length. This is measured in terms of the ratio of length to diameter (L:D). Properly, the length measurement starts at the *forward* end of the feed opening. A barrel of 20:1 length/diameter ratio is standard today for single-screw machines, and is also just about the minimum. Ratios of 24:1 are frequently found and provide higher rates by giving more plasticizing capacity—either by additional heating or by more mixing, which permits higher screw speed at good quality. These longer barrels (24:1) are not used for single-stage extrusion of rigid PVC (longer residence time promotes decomposition), but *vented two-stage* extruders of this length are being used successfully for this plastic as well as others. Vented machines are often 24:1 and sometimes longer. Very long single-stage machines (e.g., over 24:1, up to 30:1) have been made but are quite rare. Barrels for twin-screw machines are usually shorter; ratios of 10:1 to 15:1 are common. This is possible because the good mixing action allows a shorter residence time, which is one advantage cited for such machines in PVC extrusion. (A short barrel alone is no guarantee of short residence time; the channel depth, screw speed, and output rate must all be considered, and a fast-turning, long, shallow-channel single screw may yield a residence time similar to a shorter, deeper channel, slower-turning, twin-screw system.)

Barrel Material. Special alloys, produced by a single company, are almost universally used in the United States, and there are two types. The standard iron-based alloy is most common, while a complex nonferrous alloy is less hard but is preferred by some manufacturers as more corrosion-resistant—even allowing saran extrusion. These expensive materials are used as liners in a steel barrel. They are preferably centrifugally cast right into the barrel, although they can be prefabricated and pressed in. The latter method is poorer because the liner may loosen and impair heat transfer.

Barrels of nitrided steel are common in Europe, although several makers of alloys are active there too. The nitrided barrels are cheaper and are exceptionally hard, but they are reportedly less resistant to corrosion, and do not last as long. They should be satisfactory in most cases (including PVC, but not saran) provided that the nitriding goes deep enough. A .015-in. depth (0.375 mm) is possible and is adequate, and .010 in. (0.25 mm) or less is insufficient.

Other alloys including nickel, chromium and/or other metals are sometimes used for special purposes (e.g., saran) but are too expensive for general use.

Screw-flight Material. Screws are usually made of an alloy steel but the tips of the lands must receive special hardening treatment to reduce wear. Flame-hardening is the simplest, but this is not advised for high-temperature operations such as extrusion-coating. The hardening may disappear after some time at such elevated temperatures. Nitriding is also inexpensive and is a good method, yielding an extremely hard, although rather brittle, surface. It must be deep enough, however (see preceding paragraphs), and when it is done the *entire* screw must be nitrided, not just the flights, and its dimensions cannot be altered later. Where possible changes in channel depth are anticipated, nitriding should not be used.

The third common method, and the most costly, is to weld a strip of a special alloy to the flights and grind this to size. This strip is hard and wear-resistant, unchanged by heat, and permits alteration of the rest of the screw. One manufacturer even welds a strip of such an alloy to the flights of his stainless-steel screws.

Heating Systems. Before specifying a heating system, decide how much it is going to be used. Adiabatic operation or any low-temperature operation will need heaters only for start-up, and inexpensive band heaters are adequate. Metal heater blocks with the resistance elements imbedded in them ("cast-in heaters") are only a little more costly but are generally preferred, particularly for extrusions where heat is frequently added. They give a more uniform heat input because they do not warp and pull away from the barrel wall as band heaters might do. Ceramic-insulated block heaters are occasionally used. Induction heating is the newest (and most expensive) method; it is quick and clean, and provides high-power efficiency, a wide heating range, quick controller response, fast start-up, very uniform heat, and ease of cooling. The latter is perhaps the biggest advantage. With any induction heaters, it is important to know how they are replaced. In some designs much costly disassembly work is required to change a heater; this is undesirable even though replacement is less often needed than with other methods. For all types of electric heaters some attention must be paid to their output. For resistance heaters on a 2½-in. (60 mm) and 4½-in. (120 mm) barrel (20:1 L/D), 15 and 40 kilowatts, respectively, are enough. This is between 30 and 40 W/sq in. of *inner* barrel surface (= about 5 to 6 W/cm^2). The actual watt density of the heaters will be less because they have a greater contact surface than the inner barrel wall. A higher watt-density than this is often difficult to apply; it means faster heat-up and response, but also means hotter operating temperature, more chance of hot spots and warping, more heat loss, and more frequent need of replacement. Plants with available steam facilities should investigate steam-heated machines (which give very uniform, responsive heating), but *only* if the low temperature limits (around 205°C at 200 psi or 14 kg/cm^2 steam) are acceptable, and will be so for most of the extruder's expected life. Steam- and oil-heated extruders have become very rare.

Cooling Systems. For certain operations, very little extruder cooling is needed; for others, such as all rigid PVC, high-density polyethylene, or any tubular film or pipe line, considerable cooling may be needed. For less critical jobs, the conventional air fans blowing over band or cast heaters are sufficient. But there are many jobs where cooling is very important, and where lower melt temperature at the die exit means faster cooling and faster production rate. For these jobs, some system even more efficient than the simple blowers over the heaters is needed. Among the best is the one inherent in induction heating; because these heaters do not contact the barrel, air can be blown around the barrel itself. Other efficient cooling methods (see Figure 1.15, p. 18) include spraying water fog over the barrel heaters (no exposed electrical terminals, of course), heater sections with fins on their outer surfaces, and liquid evaporation. Sometimes there are channels on the *insides* of the heater sections, allowing air to flow through right near the barrel. This cools efficiently, but makes heat transfer from the heaters less uniform and more concentrated. Copper tubing (for cold water) which is wrapped around the heater sections is an inexpensive, rather efficient method. Often the tubes fit into a spiral groove in the barrel outer wall. This system has been criticized for unequal and sudden cooling, but some recent experimental work showed that good, uniform results can be gotten. The fast cooling response may impair the uniformity of the melt, unless the cooling is running continually. Metal filling-pastes can be used to improve heat transfer between both heating and cooling units and the barrel.

Controllers. Time-proportioning controllers are used for all except: (a) the simplest and the cheapest installations, where on-off control is sufficient; and (b) the most critical places, where stepless controllers are used—these vary the input voltage as needed to keep the temperature at the desired value. The stepless controllers are most often used on dies and perhaps on the last barrel zone, where temperature control is most critical. Conventional proportioning controllers were electronic instruments with tubes, but now one can buy good transistorized "solid-state" controllers at competitive (and often even lower) prices.

The temperature is sensed by bimetallic thermocouples. A system that uses thermistors instead was introduced several years ago; it works well but is still not widely used. The location of the sensing elements is important; they should not be too close to the heaters as then the control may appear much better than it is. The elements should be as close to the barrel liner (thus to the plastic) as possible, subject to the limitations of the control-time-lag; if the heater-element distance is too great, this response time is too long, and the amplitude of the temperature variations will be greater. More sensitive control (narrower control band) and/or double thermocouples (one near heater, one near plastic) can overcome this problem, but are more expensive and rarely used. Control direct from a thermocouple immersed in the plastic is possible, but the time lag between heating and response is normally too great. Such a system, which would really control the plastic, not the metal, would be welcome, if the time-lag problem were inexpensively solved. This is particularly true with the die, where temperature control is most important and where the metal can be drilled to take melt thermocouples more easily than in the barrel.

Pressure Rating. This depends on head and barrel design as well as bearing specifications. Make sure the machine can run continuously at 5000 psi (340 atm) or higher if needed. Include safety bolts or shear pins that are preset to fail at 7500 or 10,000 psi (510–680 atm) head pressure. Instead of safety bolts which fail at a preset stress, special washers can be used on the head bolts which will collapse when compressed above their limit. This causes the head to open slightly and the resultant leakage relieves the excess pressure. Some drives will stall if pressure (and thus torque needs) becomes too high, and this can act as a safety device. Still another safety system is a pre-stressed rupture disc, built into the barrel just upstream of the breaker plate. It breaks at a predetermined pressure, and the plastic leaks out through the resulting hole. The last three systems all avoid mechanical damage to the head area which may result if a conventional safety bolt or shear pin fails. It is a good idea to test a new extruder during its start-up, in the presence of the manufacturer's representative, to see if it will take the pressures and temperatures that it is supposed to take.

Valves and Pressure Gauges. A simple spherical-end plug valve set in the flow path just after the breaker plate is entirely adequate for polyolefins and polystyrenes. More complex designs are not necessary except for vinyls; in this case extreme streamlining is needed to avoid stagnation. The screw-displacement valve is such a design and is often preferred where many thermoplastic materials are likely to be extruded. Various valve designs are shown in Chapter 1, Figures 1.20, 1.21, and 1.22.

A plug valve should be designed to shut off flow completely—despite any manufacturer's fears about damage. With a proper safety system, there should be no damage. Furthermore, it is extremely difficult to machine the plug so accurately as to *really* shut off all flow, and the normal machinist's error will suffice as "safety" and should not be increased. The diameter of the passage that the valve shuts off should be about 10 to 15 per cent of the barrel diameter and *not larger*. The plug diameter should be just a little thicker than the passage.

A grease-filled pressure gauge, mounted to enter the bottom of the barrel, is commonly used. A dial that reads to 5000 psi (340 atm) is usually best, but sometimes a maximum of 10,000 psi (680 atm) is needed. Some grease gauges have a diaphragm at the point of entry; this is more expensive but allows installation in any position and keeps grease from entering the plastic stream or vice versa. The best and costliest gauge is the transducer type, in which a pressure change alters the electrical characteristics of a Wheatstone bridge circuit. This impulse can be easily indicated and recorded on conventional control instruments, and therefore this type of gauge is preferred for laboratory installations, extremely critical operations and wherever pressure is being *controlled*. This control is very rare today, but may become more common in the future; a transducer gauge would feed back to a mechanical or hydraulic system that alters the valve or screw to maintain constant pressure.

Screw Removing and Temperature Control. Mechanical screw pushers are cheap and useful, especially on big machines. They save labor and energy, and they reduce chances of equipment damage. In some designs, space must be allowed behind the extruder for the pusher, and this should be remembered when installing the machinery.

Most screws are bored through all their length for passage of a heat-transfer medium. Cooling with water used to be common, and was a successful compensation for an overly-deep screw. The cooling freezes a layer of plastic on the channel wall and, in effect, reduces the channel depth, thus producing more shear and better mixing. Unfortunately, this cooling also drastically reduces output rate and wastes energy, and if a screw is properly designed for a job and a material no cooling should be needed. The one exception is the extrusion of rigid PVC on single-screw machines, in which the preferred practice is to control the temperature of the screw (particularly the tip) with circulating hot oil at 120 to 140°C. This prevents localized heating at the tip, which could act as a nucleus of decomposition.

Where a screw is to be bored, this should be done before the flights and channels are made; if it is done later, the thermal and mechanical stresses generated during the boring may distort the screw. Most borings are simple holes of constant diameter. They detract from the torque-bearing capacity of the screw, but not much. In a few cases the bore is not a simple hole, but varies along its length (e.g., larger near the screw tip) to achieve a controlled temperature along the barrel. In most cases the "finger" that carries the control medium will extend all the way to the screw nose because it is in the last few flights that screw control is most effective.

Screen Changers. These devices are most useful where the feed is likely to contain foreign matter (e.g., scrap recovery, making color masterbatch), or where it is important to keep as much of such matter as possible from entering the die (e.g., film, monofilaments, wire covering). There are now both intermittent and continuous changers. The intermittent systems work by:

(a) sliding a fresh screen and breaker plate in place of the used screen/breaker plate combination,

(b) alternating flow between two passages, one of which can be cleaned while the other one is in use, or

(c) connecting the upstream screen surface to a passage vented to the air, through which flows the contamination-laden material.

Type (a) is most common and is illustrated in Figure 1.18. It is available from several suppliers, while the other types are specialized designs of individual firms. Recently, a continuous screen changer has been developed (see Figure 1.18) in which the screen is in strip form and is moved across the flow path at a controllable rate, carrying all foreign matter out with it. In this way, pressure can be kept constant indefinitely, even with highly contaminated feeds, and very fine screens can be used without danger of quick clogging and resultant pressure changes.

Installation and Service. Most manufacturers will send their representative, for a limited period, to supervise installation and start-up of the extruder. Where special dies are involved, some makers will run trials in their own plants, using a test extruder; in certain cases they will set the whole line up in their own plant and invite the purchaser to a trial before final purchase. Such trials are very desirable as changes are most easily made right in the manufacturer's shop. Technical service representatives of the material suppliers are also frequently invited to assist in these trials and start-ups, and it is common for the suppliers to donate free material, within reasonable limits, with the hope that the user will continue to buy that material if it works well in his extruder.

Service usually means a warranty on defective parts for a year. It is a good idea to request clear definitions of service, including the length of time, what is involved, and who pays what costs. Sometimes (but not always) start-up service is included in the purchase price.

CONCEPT OF EFFICIENCY

Earlier in this chapter typical efficiencies are given for various plastics. The extrusion *efficiency* is a measure of the power necessary to extrude a plastic and is expressed as pounds or kilograms per hour per horsepower. This is an extremely important concept. It can be used in two ways: either (a) the power is fixed, and the efficiency is used to predict top output, screw speed, and/or take-off speed; or (b) a desired output is known, and efficiency tells us what power is needed.

Calculation of the efficiency depends on the type of drive and speed variator used. For AC–DC systems, the power in watts (volts $\times$ amperes) can be measured directly at the DC motor. It can be assumed that all of the energy fed to the motor goes into the plastic, or else a loss factor (e.g., 5 to 15 per cent) can be applied, and is a function of the bearing and transmission system.

For other drive and variator systems, the volts and amperes across one phase of the 3-phase AC motor are measured. If it is assumed that the drive efficiency (percentage of energy input to motor that gets to material) is $1/\sqrt{3}$ (which is often fairly accurate), the total extrusion power is simply the product of V and A across one phase:

$$W = (3V \text{ A} \cos\theta) \times E_d$$
$$W = (V \times A\sqrt{3}) \times 1/\sqrt{3}$$
$$W = V \times A$$

where

E_d = drive efficiency,
V and A = volts and amperes across 1 phase,
W = total power to extruder, and
$\cos\theta$ = power factor, assume $1/\sqrt{3}$.

Therefore, for any of these AC drives, total power in watts can be taken as the product of the measured V and A across one phase. Then,

$$\text{Power in hp} = \text{W} \div 746$$

$$\frac{\text{Output rate}}{\text{Power in hp}} = \text{Extrusion efficiency in pounds or kilograms per hour per hp}$$

Even when all the above assumptions are not valid, the extrusion efficiency can be calculated in this way and used to compare performances of different materials on the same machine.

In addition to different materials, different conditions for the same material can alter the efficiency. For example, high pressure gives the screw more work to do and reduces efficiency. Very high temperature makes the plastic less viscous, and it raises efficiency.

In operations where much *conducted* heat is introduced into the plastic, such as extrusion coating and most twin-screw extrusion—and also wherever extensive preheating is used, the efficiency figures should be adjusted to reflect the added energy.

Efficiency should be recorded frequently, as quality control of a good extrusion operation, and it should always be measured for trials of new materials or conditions. If it deviates from its normal value, or from the typical values given earlier, there must be a reason, and it will usually pay to find it. This does not mean that the efficiency always must be high; for example, in mixing plastic with dry colors at high pressure, efficiency is quite low. However, when it is low, one should know what he is getting in return.

BALANCE OF SCREW SPEED AND DRIVE POWER

This is one of the most common single problems in extrusion. Often, a machine keeps stalling above a certain speed or with certain materials. Usually, the operator "solves" the problem by running slowly or by not using the offensive material. But these are undesirable solutions as slowed production is seldom wanted, and the material may make a better product. What, then, is the trouble?

Refer to the graph in Figure 10.1. It plots available horsepower against screw speed. Drive type A is a constant-horsepower drive and variator, such as a mechanical pulley system. Drive B is a constant-torque system, such as a magnetic clutch (eddy-current) drive and some AC–DC drives. Drive C has constant horsepower over the upper half of its speed range and constant torque over the lower half, and is representative of most AC–DC systems.

For drive type A, *all* the power is available at all speeds. This machine will never stall unless more horsepower is demanded than the motor can supply. In such a case, the machine simply doesn't have enough power to do this task. However, for drives B and C not all the power is available at all screw speeds. Suppose a material is viscous enough that all the power is needed at two-thirds of full speed. With drive B, the machine will stall, even though the motor can develop the necessary power. This stalling can be prevented, however, by changing the speed *range* and setting the top speed at or near

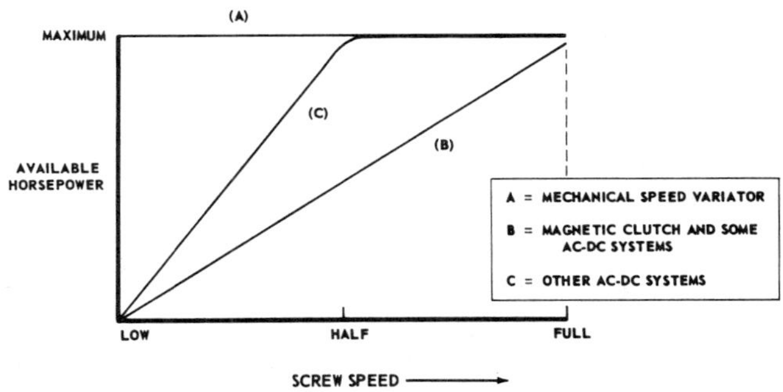

Figure 10.1. Graphs of available horsepower versus screw speed for different types of speed variators. For type (A), all horsepower is always available; types (B) and (C) have only part of their total power available at low speeds. (*Illustration by author*)

the desired speed by means of changing gears or pulleys. Now, all the horsepower is available and the machine runs! Here, then, is a good case in favor of changeable speed *ranges* in an extruder whereever the B drives (and sometimes the C drives) are used. Only one range may be needed and bought at first, but the wise purchaser will buy a machine that permits simple, low-cost changes of speed range, in case they are needed later on. One supplier even puts a handle on the machine to change range at once. This is excellent, but a simple replacement of gears or pulleys is frequently just as good since range changes usually are not made very often.

One might ask: Why not use drive A all the time? In addition to the "pros" and "cons" of the various types, discussed at the beginning of this chapter, there is another reason. Since the A drives give all horsepower all the time, this horsepower is deliverable at very low speeds. High horsepower at low speed means very high torque on the screw, which may twist and break under this load. The B and C drives are protected against such overloading because not all horsepower is available at lower speeds.

Screws made of two or more segments are at an even greater disadvantage. The weakest point is the shaft or stud that connects them. This is often even smaller than the minimum root diameter and is more liable to failure at high torque.

This balance of speed ranges, screw size, and horsepower should be calculated before indiscriminately putting a huge motor behind a small screw, and it should be looked into by any extruder purchaser and every extruder designer.

INSTALLATION

Location. A place must be picked that is free from drafts which may affect temperature control—yet not so enclosed that the operators must suffer from the heat. Sometimes the die and sizing areas are shielded with metal or wood panels, or even polyethylene film, permitting more ventilation of the remaining area. The location must be easily accessible by lift truck as well as by men. Also, there should be ample space in the front to back away the take-off and remove the screw as well as ample space in the rear to attach a mechanical screw pusher. Finally, there should be enough overhead space and support for a chain-hoist sliding on an I-beam parallel to the extruder. This is not absolutely necessary, but it is a great help in removal of heavy dies, heads, take-off rolls, etc.

Foundation. Small machines (2-in./45-mm diameter and less) can be mounted on almost any firm level floor. They are often secured by a few bolts, although a friction-lock surface is adequate. Some are even supported by lockable casters to enable their use at several locations. Bigger machines require a more solid support—preferably 2 in. or more of concrete—well supported underneath, and leveled. These, too, are just bolted to the floor at a few points. When motors are separated from the extruder, they are best mechanically insulated from the floor with rubber blocks or pads.

Supports. Most extruder barrels are supported at their front ends. This is especially important with large, long machines (over 6 ft or 2 m) or when heavy dies are to be hung from the head. Such support should permit free axial expansion of the barrel (from heat). It prevents the front end from "whipping" or vibrating excessively at high speeds, and it also prevents unequal wear back in the bearing section due to sagging of the front end.

Alignment. Misaligned drive shafts and screws can cause rapid wear and failure of both shafts and bearings. Where extruder and drive components are separate, they must be accurately aligned. It is good to check alignment again a few weeks after start-up. Flexible couplings and belt drives make alignment less critical.

Tracks. Many take-off units are mounted on tracks, which are made from angle irons. They need not be bolted firmly to the floor, as the weight of the machinery will keep them down. They must be pinned, however, to prevent sidewise motion. Usually, the tracks are parallel to the extruder. However, setting them perpendicular to the extruder allows easy use of two take-offs with one extruder. Each is rolled aside (not back) when not in use.

The take-off can also be fitted with casters which just clear the ground, while the take-off itself is resting on tracks, on grooved wheels. The casters become the only support, though, when the unit is off the tracks. This enables all movement without a lift truck. The track ends should be tapered down to floor level to enable the unit to roll on or off the tracks.

Services. Extruders need water, electricity, and drainage facilities, and sometimes compressed air for the take-off and pellet feeding devices. Extruders run on 3-phase/220 or 440 V, unless specially built. Most auxiliary machinery is available in any desired common voltage; 3-phase/440 V is the most popular. Water supply and drainage must be ample for certain take-offs and wherever a water-cooled magnetic-clutch drive is used. Calculate water consumption requirements to be certain. Water consumption for screw and barrel cooling is seldom very great.

Where take-offs frequently must be moved in and out of position quick-disconnect fittings are very helpful, as they are neat and fast and require no wrenches.

Beware of fluctuating air pressure. Where line pressure varies, and steady air pressure is needed (e.g., pipe sizing) some controlling device, or else an independent air source, must be used.

Initial Start-up. Run, if possible, only in the presence of the manufacturer's representative. Turn on temperature and pressure recorders *before* start-up, and see if they register room temperature and zero. Decide on a set of conditions and run dry first, at low speed. Then use a known material and run through entire speed range raising speed cautiously in 15-rpm steps. If a valve is present, run it up to 5000 psi (340 atm) in 1000-psi or 50-atm steps. In all cases, measure power, pressure, temperature and output. Finally, decide on the best combination of pressure and screw speed to give highest output of good quality. If take-off is the limit to output, pick the lowest pressure that still gives good quality at required take-off speed. Try to hold steady operation. Measure: power, temperature, pressure, output at *varying* screw speeds, *and* amount of fluctuation of power, temperature, pressure and output at *constant* screw speed. These data will characterize the machinery and guide its operation, and they will prove invaluable if future troubles arise.

Make sure basic operating instructions are available *at the machine* and that operators are thoroughly instructed in operation and safety hazards. No doubt, the manufacturer's and material supplier's representatives can and will provide start-up instruction in greater detail.

The above plan is given more as an example and outline, but has proven adequate by itself in actual practice.

ROUTINE START-UP

Take-off techniques vary with the product, and are covered in the individual product chapters. Following are some general rules on extruder start-up:

(1) Assemble head with clean parts and screens, with the heaviest screen nearest to the breaker plate.

(2) Make sure temperatures are all high enough.

(3) Make sure water is circulating in the feed zone of the barrel.

(4) Make sure enough material is on hand—preheated if necessary.

(5) Make sure there is grease in the entire pressure gauge.

(6) Make sure the head has been tightened uniformly *when hot.* Once a leak starts, it won't stop until the head is dismantled.

(7) Stay away from the front of the extruder when starting, and keep everyone else away.

(8) Start in low speed. Watch ammeter or wattmeter and pressure gauge, and slowly raise speed to desired value.

(9) If feeding is erratic, feed a little at a time (starve-feed) until operation smooths out.

(10) Inspect controllers—see if all are working properly.

OPERATION

Operation techniques vary with the product, and are covered in the individual product chapters. Following are some topics relevant to all extrusions:

Feeding Systems. The simplest is still a man with a scoop. In larger operations an air-loader blows the pellets up into the hopper (and sometimes dries them first). These loaders either have a bin into which bags or drums are poured, or a "snorkel" tube, which fits into a drum or large shipping carton placed under the loader. The most elaborate installations have push-button conveying systems—usually air-powered—that unload bins, cartons, rubber bags (5-ton capacity), silos or even railroad cars. A unique system that works without air is a simple grain conveyor, i.e., a screw inside a metal tube, commonly used on farms. It is fast and efficient and uses no air, but it is noisy, hard to clean, requires some manual operation, and has had to be custom-fitted to each installation. The latter disadvantage has been recently overcome by a few suppliers who now build them specially for plastics use. Recently *flexible* tube conveyors have been developed for moving plastic—both powder and granules.

Feeding Problems. Occasionally, material will not feed smoothly through a normal hopper. Possible causes include:

(1) Material too light and fluffy for gravity feed (e.g., chopped film scrap).
(2) Material wet.
(3) Dryer temperature too hot (causing bridging).
(4) Screw channels in feed zone not deep enough.
(5) Too much external lubricant (there shouldn't be any at all, but a little—up to .05 per cent—is harmless).

Mechanical stimulation of the hopper is often employed in addition to the obvious remedies for the above causes. Although the cheapest solution is often a man beating the hopper with a plank, an electric vibrator affixed to the hopper wall is preferred, except for powders, where a screw conveyor in the hopper is commonly used. Sometimes a magnetic metal detector also is used in the hopper to keep out nuts, bolts and granulator blade chips that may have gotten into the material accidentally.

Typical Conditions. The table on pages 300–301 shows typical conditions for various products and materials. There is no one set of exact extrusion conditions for each material or process. Equipment, products, and people themselves vary so much that a wide range of conditions are encountered. However, one can seldom go far wrong using these conditions, although it may be found desirable to alter them to fit special needs once experience is gained.

Heating and Cooling. With automatic control, little or no attention is needed. However, the controls may not be able to prevent overheating. The operator should watch for this (a thermocouple in the melt stream is invaluable here) and he should be ready to add more cooling if necessary. In some machines this means manually opening vents and starting blowers, or opening valves to admit or increase water flow. If full cooling capacity cannot bring down temperature enough to permit quality product, one can:

(1) Reduce die temperature, turning off heaters if necessary. This reduces melt temperature at exit, and makes overheated melt more manageable.

(2) Cool the screw with water or oil. Use only a trickle of water—at least at the beginning. Oil is recirculated through a controller: start at about 130°C and work down. Warning: cooling of shallow-channel screws may sometimes *raise* compound temperature, as channels get so "shallow" that intense frictional heat is produced. Also, screw cooling always means lower output, sometimes as much as 50 per cent less.

(3) Reduce pressure (open the valve), as long as product quality can be maintained.

(4) Reduce screw speed, which, of course, lowers output too.

(5) Change to a less viscous material, or a deeper-channel screw.

Sometimes output is very low with crystalline materials. They take much heat to melt, and the feed zone may be too small to melt what the screw could otherwise pump out. Output can be raised:

(1) By overheating the feed zone—beware of bridging and sticking in the flights, stopping almost *all* the feed.

(2) By *sparingly* adding external lubricant, which packs pellets a little closer—.05 per cent by weight is plenty.

(3) By changing the screw dimensions, making a bigger feed zone.

(4) By preheating the feed.

Vents and Valves. On a vented (two-stage) extruder the stages are designed so that output is independent of the pressure at the head—up to that pressure above which the material will extrude out of the vent. This is usually around 2500 psi (170 atm) unless a special screw design has been used. Therefore, some care is required when altering pressure on such a machine. For best quality (best mixing) the pressure should be run just short of this critical point. Start always at open valve; go up to full speed, and increase pressure slowly to desired value. Then open the vacuum line to the vent.

Valve Adjustment. Turn a valve slowly, especially when near its bottoming point. A little turn can mean much pressure increase near this end. Always watch pressure gauge while increasing pressure.

TABLE 10.1. Typical Extrusion Conditions

(all temperatures in °C, pressures in atmospheres [1])

Application and Material	Extruder hp	Extruder Barrel (in./mm)	Barrel Rear Temp.	Barrel Front Temp.	Die Temp.	Typical Pressure	Typical Mat'l Temp.	Some Typical Take-off Conditions
Pipe								
Polyethylene	40	3½/90	150	160	165	100	165	Cooling water room temp. or less.
Rigid PVC	50	3½/90	140	160	170	175	175 [2]	Cooling water room temp. or less.
ABS	50	3½/90	170	195	200	175	200	Cooling water room temp. or less.
Sheet								
Polystyrene	100	4½/120	180	200	210	70	210	Rolls—50–90°, depends on rate.
Linear polyethylene	100	4½/120	230	205	200	175	220	Rolls—70–125°, depends on rate.
Film								
Tubular polyethylene	40	3½/90	150	160	165	100	165	Blow-up ratio 2:1.
Flat polyethylene	40	3½/90	200	240	250	65	250	Chill roll—50°, Gap ½ in. (12 mm).
PVC, rigid	25	2½/60	140	160	170	150	175	Horizontal tubular bubble.
PVC, plasticized	50	3½/90	140	170	175	100	180	Vertical bubble, blow ratio 2½:1.
Coating								
Polyethylene	100	4½/120	250	315	325	75	320	Preheat substrate 100° chill roll 50°.
Wire								
Polyethylene	50	3½/90	220	240	240	200	240	Preheat copper 150°, water baths 70°, 40°, 15°.
Plasticized PVC	25	2½/60	130	165	170	100	170	Preheat copper 150°, bath 20°.

Nylon	15	2/45	260	295	300	60	300	Preheat copper 150°, baths 70°, 20°.
Monofilaments								
Polypropylene	25	2½/60	190	240	250	3	250	Quench 50°, air oven 200°.
Nylon	25	2½/60	260	290	300	3	300	Quench 90°, oven 260°.
Polystyrene	20	2½/60	180	205	210	100	210	No quench, glycol bath 125°.
Contours								
Polystyrene	20	2½/60	175	200	205	100	200	Flame-polish with 400° flame. Air-cool if no polishing.
Rigid PVC	25	2½/60	140	165	170	175	170	Vacuum sleeve cooled with tap water.
Plasticized PVC	20	2½/60	140	165	165	75	165	Flame-polish with 400° flame; water bath at room temp.

1 Multiply by 15 to get approximate psi.
2 Some new systems run at high speed at over 200°C.
3 Positive displacement pump in head provides very constant feed to die.

Prepare to change take-off speed to match rise or fall in output because of pressure change. Never unscrew more than necessary, and mark stem to ensure this. Otherwise, plastic gets up into the valve-stem hole, clogs the hole, makes valving difficult, and adds a stagnation point.

Pressure Gauges. The common grease-filled type needs little attention other than seeing that it is full of grease before operation. If it is suspected of sticking during extrusion, give it more grease; when it is full, the grease will come out of the die in slippery streaks, distorting the cross section for a little while. Also, the impulse of the grease gun will show on the pressure gauge. A high-temperature grease, such as a silicone, must be used.

When the gauge remains stuck, there is usually a solidified plug of plastic part way down the connecting tube. The gauge must then be disassembled and cleaned. Having a clean spare connecting tube on hand will allow such repair with just a brief stoppage of extrusion. The diaphragm-grease gauges prevent the interchange of plastic and grease, and thus make maintenance much easier.

The transducer gauges require more care, but they are much more accurate. However, this accuracy is not always needed; the grease gauge is adequate where no control or record impulse is wanted. Some transducer gauges must be cooled, and lack of cooling will damage the sensitive element. However, these gauges are now made without the need for such cooling. All transducer gauges should be calibrated upon installation and at regular intervals thereafter (e.g., every six months) to ensure accurate readings.

Leakage of Molten Material. Most leakage is at the head of the extruder. It is caused by high pressure, uneven tightening of head assembly, or both. At pressures of 200 to 300 atm, often used in modern extrusions, some machines do not seal smoothly and tightly enough to stop leaks. If the head actually bends, it will not be possible to run at these pressures at all. In such a lightly built machine, the drive is also usually unsuited for these pressures. But sometimes the sealing surfaces of a strong machine become nicked or worn and thus leaks can develop, even when the head is properly and evenly tightened. These leaks can be stopped with a brass gasket cut to fit the leaking seal area. Gaskets made from an asbestos/aluminum combination have also worked well. The gasket should be as thin as possible (e.g., 0.010 in. = 0.25 mm) to inhibit its own failure. It should be carefully cut to leave *no notches* at all on the inside edge, which could otherwise act as stress-concentrators. Of course, regrinding the leaking sealing surfaces is also an acceptable remedy.

When tightening the head it is wise to go around the bolt circle several times, turning the bolts a little at a time to make them all equally tight. A torque wrench is a useful tool here. The head should be re-tightened *while hot,* just before starting. Heads with a single tightening bolt escape this problem of nonuniformity, but may not seal so tightly. A breech-block (wedge) head closing system is about the best of all.

Once a leak starts, it will continue. Reducing pressure will slow the leak, but it will not be stopped until the system is disassembled and cleaned.

Timers. A timer is a useful device which automatically heats up the machine several hours before start-up, thus saving some labor cost. It should have at least a four-day span to allow presetting across holiday periods. It should be connected only to the heaters, never to the drive. Finally, it should not be set to turn on the heaters too far in advance, especially with less stable materials which might decompose and corrode the internal surfaces and/or produce internal hard particles that cause "die lines."

Recorders. Any electrical impulse such as temperature, power, and some pressure gauges can be recorded on conventional instruments. This record can be a useful analytical tool at installation or whenever troubles arise. It can also serve as a check on consistency of product conditions and perhaps be linked to quality control tests. It is advisable to have such a portable recorder on hand, provided that there are personnel with the time and ability to use it, understand it, and apply the results.

Output Calculation. The best method is to collect the extruded output for 1 min, weigh it, and multiply by 60 to get pounds or kilograms per hour. When output cannot be easily weighed, as in a running roll of sheet or a coil of wire, it can be calculated by measuring dimensions of a 1-min piece. As an example, the output for high-impact polystyrene sheet 36 in. wide and .030 in. thick, running at 20 ft/min is:

$$20 \times \frac{36}{12} \times \frac{.030}{12} \times 60 = 9 \text{ cu ft/hr}$$
$$9 \times (65 \text{ lb/ft}^3) = 585 \text{ lb/hr}$$

In metric units, if the sheet is 1 meter wide and 0.5 mm thick, running 6 meters/min, the output is:

$$6 \times 1 \times \frac{0.5}{1000} \times 60 = 0.18 \text{ cu meters/hr}$$
$$0.18 \times 1.05 \text{ g/cc} \times 1000 \text{ (conversion factor)} = 189 \text{ kg/hr}$$

Pocket "slide-rules" for output calculation are occasionally available from machine makers, material suppliers, and extruders, printed specifically for different extruded products. These eliminate all paper work.

Do not waste time weighing rates with a very good balance. Weights to 1 or 2 per cent are adequate as output sampling is seldom that accurate. Furthermore, changes in material viscosity and machine temperatures will introduce a variation, too. For time-study and scheduling purposes, weight of good product per shift is a better measure than actual extruded weight per hour at any one time.

Labor Requirements. Where operation is steady and only occasional adjustments are needed, one man can run two or three machines. He requires the assistance of automatic loading machinery and of men who remove scrap product, granulate it, and return it with the virgin feed. Thus, in a plant with many machines, the manpower requirement *totals* about one and one-half men per machine, including direct supervision and product removal. With frequent changes of product or equipment, each machine must have an operator, and the total labor requirements are two men per machine.

In continuous sheet-extrusion/forming lines, one man is able to run one line, but for long runs only. Short runs need more men, and any intricate job will be special. Many shops use a "set-up man" technique; i.e., one semiskilled operator is assigned to each machine and one experienced man travels around the shop, assisting in start-ups, purges, die changes, etc.

Extrusion is usually quite highly automated. However, in areas of low-cost labor, automation is often minimized because it is less costly to have men do the jobs. The labor requirements then become greater than the typical examples in the above paragraphs.

Morale. This is not really a technological topic, but it affects operation so much that it merits inclusion. The operator is at the machine all the time, and his attitude can make or break the success of an operation. Attention should therefore be paid to items such as:

(1) Ventilation (but avoid drafts blowing on dies).

(2) Ease of readings, recordings, and adjustments (if difficult, operators may not want to bother with them).

(3) Proper tools (without them, cruder tools will be "invented," and machinery may be damaged).

(4) Safety (provide gloves, groundings for electrical equipment, hoists, etc.).

(5) Education. Tell operators what is being made and how it is done. Show simply how and why the extruder works. Plastics, because of the variety and novelty of their uses, can easily gain an employee's interest.

SHUTDOWN

It is common to run an extruder dry when shutting down; it is kept turning around after the feed is stopped until no more plastic is extruded. This ensures against start-up with cold material, which can seriously damage the equipment. However some extruders (film, coating) prefer to shut down with the screw full, taking care that the machine is fully heated before starting again. In this way, air is prevented from getting down into the screw and oxidizing the bits of plastic that remain there. These oxidized bits would otherwise cause breaks and pinholes in the extruded film of coating.

To help minimize decomposition of plastic in the head and die, these parts should be cooled as fast as possible. Air hoses or fans are commonly employed for polyethylene and PVC. Polystyrene is less critical. The cooling time can be shortened further by anticipating shutdown, which is done by starting to reduce temperatures some time before actual extrusion ends.

Ease in disassembling the head and die should be considered when designing them. Sometimes the die need not be disassembled at every shutdown, but usually the head is opened and the screens changed.

Some materials are best purged before shutdown. High-impact polystyrene is commonly left in the die, but a general-purpose polystyrene is sometimes used as a purge; such a purge is strongly recommended for styrene-acrylonitrile copolymer, which easily decomposes on standing. Producers of polyethylene film commonly use a special purge compound—a highly-stabilized polyethylene with an abrasive inorganic filler—to clean internal surfaces before shutdown. Some vinyl is self-cleaning, but must never be left standing in a hot die, as it will decompose. A plasticized vinyl purge compound, also highly filled, is recommended to clean barrel and screw for all materials. To use it, the head is opened, the extruder is started again and a few pounds of this vinyl compound are run through.

After cleaning the machine, the main electrical switches should be turned off. If "start" buttons are pressed after this, nothing should happen. The timer, if present, should then be set to turn on the heaters an hour or two before the next scheduled start-up.

CLEANING

The tools of the extruder cleaner are the wire brush, the air hose, the brass blade and the cleanser can. With these, the most tenacious melt can be removed from the metal surfaces of the machine.

The screens and breaker plate are the first to be exposed when the head is opened. The screens are thrown away, and the breaker plate must be cleaned quickly while it stays hot. It should be clamped in a vise between blocks of wood (insulators as well as protectors), and the plastic can then be removed as cleanly as possible with an air hose and pliers. See Figure 10.2. In this technique, a stream of air is directed at the point of contact of plastic and metal. The plastic cools at the surface, and shrinks away from the metal. This leaves a clean surface and also gives the plastic enough tensile strength to be continuously pulled away. Care must be taken to avoid blowing bits of plastic around. Goggles are recommended, and the air hose should never be directed at bare skin. All extruder operators should learn this technique; it takes a little practice to do well, but is worth the effort.

Difficult spots are accessible with a pointed brass blade or with small brass-wire brushes. The big steel-wire brush is used for the out-

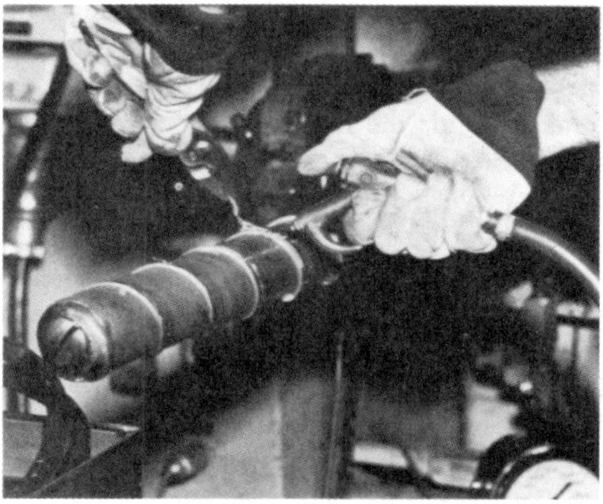

Figure 10.2. Using an air-hose and pliers, the operator is removing hot plastic from the metal surface of the screw. Cooling the "skin" of the plastic with air makes it peel off cleanly and completely. (*Courtesy of Union Carbide and the author*)

side of the head, the head bolts—anywhere except on an internal or sealing surface.

After cleaning the breaker plate, get at the head, and use the same air hose and blade techniques. If the head appears to be cooling off too much before the breaker plate is finished, put the breaker plate on a hot-plate to keep the plastic molten while the head is being cleaned. Use aluminum foil between the breaker plate and the hot-plate.

Then, if desired, the screw is purged with vinyl "cleaner" compound. The screw comes out of the front (nose) end in most machines; in a few, the barrel moves and comes off the screw(s). A mechanical screw-pusher is preferable to a sledge-hammer or its equivalent. A brake-drum puller can also be used. After the screw is out, it can be cleaned. Clean with the air hose, brass blades, and wire brush.

The air-hose technique is most valuable in cleaning dies. Use it wherever possible, and use appropriate wire brushes to finish up. Clean bolts upon removal so that they can be easily replaced and put them into a container. Before reassembly, coat their threads with molybdenum sulfide or graphite lubricant for ease in future removal.

A hoist is a major help in handling large dies such as those for sheet and film. Another convenience is a sheet of asbestos board on which the hot die parts can be laid out safely. If the die is to be put away, it should be reassembled; the lips should be covered with thick tape, and the die should be placed where nothing can damage it. It is often big and heavy, but nevertheless delicate and expensive.

Solvents are seldom used in die cleaning. Their vapors are harmful, and some are dangerously flammable. Also, the safest ones are rather expensive. About the best one is tetralin (tetrahydronaphthalene), which attacks just about every thermoplastic when hot. A better cleaning aid is an ordinary kitchen cleanser used *dry* or slightly damp. Its abrasive action is very fine, and will not damage most finishes; it can be tested on a noncritical point if in doubt. This cleanser is excellent for metal rolls. Another roll-cleaning technique is to use a clean rag and cold water. The water shrinks the plastic away from the roll where it may be wiped away. This has worked well for linear polyethylene when all else has failed.

Dies, breaker plates and other small parts can be cleaned by dipping them into a molten-salt bath. This is particularly useful to clean out tiny holes and crevices.

More aids in cleaning: Before extrusion, put a film of silicone or fluoroplastic on all outside surfaces that are likely to be covered with molten resin (e.g., fronts of sheet and film dies). Mold-release sprays can be used, as well as prepared silicone-grease compounds. These surfaces become much easier to clean later because nothing will have stuck to them. Also, the wire brush mentioned above can be mounted on the end of a flexible shaft, driven by a small motor on a cart, and the whole assembly wheeled around like a small, portable vacuum cleaner. This "power brush" is inexpensive and saves a lot of time and labor. It should be built to take several sizes of standard brass and steel brushes.

MAINTENANCE AND REPAIR

A strict maintenance schedule should be followed to keep extruders in good working condition. Actual needs and periods vary with machine manufacturers and should be determined in discussion with them. A checklist should be kept and followed, with preset periods of checking and dated entries. (See *SPE Journal*, May 1960, for further instructions.) Some possible items on this list include:

(1) Flush and change oil in drive gears.
(2) Scrub down machine and clean area around it.
(3) Calibrate thermocouples and other gauges.
(4) Visually check contact of band heaters.
(5) Measure wear of screw flights.
(6) Check condition of belts in belt drives.
(7) Grind down breaker plate surfaces to effect better seal.
(8) Inspect thrust bearing for wear.
(9) Test die and extruder heaters for "burn-out."
(10) Re-tape frayed electric wires.
(11) Check tightness of all foundation and mounting bolts.
(12) Flush and clean insides of take-off rolls.
(13) Check dies and sizing devices for wear.

Some spare parts are always desirable, especially if the machine is made at a distant place. Examples are: heaters, thermocouples and thermocouple wire, pressure gauge, breaker plate, and V-belts. A portable spare controller is also desirable.

Repairs on an extruder should be made wherever possible with the advice and/or assistance of the manufacturer. The skills of electrical engineer, mechanical engineer, and instrument mechanic will be the ones most needed by the responsible repairmen. Examples of possible repairs are:

(1) Replacement of burnt-out heater.

(2) Repair of malfunctioning controller (sticking solenoid, severed wire, blown tube).

(3) Replacement of V-belt in drive.

(4) Replacement of thrust bearing.

(5) Regrinding and refinishing of rolls.

(6) Regrinding and refinishing of die lips (sheet and film, especially).

(7) Repair of malfunctioning water-circulating unit.

ECONOMICS

Brief mention of the contributing costs of producing an extruded item are given here. Many methods of accounting are used, and no attempt is made here to show them all. Rather, a typical example is given which extruders can use as a guide while fitting their own systems and costs into their calculations. See *SPE Journal,* March 1966, for further discussion of this topic.

TYPICAL EXAMPLES:

(1) Amortized costs

Machine and take-off	\$50,000
Auxiliary machinery, dies, supplies, tools, installation	10,000
Total initial fixed investment	\$60,000

Assume payoff in five years,
straight-line system of depreciation = \$12,000/year
Assume 6000 hr operation/per year of good product = *\$2.00/hr amortization* costs

(2) Material costs

Assume 400 lb/hr good product @ 30¢/lb raw material cost, and average of 5 lb/hr lost as scrap, worth 10¢/lb = *Total material cost* = (405 lb @ 30¢) − (5 lb @ 10¢) = \$121.00/hr.

(3) Labor and overhead costs

1 man to run machine
⅓ man to supply material, work at shipping dock, etc.
⅓ man direct supervision
⅓ man engineering/maintenance supervision

Total = 2 men at \$5.00/hr gross cost = \$10.00/hr
Overhead at \$0.015/lb (standard factor) = another \$6.00/hr

(4) Services cost

Electricity—50 kw at \$0.015	= \$0.75/hr
Water, air, miscellaneous	= \$0.25/hr
Total	= \$1.00/hr

Ex-factory costs = 2.00 + 121.00 + 16.00 + 1.00
= \$140.00/hr = \$0.35/lb good product

Other items involved in total costs, but not directly concerned with extrusion, include selling and product development expense, packaging, shipping, commissions, and royalties. These vary widely with the product.

The scrap value and overhead factor are assumptions which should be based on experience with the particular extruded product. They should be checked periodically against actual costs to make sure they are accurate, and they should be revised when necessary.

Technical improvements in machinery are being developed very fast. Often, if an extrusion plant manager wants to keep up with his newer competitors, it might be advisable to allow some development money and engineering time just for this purpose in the calculation of product costs. For the above example this might be \$8000/year in money and time, which would add \$.003/lb to the product costs. This does not include cost of the improvements themselves, which would be an amortized expense. Note that the material costs are by far the predominant factor in the above example. Power is relatively unimportant. Thus, improvements in materials or design that allow less material to be used have a major effect on costs.

The example is representative of a large, automated line, with an *average* output of 400 lb (180 kg) per hour of salable product. Lower output means higher costs per pound or kilogram because the fixed hourly costs must be distributed over less product. But even more important, less output per hour means less *profit* per hour and less return on investment—even if cost per pound or kilogram were to remain the same. This explains the stress placed on output by most extrusion companies. It is still usually valid today to say: "If you can keep a good extruder going *and* if you can sell all you make, you will make money." The key is in the selling.

SAFETY

Extrusion equipment has four major types of hazards: heat, weight, moving parts, and electricity.

Heat. Dies, rolls, and extruders are all hot, and can give bad burns. Keep asbestos or heavy cotton gloves on hand, as well as proper first-aid supplies for burn treatment. Make some metal "HOT—DANGER" signs to hang on hot equipment or to place near hot dies on the cleaning table. Insulate all steam or hot oil lines. Insulate dies wherever possible. Insulate the extruder cover, if possible. Some have no cover—the cast iron body is both support and outside. Such machines are seldom insulated and can get quite hot. Beware of burns when adjusting die bolts, and use gloves if possible. Finally, try to direct heat from the extruder up through a hood, or to any place other than around the operator—but avoid causing uneven drafts around the die. The die sections of film, filament and even sheet lines are occasionally enclosed to prevent drafts, thus keeping more constant ambient conditions around these dies.

Weight. Many dies are heavy, as are extruder screws. A 4½-in. (120 mm) screw can weigh more than one-quarter ton! Both dies and screws are frequently removed and installed. A well-placed hoist is almost indispensable for this handling as well as for roll replacement. Put "stops" on the ends of the hoist beam to prevent the hoist itself from rolling off. Any lifting, of course, should be properly done to avoid hernia. Safety shoes are another guard against foot injuries caused by heavy objects.

Moving Parts. All chains and belts should have covers or guards. All rolls should have safety cords across their bites in such a position that an operator can strike one, with his head if necessary, if he gets caught in the rolls. The cord should *open* the rolls as well as stop them. Gloves worn by workers near rolls should have slit backs so that the hand can quickly slip out if the glove is caught. Shears and saws should be adequately guarded. The screw feed opening should be deep enough to prevent a hand from entering or should be barred by a thin rod welded across the opening. No one should wear a necktie or loose coat while working around an extruder. In high-speed wire coating, the fast-moving wire itself may be dangerous, and should be marked, shielded, or otherwise guarded.

Electricity. All electrical equipment *must* be grounded. This includes single-phase 110- and 220-V equipment, such as heaters, which may come with two-prong plugs. Include a third lead and ground the equipment. All electricity should be shut off during cleaning and disassembly; a lockout procedure is advisable. A problem arises because the screw must be turned around during cleaning, and sometimes during removal. Perform these tasks with appropriate caution. More precautions are: keep water off the floor, do not allow leaking plastic around heaters, and wear insulated shoes. Static electricity may be a problem with rolls of sheet or film. Commercial static eliminators are available, or a chain, wire, or "tinsel" may be hung in contact with the roll during windup to drain off the charge. A big charge can otherwise build up, especially at high speeds and on dry days.

Miscellaneous Safety Measures. Oil can leak from motors or gearboxes and make slippery spots on the floor. Have some sawdust or its equivalent in the plant to spread as soon as this is noticed. Cork-soled safety shoes also help.

An air hose can be dangerous. Turn air on and off at the source; otherwise, if the nozzle breaks under the pressure, the hose will fly about like a heavy whip. Also, never direct a compressed-air stream at bare skin. If it strikes a cut or other skin break, it may force air into the blood stream, which is dangerous and possibly even fatal. Clean off plastic dust with brushes or a vacuum cleaner, even if it is slower.

Chemical hazards are rare in extrusion shops unless special chemicals are used in plastic formulations. Most plastics do burn, so fire extinguishers should be at hand; the carbon dioxide type is most common. Smoking is permitted in most extrusion plants but is seldom advisable because ashes and butts somehow find their way into the reworked scrap, and there is always the possibility of fire. Where solvents, flammable powders or flammable plasticizers are used, smoking must be prohibited.

Eye protection—preferably a face shield—is recommended whenever using the air hose.

Never stand in front of a machine during start-up, and always watch the power indicator and pressure gauge. One never knows what might come out.

Safety also implies safety to machinery. In this regard the following rules apply:

(1) Protect chromed rolls, when not in use, with sponge rubber or equivalent.

(2) Keep dies where they cannot be scratched or nicked.

(3) Put "stops" on all sliding tracks.

(4) Never use steel tools on rolls, dies or die lips. Use brass or wood.

(5) Use brass blades or wooden shims to set gaps in rolls or sheet/film dies.

(6) Run the regrind, and preferably all feed material, through a metal detector (which is best located in the hopper).

(7) Do not continually run over rated motor loads, even if the system doesn't stall.

(8) Keep use of hammers (including sledgehammers) to a minimum. Use rawhide or rubber mallets wherever possible.

REFERENCES

Many of the references at the ends of the other chapters, particularly Chapter 1, also include information on selection and general operation of extrusion machinery.

Selection

"Single and twin-screw extruders—a technical comparison," F. Martelli (Gerd Lester Corporation—agent of LMP/Colombo), *SPE Journal,* June 1967. Principles of both types are analyzed, pointing out positive-displacement nature of twin-screw machines.

"What you can expect from an extruder manufacturer today and tomorrow," N. V. Fay (Davis-Standard), *SPE RETEC,* Boston, March 1, 1967. Some comment on buying extruders, and trends for the future.

Drive Selection

"Reliability, economy, fast response offered by SCR drives," W. H. Welsh (General Electric), *Plastics World,* June 1967. Advantages of SCR drives and predictions that they will monopolize plastic industry soon. Adapted from paper presented at *SPE RETEC,* Boston, March 1, 1967.

"Application of coordinated drive systems for plastics extrusion lines," R. H. Soderberg (Louis Allis), *SPE RETEC,* Boston, March 1, 1967. Comments on various drives: advantages and disadvantages, feed-back, regulation, etc., especially concerned with take-off machinery.

"Extruder drives," P. J. Moroz (du Pont), *SPE Journal,* March 1963. Long and well-illustrated presentation of different types of drives, showing what is used, how they work, and criteria for choice. Written before recent development and growth of solid-state DC drives.

"Drives for extruders," G. D. Campbell (Reliance), *SPE Journal,* November 1958. Principles and advantages of the three main types of continuous speed variators.

Measurement and Control

"Messung der Masse-Temperatur bei der Kunststoff-Verarbeitung" (Measurement of melt temperature in plastic processing), W. Knappe, paper delivered at Deutsche Kunststoff-Tagung, Munich, April 1967.

"A pressure measuring device for plastic melts," J. Schijf, *Plastica,* April 1967 (Dutch). Description of measurement technique up to 250°C.

"Pneumatischer Druckmessumformer für Kunststoff-Schmelzen" (Pneumatic pressure measuring device for plastic melts), H. Schink, *Kunststoff-Rundschau,* February 1967.

"Die Uberwachung und Regelung von Spinn- und Folien-Extrudern" (The regulation and control of filament and film extruders), E. Liebler, *Kunststoff-Rundschau,* February 1967. Temperature and pressure regulation through computer control on extruders with electric or diphenyl heating.

"Pressure measurements in twin-screw extruders," H. Marhenkel (Schloemann), *Plastics,* August 1965, and *Kunststoffe,* May 1965. Brosa electric gauges are used, in extrusion of PVC pipe.

"Torque measurements on an extruder," W. J. Allen and D. A. Hillman (ICI), *Plastics,* October 1957. Description of test apparatus and presentation of data.

Economics

"Economics of extrusion," Allan L. Griff (Edison Technical Services, Inc.), *SPE Journal,* March 1966. Discussion of economic factors in extrusion, with practical examples and suggestions.

"Uber die Mengenleistung von Einschnecken-Extrudern" (Concerning the output of single-screw extruders), F. Hensen and H. Vinkeloe, *Der Plastverarbeiter,* January 1965. Output predictions and economics of single-screw extruders.

"What price output?" R. K. Senn (NRM), *SPE Journal,* March 1961. Comments about the relation of output to equipment cost.

Operating Techniques

"Purging of extrusion dies," P. N. Richardson (du Pont), *SPE Journal,* January 1962. Short mathematical study of the theory of purging, leading to the recommendation that cleaning is better than purging.

"Try a rear-brake drum puller to free that 'stuck' extruder screw," S. D. Nesmith (U.S.I.), *Plastics Technology,* May 1967. One-page illustrated description of this technique of screw removal.

"Der Einfluss der mehrmaligen Extruder-Verarbeitung auf einige Eigenschaften von Polyäthylen (The influence of repeated reworking on some properties of polyethylene), J. Müller, *Kunststoff & Gummi,* February 1965. Well-illustrated study of effect of reworking on properties of polyethylene.

"The colouring of thermoplastics," R. Mather (Monsanto), E. J. G. Balley (ICI), and the *British Plastics* staff, *British Plastics,* April 1961. A 15-page survey of this subject, including techniques, equipment, problems, color photographs and special discussion of coloring in an extruder.

"Extrusion-coloring natural polyethylene," H. Segal (Celanese), *SPE Journal,* December 1957. Comments on colorants, conditions, and results.

"Preheating polyethylene and impact polystyrene prior to extrusion," P. Wilton and F. D. Marrow (Union Carbide), *Plastics Technology,* March 1956. Advantages of preheating, including higher output and better surface.

Maintenance

"Cut extrusion die cleaning time with salt baths," *Plastics World,* December 1966. Brief, adequate review of this technique. Further information in *Kunststoffe,* August 1962, discussing "Kolene" salt baths.

"Eliminating PVC die corrosion," William Suski (Electro-Coatings, Inc.), *Plastics World,* December 1966. Description of properties and advantages of electroless nickel coatings on internal die surfaces.

"Servicing electronic drives," J. M. Lavoie (Louis Allis), *Plastics Technology,* December 1966. Explains in detail how magnetic eddy-current types of drives work. What can go wrong and how to find it.

Discussion of drive, maintenance and repair of extruders, R. Marcu, *Materiale Plastice,* May 1966 (Rumanian).

"Maintenance of extrusion equipment," N. V. Fay (Davis-Standard), *SPE Journal,* May 1960. Installation and operation hints, including check-list of things to be done daily and periodically.

"The effect of wear on delivery capacity of extruder screws," B. H. Maddock (Union Carbide), *SPE Journal,* May 1959. Screw should be replaced or reflighted when clearance to barrel wall is 15% of channel depth.

ELEVEN
basic extrusion fundamentals

PRINCIPLE

An extruder is a combination of screw conveyor and compressor. As a conveyor, it can be visualized as screwing itself backwards out of the mass of material. Since the screw itself stays in one place, the material does the moving, and is forced out of the delivery end. The plastic is viscous and offers frictional resistance to flow. Therefore, power is required to turn the screw.

The screw acts as a compressor also, as the feed is lighter in bulk than the melt. Actually, the solid feed pellets are *denser* than the hot melt—especially with crystalline plastics—but they must always include spaces between them which give them a much lower bulk density. Typical values for polyethylene are:

Solid density	= 0.96 gram/cc
Bulk density of feed	= 0.55 gram/cc
Melt density at 205°C	= 0.78 gram/cc

Thus, the screw compresses the pellets together, while the air is pushed back out of the feed opening and the plastic is conveyed forward.

THE EXTRUDER SCREW

In a single-screw extruder, there are usually three zones: feed, compression, and metering. See Figure 11.1. The pellets fall by gravity into the *feed zone,* which conveys them some distance down the barrel while they are heated by conduction and mechanical friction until they are quite soft. For powders and other bulky feed materials, a mechanical aid such as a screw is recommended to force the plastic uniformly into the feed zone. In this zone, the screw channels usually have a constant cross section; neither the pitch nor the depth changes. This zone occupies anywhere from zero to 75 per cent of the screw length.

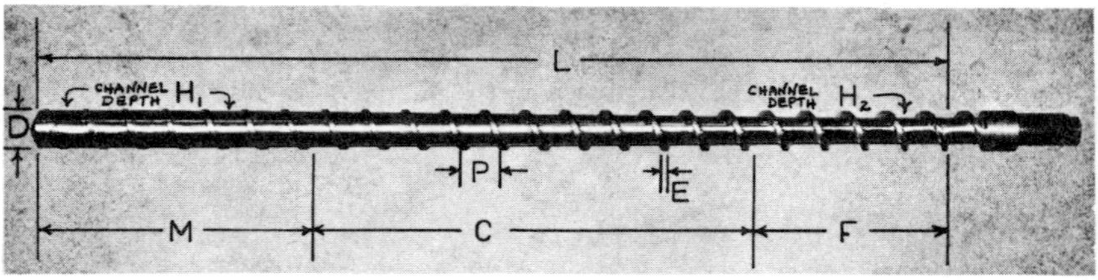

Figure 11.1. An extruder screw, showing important dimensions. The symbols correspond with those in Table 11.1, page 323. (*Courtesy of Union Carbide*)

It is longest for crystalline polymers which require much heat to melt, and shortest for soft amorphous plastics like plasticized vinyls, which can be compressed immediately upon entry into the extruder. Other, more rigid plastics require the "soaking" period that a long feed zone provides, to heat and soften them before compression.

The *compression zone* contains the transition from softened, discrete pellets into continuous melt. This is usually done by reducing the channel *depth* as the plastic moves down the barrel. Heat is produced by mechanical working and more is added by conduction through the barrel walls if necessary. This zone is commonly about 50 per cent of the screw length, but can be as high as 100 per cent (for plasticized vinyls) or as low as 5 per cent (nylon screw, with all compression in one turn). Materials like nylon need a smaller compression zone since melting takes place very quickly.

The *metering zone* has constant cross section, like the feed zone, but it has a smaller channel depth. In it, the plastic is smeared, mixed, and sheared to give a melt of uniform composition and temperature. Most of the shearing in the machine takes place in the metering zone. Its constant cross section tends to damp out any variation in flow rate produced in earlier zones; this zone thus acts as a constant-volume *metering* pump—hence its name.

The length of the metering zone varies widely. Some plastics are heat-sensitive (rigid vinyls) and cannot take the heating that results from the high shear in the metering zone. These must be removed from the extruder as soon as they are melted, and therefore there is no metering zone. In some rigid vinyl screws, there is a metering zone, but its channels are deep enough to avoid excessive shear. In a few others, the screw runs quite fast, and residence time is reduced enough to permit a short, shallow-channel metering zone. Less sensitive materials (polyolefins, polystyrenes) can profit from the good mixing in this zone, and for them a common metering length is 20 to 25 per cent of total screw length. Some operations gain advantages of more uniform melt and better mixing at high speeds with a metering zone 50 per cent of the screw length. Such screws are finding increasing use, especially for blending different grades of polystyrenes and polyethylenes. High temperatures are produced in these screws by the intense mechanical working, and only stable polymers can be employed.

FACTORS AFFECTING OUTPUT

It would appear that the output should be directly proportional to the volume of the last flight and to the screw speed. This is not quite true, as this theoretical output is modified in four ways:

(1) Some screws cannot melt enough polymer to feed the metering zone as fast as that zone wants to pump it out. This is rare.

(2) A pressure peak is built up in most screws, several flights before the end, and this raises the output above the theoretical.

(3) The die offers restriction to flow, and this back pressure opposes the forward flow of material.

(4) Finally, if the screw does not fit the barrel tightly, there will be some leakage backwards over the screw flights. This is negligible, except on severely worn or mismatched equipment.

The real output for a given screw can be described as follows, assuming that the feed zone can melt enough plastic, and that leakage over the flights is negligible:

$$Q = AN - \frac{BP}{\mu} + C$$

where

Q = real output, lb or kg/hr
N = screw speed, rpm
P = pressure of the melt at nose of screw, psi or kg/cm^2
μ = viscosity of plastic
A = constant, based on geometry of last screw flight
B = constant, based on geometry of metering zone
C = a term expressing the effect of the pressure peak, if there is one.

The above equation shows that higher pressure means lower rate. This effect of pressure varies greatly with screw design. A shallow channel is least affected by pressure change because the screw and barrel surfaces are close together and provide more positive conveying. However, a deep channel is very sensitive to pressure, and cannot be used where pressures are frequently changing or are always high. These conclusions are illustrated in Figure 11.2.

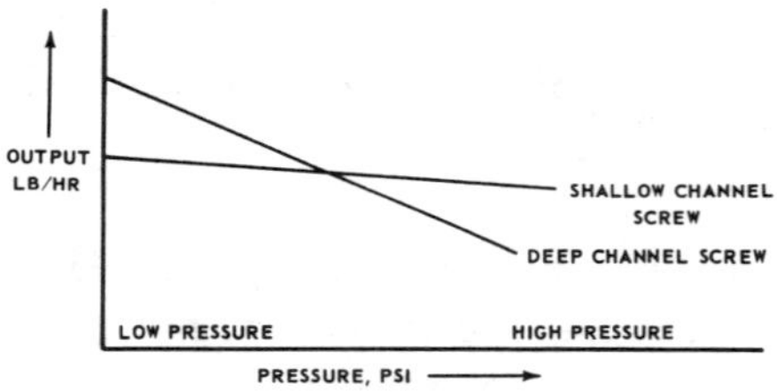

Figure 11.2. Screw characteristics for deep and shallow-channel screws. The deep channel is far more sensitive to changes in pressure than the shallow channel. (*Illustration by author*)

FLOW IN CHANNEL

A motion picture has been made showing the behavior of plastic in a screw channel. A transparent glass barrel was used, with a clear viscous liquid as the "plastic." Ink was injected at various levels in the channel, and the flow patterns became visible. They show that the plastic never moves backward relative to the barrel; rather, it moves in a rolling manner in the channel as well as forward in the barrel.

This motion, as well as the whole subject of flow of *molten* plastic in extruder channels, has been treated in many excellent papers in recent years, listed in the bibliography at the end of this chapter.

SINGLE-STAGE SCREW DESIGN

Screw design for a single-screw extruder must consider many factors, such as:

Extruder barrel length	Desired output
Proposed material—stability	Available power
Proposed material—viscosity	Barrel diameter
Degree of mixing wanted	Size and shape of feed

The screw designer first determines as much of this information as he can and then applies rules such as these:

(1) More stable polymers can take shallower channels.

(2) More viscous polymers can take deeper channels.

(3) Shallower channels mean better mixing, especially in metering zone, but thus generate more frictional heat and yield higher melt temperatures.

(4) Deeper channels mean more output at low pressure, and less output at high pressure.

(5) Deeper channels mean more output change with pressure change.

(6) Larger diameters can allow deeper channels.

(7) Feed zone channel depth must be big enough to take feed particles without jamming or bridging.

(8) Feed, compression, and metering zones should be distributed in line with the needs of the material (see above discussion of zones).

(9) Most designs have a pitch equal to the diameter, and the thread width equal to about $1/10$ of the diameter.

(10) Clearances with barrel walls should be about 0.0015 in. (.04 mm) for each inch of barrel diameter. One investigation, however, claims that the clearance should be about 4–8 mils (.1–.2 mm), independent of diameter.

(11) Minimum root diameter must be great enough to resist expected torques.

The designer has the basic tasks of (a) dividing the total length into the three zones mentioned earlier, and (b) determining the channel depths, particularly in the metering zone. Despite attempts to specify screws by channel depth, the concept of "compression ratio" is still used sometimes. It refers to the ratio of volumes of first to last flights, and is approximated by the ratio of channel depths at the two ends. It has value in itself, but it doesn't tell the whole story and is too often the only thing an extruder knows about his screw. Two screws can have the same compression ratios and yet be totally different in performance. For this reason, it is now considered sound practice to "name" a screw by its channel depths rather than (or in addition to) its compression ratio.

Many firms now run simulated trials of their proposed designs on a computer, and the computer is even available to small companies on a time-sharing basis. Not only does this save the time and expense of making and testing real screws, but it also changes the whole nature of screw design, making it much more of a science by encouraging experiment.

Table 11.1 shows dimensions of typical screws. There is no one universal screw for all products; however, the screw listed in Table 11.1 for low-density polyethylene will do a good job on impact polystyrene, plasticized vinyls and the lower-viscosity high-density polyethylenes, and is as "general-purpose" a screw as is currently known.

To aid in scaling up or down, resistance to flow varies approximately as the cube of channel depth. Also, the following depths are approximately equivalent for the low-density polyethylene screw:

Barrel diameter,	in.	1½	2½	3½	4½	6
	mm *	35	60	90	120	150
Metering-zone depth,	in.	.075	.095	.110	.125	.150
	mm *	1.8	2.3	2.8	3.2	3.8

Another important concept in scale-up is *peripheral speed*—the speed of the flight tips relative to the barrel. This is useful in determining mixing action, for mixing depends on shear, and shear rate is proportional to the ratio of peripheral speed to channel depth. Thus, for equivalent performance, small machines must run faster than larger ones because the distance traveled per revolution is less on the smaller machine. This is something like a small boy having to take many more (and smaller) steps per minute to keep up with an adult. Today, modern machinery goes up to 150 ft/min (46 meters) peripheral speed, and anything over 100 ft/min (30 meters) is considered quite high. To calculate peripheral speed, multiply:

$$(\text{Screw diameter}) \times (\pi) \times (\text{screw speed}) = P$$

* Metric numbers are not exact equivalents of others, but they all represent common extruder sizes.

TABLE 11.1. Typical Screw Designs *

Letter	Dimension	Rigid PVC	Impact Polystyrene	Low-density Polyethylene	High-density Polyethylene	Nylon	Cellulose Acet/Butyrate
D	diameter	4½ (90)	4½ (90)	4½ (90)	4½ (90)	4½ (90)	4½ (90)
L	total length **	90 (1800)	90 (1800)	90 (1800)	90 (1800)	90 (1800)	90 (1800)
F	feed zone	13½ (270)	27 (540)	22½ (450)	36 (720)	67½ (1350)	0
C	compression zone	76½ (1530)	18 (360)	45 (900)	18 (360)	4½ (90)	90 (1800)
M	metering zone	0	45 (900)	22½ (450)	36 (720)	18 (360)	0
P	pitch	4½ (90)	4½ (90)	4½ (90)	4½ (90)	4½ (90)	4½ (90)
H_1	depth in M	.200 (5.0)	.140 (3.5)	.125 (3.1)	.155 (3.9)	.125 (3.1)	.125 (3.1)
H_2	depth in F	.600 (15)	.600 (15)	.600 (15)	.650 (16)	.650 (16)	.600 (15)
E	thread width	½ (12½)	½ (12½)	½ (12½)	½ (12½)	½ (12½)	½ (12½)

* All dimensions in inches; dimensions in parentheses are millimeters. Metric equivalent of extruder diameter, lengths, and pitch are not exact; rather, they represent common commercial usage.

** Total length starts at the *forward* end of the feed opening.

TWO-STAGE SCREW DESIGN

In a vented two-stage single screw the design is somewhat different. This is really two screws in succession; the first one is a melter and delivers molten plastic to the second, the pump. Both provide mixing. The first (melting) screw operates against no pressure if the vent is open, while the second (pumping) screw normally operates against the back pressure of the die. The pumping screw must have a higher output than the melter under no-pressure conditions. In other words, the channels in the metering zone must be deeper in the pump than in the melter. Thus, when die back-pressure is applied, the pump can lose in output capacity and *still* be able to handle the constant output of the melter, which is oblivious to all that goes on at the front end. For example, the metering zone in the pump may be 50 per cent deeper than that in the melter. These zones have their limits, however. The pump metering zone cannot be made too deep, for then it will be very pressure-sensitive (see Figure 11.1); a high-resistance die will cut its output capacity and cause the melter to extrude out of the vent. Nor can this depth be made too shallow, as the melter must always be shallower still. At no pressure in vent zone (normal operation) shallow melter channels mean lower output (see Figure 11.1) which is undesirable. Moreover, a very shallow melter may result in excessive mechanical work and consequent excessive material temperature.

A good compromise is to make the pump as deep as will allow reasonable die pressures (for example, about 2500 psi [170 atm] maximum) and then make the melter about two-thirds of this depth. Typical values for polystyrene are .150 in. and .100 in. in a 4½-in. screw (3.8 and 2.5 mm for 120-mm screw) with 24:1 length/diameter ratio. The melting screw is 16 flights long; the pump screw, including vent zone, is 8 flights long. A few more elaborate designs include a valve at the end of the melter, just before the vent. One such design uses a bypass, in which the melt passes through a simple valve outside the barrel and then returns to be vented and pumped out. Another design has a bulge on the screw corresponding to a ridge or groove on the barrel wall; axial screw motion changes the clearance and thus offers valving. These devices make a given screw design more versatile, allowing the use of high-pressure dies, but add an additional variable to control.

TWIN-SCREW DESIGN

For twin-screw extruders, design is even more complex, and falls beyond the scope of this book. Each maker has his preferred design, and some make their screws with interchangeable segments. This permits "making up" of different screws from the same segments. Several makers even offer a two-stage twin-screw machine; i.e., one twin-screw "melter" combination feeding melt to another "pump" twin-screw combination.

In a twin-screw extruder, the two screws operate as a positive-displacement pump, forcing the plastic forward at relatively low pressure while compressing and mixing it as it passes between them. The latter action is similar to a roll mill. The positive conveying allows very viscous melts to be moved through narrow channels at the relatively low pressure. Thus, screw pitch can be short and residence time long, and heat-sensitive materials like PVC can be extruded at low temperatures, despite their high melt viscosity at these temperatures. They can pass through the twin-screw extruder at low total shear per minute, thus building up little heat and avoiding decomposition.

Some twin-screw machines have their screws turn in the same direction; others (the majority) use opposing directions. All American and most all other *single* screws are right-handed screws and turn clockwise—looking at the screw from the discharge end.

HEAT AND ENERGY BALANCE

Most heat is added to the plastic by mechanical working, especially in the metering zone. Some is always added by conduction in the feed zone, and some may be added through a hopper-preheater (dryer). Heat is further added by conduction in the compression and metering zones if necessary (e.g., in extrusion-coating). On the other hand, heat is removed as sensible heat in the plastic, as radiation from the extruder, through cooling systems and occasionally through screw cooling. Further energy is lost through inefficiencies in the motor and drive; this depends greatly on the type of drive. Thus the energy balance on an extruder is:

$$Q + R + E + C = P + W + H$$

where

Q = heat gain of plastic,
R = radiation loss,
E = efficiency loss in drive (dissipated heat),
C = heat lost through cooling systems,
P = total energy consumption of drive,
W = total heater consumption of energy, and
H = heat added in hopper-dryer.

Preheating is often done to dry the feed, but it also can be a significant means of energy input to the system. If preheating is properly used, motor power requirements and load on the mechanical system can be reduced, and output of an existing extruder can be increased. Screw design would have to take preheating into account as less melting (feed) zone would be needed. Another advantage of preheating: feed temperature is controllable and can be kept constant, eliminating one source of dimensional and property variation. Force-feeding may

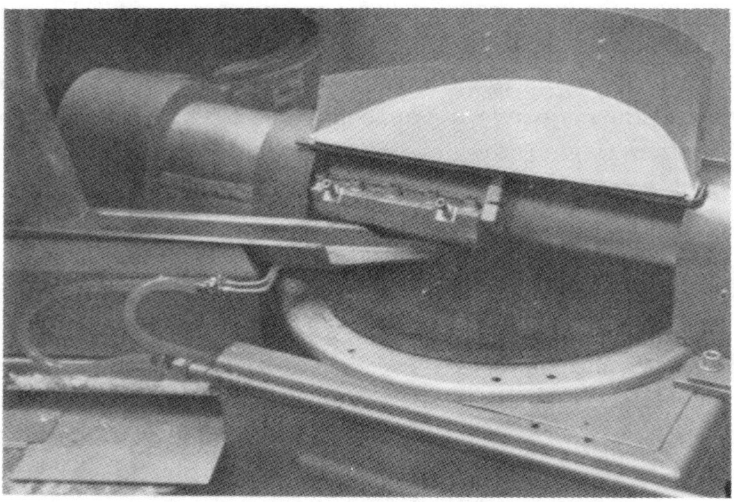

Figure 11.3. The Rotomeltor (Rotessor) extruder, in which the feed falls on a slowly turning heated disc (black, at right), melts there, and is then picked up by a conveying screw (enclosed, upper left) and extruded out. (*Photograph by author, machine made by Baker-Perkins-Granbull and T. Engel*)

be necessary to prevent surface sticking and consequent erratic feed, but techniques for such feeding have been developed already for other hard-to-feed materials such as chopped film and powder. Most of these involve a screw or other mechanical device in the hopper. One unusual solution (the Rotomeltor, Figure 11.3) drops the feed on a slowly-moving heated disc (6 rpm); it melts there and is continually scraped off and into a short extruder which delivers it to a die.

SCREWLESS EXTRUDERS

Several extruders have been developed that do not use screws at all, but rely on other methods to melt and pump the plastic. Probably the best-known of these is the elastic-melt extruder (Figure 11.4), which was studied both in the United States and the Soviet Union, and is now commercially available. The operation is based on the elasticity of melted plastic. The machine consists of a thick disc or cylinder rotating in a heated chamber, with a narrow gap between the rotor and chamber wall. The plastic is fed onto the periphery; it melts there and then moves through the gap toward the central axis because of its elasticity. This can be represented by an arc of a circle contracting into a chord. The system works best for melts with a high elastic component (e.g., polypropylene, PVC). There is melt already

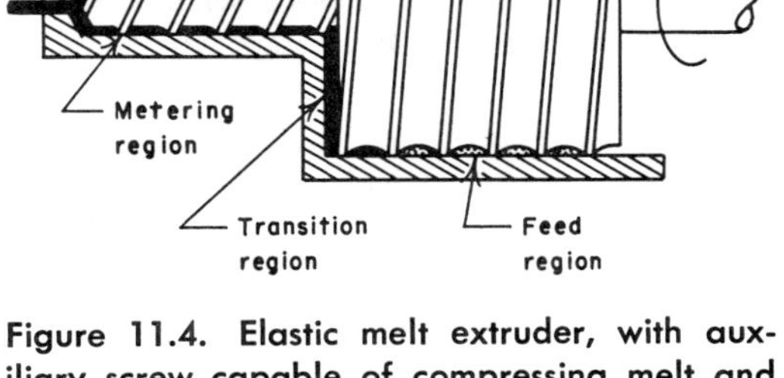

Figure 11.4. Elastic melt extruder, with auxiliary screw capable of compressing melt and forcing it out through normal extrusion dies. *(Courtesy of KPT Manufacturing Co. and Plastics Technology)*

in the gap but the elastic force of the new melt pushes the old melt toward the axis of the system, and then out through a hole in the center of the wall.

This system is small and compact, simple and inexpensive to build, and operates with a short residence time, low power and good mixing ability. Its major disadvantage is its inability to work against high pressures such as those generated by extrusion dies. A major United States processor studied this machine and built large models, and finally developed a system in which the melt is extruded at low pressure into the feed end of a short conveying screw, with the die attached to the other end. The commercial version now offered uses this combination claiming maxima of 300 lb/hr (136 kg/hr) and 200 atm pressure. Other screwless extruders include (a) another rotating-disc machine with radial ridges on one disc: the melt is compressed as the ridges push it around and out through holes in one of the discs; and (b) connection of two ram (piston) extruders to a single die, with one or the other ram in operation at any one time.

REFERENCES

The following pages present some of the important literature references in recent years dealing with theory of extrusion, rheology, screw and die design, and related topics. Further information can also be found in the Supplementary Bibliography immediately following this chapter.

Rheology, Flow Experiments and Studies of Extruder Performance

"Introduction to melt rheology and application to extrusion," R. C. Kowalski (Esso), *SPE ANTEC* 1967. Discusses rheological data, curves of stress versus shear rate, elasticity and melt fracture, measurements, swelling, brief practical application.

"Analysis of polymer melt flow in capillaries, including pressure effects," I. Duvdevani and I. Klein (Western Electric), *SPE ANTEC* 1967. Study of polyethylene at constant diameter and temperature, varying the pressure and the capillary length.

"Flow in an extruder," S. V. Nasirova and I. N. Kaufman (USSR), *Mechanika Polimerov,* No. 6, 1966. Mathematical description of rheology in an extruder.

"Comparison of flow characteristics and elastic properties of polyethylene melts," H. Horak (BASF), *Kunststoffe,* March 1965. Shows change of melt index with shear rate, and the corresponding effect on surface.

"Plastifying extrusion," L. G. Street (Welding Engineers), *International Plastics Engineering,* July 1961, *SPE ANTEC* 1961, and *Plastics Technology,* August 1959. Illustrated analysis of flow in *melting* zone.

"The performance of single-screw extruders in the processing of polyolefins and PVC," K. O. Schiedrum and H. Domininghaus (Hoechst), *Plastics,* February and March 1961. French version in *Industrie des Plastiques Modernes,* March, April and June 1961; German (original) in *Der Plastverarbeiter,* November and December 1960 and January 1961. Extrusion theory, prediction and measurement of screw performance at various die resistances.

"Flow patterns in a non-Newtonian fluid in a single-screw extruder," W. D. Mohr *et al.* (du Pont), *SPE Journal,* September 1960, *SPE ANTEC* 1960, *Industrial and Engineering Chemistry,* May 1958. Classic experiments with transparent barrel and corn oil (simulating melt), with colored ink injected to show flow paths.

"Polymeric melts," A. B. Metzner *et al.* (University of Delaware), *Modern Plastics,* July 1960. Capillary experiments with varying lengths and entrance angles, concluding that a 20° maximum angle will avoid most melt fracture problems.

"Pressure development in extruder screws," B. H. Maddock (Union Carbide), *SPE Journal,* March 1960, and *SPE ANTEC* 1960. Study of pressures along the barrel; explains pressure peaks and recommends long metering zones.

"Extruder scale-up theory," B. H. Maddock (Union Carbide), *SPE Journal,* November 1959. Hypothetical applications of theory; four examples.

"Extrusion studies on thermoplastics," D. A. Lannon and G. C. Karas (British Resin Products), *Plastics Progress,* 1959. Relationship of power, output, temperature, viscosity, of high-impact polystyrene; small extruder (1½ inch = 35 mm) and tape die.

"Modellgesetze für Kunststoff-Schneckenpressen" (Basic laws for plastics extruders), G. Schenkel (Troester), *Kunststoffe,* January, February and March 1959. How a single-screw extruder works, including power balance and channel geometry. Derivation of flow equations. Basis of part of the author's book on extrusion.

"The significant flow properties of thermoplastics," L. B. Ryder (Celanese), *SPE ANTEC* 1959. Theory and practice with extrusion-plastometer, a ram-type device to measure viscosity.

"A visual analysis of flow and mixing in extruder screws," B. H. Maddock (Union Carbide), *SPE ANTEC* 1959. Mixtures of colored pellets are extruded, the extruder is stopped, the screw is removed while still full, and the contents photographed and analyzed.

"Screw-extruder pumping efficiency," P. H. Squires (du Pont), *SPE Journal,* May 1958, and *SPE ANTEC* 1958. Geometrical analysis of flow in screw channels, and development of basic equations which consider the curvature of the channel.

"How to predict extruder performance," B. H. Maddock (Union Carbide), *Plastics Technology*, May 1957. Shows effect of varying screw dimensions on polyethylene output, including corroborating experiments.

"Factors affecting quality in polyethylene extrusion," B. H. Maddock (Union Carbide), *Modern Plastics*, April 1957, and *SPE ANTEC* 1957. Effects of screw design and extrusion conditions on uniformity and mixing. Use of a valve is recommended.

"Viscosity data for extruder flow equation," R. D. Sackett (Monsanto), *SPE ANTEC* 1956. Describes 1½-inch (35 mm) extruder set up as a rheometer, and gives temperature-viscosity data collected for six plastics.

Screw Design: Single-Screw Extrusion

"An improved-mixing screw design," B. H. Maddock (Union Carbide), *SPE Journal*, July 1967, and *SPE ANTEC* 1967. New design which includes a short length with no flights, but with grooves parallel to the screw axis; plastic enters one set of grooves and flows into the other.

"Extruder screw design developments," D. J. Schmidt (Davis-Standard), *SPE RETEC*, Boston, March 1, 1967. Comments on NRM, Maillefer, Davis-Standard and other mixing screws, and on operating conditions, particularly screw temperature.

"Screw requirements," J. C. Houston (du Pont), *SPE RETEC*, Boston, March 1, 1967. Discussion of reciprocating screws for injection molding: design, special features, valves, etc.

"Melting in plasticating extruders—theory and experiments," Z. Tadmor, I. Duvdevani and I. Klein (Western Electric), *SPE ANTEC* 1967. Latest developments in Western Electric's work to develop computerized screw design. Earlier work included articles by the above authors and D. I. Marshall at the *SPE ANTEC* 1966, and a presentation by these authors at the *SPE ANTEC* 1965, which later appeared in the *SPE Journal*, October, November and December 1965.

"Effects of radial screw clearance on extruder performance," R. A. Barr and C. I. Chung (Midland-Ross, Waldron-Hartig division), *SPE Journal*, June 1966, and *SPE ANTEC* 1966. Develops and justifies optimum clearances for different plastics. Optimum clearance is relatively unaffected by machine size.

"Notched extruder screws," by Johnson Manufacturing Company, *Plastics Technology*, May 1966. Simple idea, cut slashes into flights in metering zone.

"Extruder screws," L. L. Scheiner (Journal staff), *Plastics Technology*, November 1965. Some new designs, and comments on old ones.

"QLT extrusion," a review of articles by R. B. Gregory (NRM), *International Plastics Engineering*, September 1965. Explains quality low-temperature extrusion using a three-section screw with one reverse-flighted section. No illustration. Experiments with this screw are reported in *SPE Journal*, August 1964, *Plastics Technology*, February 1964, and *SPE ANTEC* 1964. High-density polyethylene is used.

"Extruder Bauelemente" (Extruder building elements), W. Mink (Battenfeld), *Der Plastverarbeiter*, November 1963. Screw designs for various plastics.

"Vented single-screw extruders," G. Schenkel (Troester, now at Stuttgart Technical College), *Der Plastverarbeiter*, November and December 1962 and January 1963; also *International Plastics Engineering*, September through November 1962. Thorough discussion of theory and experience with vented extruders of various designs.

"Grundlagen der Extrudertechnik" (Basic principles of extrusion), G. Schenkel (Troester, now at Stuttgart Technical College), *Kunststoff-Rundschau*, September and October 1961. Extrusion principles discussed, mainly screw design, with data on dimensions, power balance, valving.

"Design of extruder screws for single-screw machines," K. O. Schiedrum and H. Domininghaus (Hoechst), *Industrie des Plastiques Modernes*, February and March 1961.

"A feedscrew design for extruding rigid polyethylene," W. Mundy (NRM) and A. L. Walters (Phillips Petroleum), *Plastics Technology*, July 1960. Development of design: 0.150-inch (3.75 mm) metering depth on a 4½-inch (120-mm) extruder.

"Feedscrew selection," J. Badonsky (Waldron-Hartig division of Midland-Ross), *SPE Journal*, February 1960. Some basic rules and necessary information. Additional similar information by R. D. Sackett (Monsanto, now Waldron-Hartig) is in *SPE Journal*, June 1957, and in *SPE ANTEC* 1957.

"The mixing torpedo," F. H. Collins (Dow), *SPE Journal*, May 1957. Describes the Dulmage screw, formerly widely used for high-impact polystyrene. No data given.

Twin-Screw Extruders

"Maschinen & verfahrenstechnische Fortschritte beim Bau von Doppelschneckenextrudern" (Machinery and processing progress in the construction of twin-screw extruders), H. Domininghaus (Hoechst), *Kunststoffe*, October 1966.

"Pressure measurements in twin-screw extruders," H. Marhenkel (Schloemann), *Plastics*, August 1965, and *Kunststoffe*, May 1965. Brosa electric gauges are used, in extrusion of PVC pipe.

"Theoretisch und wirklich Ausstossleistung der Doppelschneckenextruder" (Theoretical and actual output of the twin-screw extruder), Z. Doboczky, *Der Plastverarbeiter*, July 1965. Output is much greater per revolution, compared to a single screw of same diameter. Further comparisons by same author are in 11-page article, well illustrated, in *Der Plastverarbeiter*, February 1965.

"Observations on machine construction in the development of twin-screw extruders," H. Mäkelt (Derex), *International Plastics Engineering*, January 1964. Theory and experiments based on Derex machine.

Heating and Cooling

"Refined screw cooling—its advantages with the newer resins," R. E. Lowey (Gene Lowey, Inc.), *SPE Journal*, October 1966. Describes various methods used.

"Estimating polyethylene extruder cooling capacity," R. A. Barr and C. I. Chung (Waldron-Hartig division of Midland-Ross), *Modern Plastics*, October 1966. Based on material presented at *SPE ANTEC* 1966 by the same authors, who recommend an air-water mixture as the cooling medium.

"Heating of thermoplastics in extruders," V. Jinescu, *Materiale Plastice* (Rumanian), April 1966. Discusses variables, heating methods, temperature measurement.

"Temperature profile of molten plastic flowing in a cylindrical duct," H. Schott (Lever Brothers) and W. S. Kaghan (Olin), *SPE Journal*, February 1964. Wall and center temperatures vary considerably.

"Heating capacity limitations of extruder screws," B. H. Maddock (Union Carbide), *SPE Journal*, April 1961, and *SPE ANTEC* 1961. Experimental study of effects of barrel temperature and other variables on melt temperature. Related information is in another paper by Maddock at the same *ANTEC:* "Power and heat energy relations in polyethylene extrusion."

"Autogene und konventionelle Einschneckenpressen für die Kunststoffverarbeitung" (Autogenous and conventional single-screw extruders for plastics processing), G. Schenkel (Troester, now at Stuttgart Technical College), *Kunststoffe*, November and December 1958. Theoretical analysis of autogenous (often called "adiabatic") extrusion.

Die Design—General References

Articles on design of dies for specific products such as film or pipe are listed at the ends of the appropriate chapters. More general information is in the following articles:

"Calculation of circular and wide flat dies," L. Kreft and Z. Doboczky, *Kunststoff & Gummi*, February 1967. Part of a long series on extrusion technology.

"Calculation and construction of extruder dies," five-page article by E. Gluckov *et al.*, *Plasticheskie Massi*, August 1966. Another related article, "Flow of polymer melts in extruder head," by M. G. Blumental *et al.* is in *Plasticheskie Massi*, July 1966, and discusses control of stresses on the melt.

Finally, in the same journal, January 1966, there is an article entitled "Entrance profile of extruder head and its influence on extrusion conditions," based on large polypropylene pipe, by S. I. Klaz and I. F. Kanavetz.

"Flow in various extrusion processes," P. L. Clegg (ICI), *British Plastics,* February 1966. Discusses die swell, dimensions, elastic recovery, entrance angles, for polyethylene, polypropylene, and PVC. Shear stress/shear rate data are included; also effects of conditions on optical properties of film, blown bottle surface, and extruder surging.

"An introduction to extrusion die design," K. J. Braun (Bone Brothers), *Plastics,* December 1964 and January 1965. Theory and practical examples.

"Planning and design of entry zone of an extrusion die," Ya. V. Kaplun and A. N. Levin, *Plasticheskie Massi,* January 1964. Study of shape transitions in entry zone, especially for non-circular extrusions. Same authors have discussed exit zone and die swell (see next subdivision).

"Problems of flow in extrusion dies," J. F. Carley (Prodex, now at University of Colorado), *SPE Journal,* December 1963. Sample calculations of dies, based on principles presented by the same author in *SPE Journal,* September 1963, and at an *SPE RETEC,* November 1962, in Newark.

"Calculations for and construction of extrusion dies," L. Kreft and Z. Doboczky, *Kunststoff & Gummi,* January 1962.

"End-correction implications in die design and polymer processing," L. B. Ryder (Celanese), *SPE Journal,* December 1961, based on presentation at *SPE ANTEC* 1961. Discussion of exit-end effect on flow, with correction factors, data and examples.

"Design of extruder dies," R. Anguita, *Revista de Plásticos,* January and February 1961. Covers dies for film, sheet and wire covering.

"Polymeric melts," A. B. Metzner *et al.* (University of Delaware), *Modern Plastics,* July 1960. Experiments with varying entrance angles and land lengths, showing an angle of 20° or less to be desirable, to avoid melt fracture.

"Die design in extrusion," P. L. Clegg (ICI), *Plastics Institute Transactions,* Winter 1960. Practical study of die design, especially for wire covering and film.

"Rheologische Formgebung einfacher Spritzwerkzeuge für Extruder" (Rheological bases for simple extrusion dies), G. Schenkel (Troester, now at Stuttgart Technical College), *Kunststoffe,* April and May, 1959. Theoretical and practical aspects, stressing crosshead and mandrel dies.

"Streamlining extrusion dies," P. N. Richardson (du Pont), *SPE Journal,* August 1958. How to design head and neck assemblies.

"Some aids to the design of dies for plastics extrusion," D. J. Weeks (ICI), *British Plastics,* April and May 1958. Mathematical analysis of flow in dies for rods, slits and annuli.

"Extrusion dies and take-off equipment," K. O. Robbins (Robbins), *Modern Plastics,* December 1955. Early work by a leading die maker, including practical discussion of irregular cross-sections.

Die Design—Swelling

"Comparative swelling behavior of various thermoplastic polymers," A. P. Metzger and J. D. Matlack (Continental Can), *SPE ANTEC* 1967. Viscosity alone cannot determine swelling behavior. Data are given for effects of shear on viscosity and on swelling.

"Experimental method to determine swell at die in extrusion," Ya. V. Kaplun and A. N. Levin, *Plasticheskie Massi,* February 1965. Includes mathematical analysis.

"Post-extrusion swelling of polyethylene," E. B. Bagley *et al., Journal of Applied Polymer Science,* Vol. 7, No. 5, September 1963. Study of length-diameter ratio, shear stress and strain, and melt fracture, in a capillary tube.

"Swelling behavior in extrusion," J. C. Miller (Union Carbide), *SPE Transactions,* April 1963.

Miscellaneous Topics

Elastic Melt Extruder: Developed by Prof. Bryce Maxwell (Princeton University) and first described by him in *Modern Plastics,* October 1959; further developed at Owens-Illinois, whose personnel (J. E. Henry and C. E. Plymale) presented their findings at the *SPE ANTEC* 1964 and *SPE Journal,* April 1965. The machine was then made commercially by KPT Manufacturing Company, and this version was described in *Plastics World,* May 1965, and *Plastics,* October 1965. Finally, technical workers in the USSR have also experimented with this principle, as presented in *Plasticheskie Massi,* April 1962, by V. I. Morozov, B. P. Shtarkman and E. E. Rylov.

Pulsating Extrusion: Discussions in *Gummi, Asbest, Kunststoff,* Vol. 17, No. 2, by W. L. Krueger and R. B. Kirby, presenting theory and results obtained in a glass-walled extruder, with corn syrup and polyethylene as the extruded materials.

Resin Structure: "How resin structure affects extrudability," R. F. Heitmiller (Cabot), *Modern Plastics,* January 1963, and *SPE RETEC,* Newark, November 1962. Effects of molecular weight and its distribution, crystallinity, and chemical nature on plastication and product quality—especially for polyethylenes.

Supplementary Bibliography

Descriptions of Extruders and Extrusion Lines

Collective reviews of commercial extruders have appeared in journals as follows:

Extruders of the World—*Plastics,* March 1965;

United Kingdom extruders—*International Plastics Engineering,* November and December 1964;

West German extruders—*International Plastics Engineering,* August and September 1964;

French extruders—*Plastiques Modernes et Elastomeres,* June 1966;

Non-German extruders—*Kunststoffe,* June 1964.

Reviews of individual machines and lines are given in the following list. Many of these descriptions include detailed and/or cutaway drawings of the machines, especially in the British journals. It should be remembered that many of the older designs (e.g., 1961–1962–1963) have since been changed.

Alpine high-speed autothermal units; *International Plastics Engineering,* July 1962; *Kunststoffe,* July 1959, July 1957, January 1956; *SPE Journal,* June 1958; *British Plastics,* October 1957.

Amigo hydraulic: *International Plastics Engineering,* February 1962.

AGM "Cotruder" with twin conical screws: *Plastics Technology,* April 1966; *International Plastics Engineering,* March 1966; *Plastics,* September 1965; *Kunststoff und Gummi,* June 1965.

Amut standard extruder: *International Plastics Engineering,* September 1961.

Anger standard twin-screw machine: *International Plastics Engineering,* March 1962; *Industrie des Plastiques Modernes,* December 1961. Four-screw (one twin feeding another twin); *Plasticheskie Massi,* November 1966; *British Plastics,* October 1966; *Kunststoffe, British Plastics* and *Plastics Technology,* all April 1966.

Aragon, see **Egar.**

Baker-Perkins Rotessor, see **Engel** Rotomeltor.

Barand (Windsor-owned), standard line: *Plastics,* August 1963. Laboratory coating unit; *International Plastics Engineering,* November 1962 and *Plastics,* September 1963.

Barmag for low-bulk-density feeds: *Industrie des Plastiques Modernes,* August 1964. Giant (250-mm diameter) extruder; *International Plastics Engineering,* August 1963. Standard line; *International Plastics Engineering,* October 1961.

Bell Laboratories, Westover stepped-disk machine: *Modern Plastics,* November 1962; *SPE Journal,* December 1962; *International Plastics Engineering,* June 1963. Alternating twin-ram machine: *Modern Plastics,* March 1963. Both types, *SPE Journal,* May 1965.

Berstorff with short second screw in feed zone: *Kunststoffe,* September 1963; *International Plastics Engineering,* May 1963; *British Plastics,* September 1960.

Bielloni film and bag line: *International Plastics Engineering,* February 1963.

Black-Clawson 4½ in. induction heated: *International Plastics Engineering,* September 1963.

Bone Brothers twin-screw machine: *Plastics,* June 1965.

Davis-Standard: *International Plastics Engineering,* November 1963, and March 1961.

Derex twin-screw extruder: *International Plastics Engineering,* August 1963.

Egar, vertical extruder, front-end drive: *Modern Plastics,* March 1966.

Engel Rotomeltor: *SPE Journal,* May 1965; *International Plastics Engineering,* April 1965; *Plastics,* May 1963; *Kunststoffe,* April 1963, August 1959, and July 1959.

Finton, Canadian, with hydraulic drive: *International Plastics Engineering,* September 1965.

Fraser and Glass high-speed autothermal extruder: *International Plastics Engineering,* June 1961 and January 1962.

Frieseke and Hoepfner extra-short 150-mm extruder, output like 60-mm, for large cross-sections: *Kunststoffe,* April 1960; *Plastics,* February 1960.

Humboldt standard line: *International Plastics Engineering,* October 1963.

Ikegai screw-die system for PVC sheet: *Japan Plastics Age News,* June 1963.

Japan Steel Works, circular die for PVC sheet, and wire covering line: both *Japan Plastics Age News,* June 1963.

Jumex four-screw machine: *International Plastics Engineering,* February 1963.

Kawasaki three-screw machine: *Modern Plastics,* January 1965.

Kestermann general line: *Plastics Technology,* November 1966; *British Plastics,* June 1966; *Plastics,* February 1966. Pipe line: *Der Plastverarbeiter,* October 1964.

Kleinewefers small machines: *Kunststoff & Gummi,* September 1965 and September 1964. Calandrette line for PVC and other materials: *Plastics,* August 1963.

KPT elastic melt extruder: *Plastics,* October 1965; *SPE Journal* and *Plastics World,* May 1965. Original Maxwell work in *Modern Plastics,* October 1959.

Lauffer and Butscher rotating extruder: *SPE Journal,* February 1962.

Leonard, Italian, both twin and single screw: *International Plastics Engineering*, June 1965.

LMP/Colombo for PVC powder feed: *Kunststoff & Gummi*, September 1963; *International Plastics Engineering*, June 1962; *Kunststoffe*, April 1958.

Maillefer BM double-channel extruder: *SPE Journal*, May 1965; *Modern Plastics*, January 1963; *International Plastics Engineering*, January 1962; *Kunststoffe Plastics*, No. 3, 1961.

Mapré twin-screw machine: *International Plastics Engineering*, June 1965; *Kunststoffe*, March 1962.

Maxwell elastic melt extruder, see **KPT.**

Moi twin-screw machine: *International Plastics Engineering*, December 1962.

Müller extruders: *International Plastics Engineering*, February 1961. Flooring line: *International Plastics Engineering*, April 1963.

Negri Bossi standard extruders: *Materie Plastiche*, October 1965; *Kunststoff & Gummi*, June 1964; *Kunststoffe Plastics*, Vol. 11, No. 2.

Oerlikon "Rotatruder" for film: *Kunststoff & Gummi*, November 1963; *International Plastics Engineering*, September 1962; *Kunststoffe*, August 1962; *SPE Journal*, February 1962.

Pasquetti, see **Schloemann.**

Pastorello mixer-extruder combination: *International Plastics Engineering*, August 1963.

Peco 60-mm extruder: *International Plastics Engineering*, June 1965.

Plamvo vertical machine: *Verpackungs-Rundschau*, December 1965; *Kunststoffe*, September 1964.

Polish extruders, 90 to 130 mm: *Plastics*, February 1966.

Reifenhäuser tubular film line: *British Plastics*, December 1966.

Samafor standard extruders: *International Plastics Engineering*, June 1961.

Schalker planetary gear extruder: *British Plastics*, April 1967.

Schloemann twin-screw extruders, Pasquetti system: *Kunststoff & Gummi*, August 1965; *International Plastics Engineering*, March 1965; *Kunststoff & Gummi*, November 1963; *International Plastics Engineering*, May 1963.

USSR elastic melt extruder: *Plasticheskie Massi*, April 1962. High-speed autothermal extruder for nylon: *Plasticheskie Massi*, January 1961.

Werner and Pfleiderer twin-screw compounders: *International Plastics Engineering*, January 1965; *Kunststoff & Gummi*, June 1964; *Plastics*, December 1962.

Windsor twin-screw extruders (LMP system, see also **LMP/Colombo**): *International Plastics Engineering*, January 1965. Screw-die PVC sheet line, see **Ikegai.**

EXTRUSION TECHNOLOGY FOR SPECIFIC PLASTICS

Information on how to extrude various polymers is also included in the many references listed at the end of each chapter. The references that follow here are those which are not restricted to a single application. Brochures and booklets on how to extrude different plastics are also available from the material suppliers.

General

Plastics Design and Processing, July 1966, brief review of conditions for each plastic.

ABS

Plastics, April and May 1965, (Marbon); *International Plastics Engineering,* October 1964; *Industrie des Plastiques Modernes,* August 1964, both using CAFL twin-screw machines; *SPE Journal,* January 1964, (U.S. Rubber).

Acetals

British Plastics, October 1965, H. Schmidt (Hoechst); *SPE Journal,* November 1962, A. Ciarlone (Celanese); *Der Plastverarbeiter,* April 1961 (du Pont); *SPE Journal,* December 1960; *Plastics,* April 1960.

Acrylics

British Plastics, February 1966, staff written; *SPE Journal,* 1962; and *SPE ANTEC* 1962, J. F. Garber and R. T. Cassidy (Rohm and Haas), discussion of screw design and conditions; *Plastics Progress* (Papers at British Plastics Exhibition) 1959, L. Griffiths (ICI); *Kunststoffe,* April 1957, (Rohm and Haas).

Cellulosics

Kunststoffe, May 1962.

Ethylene/Vinyl-Acetate Copolymers (EVA)

Plastics Technology, February 1965, B. H. Krevsky (Union Carbide).

Fluoroplastics

SPE ANTEC 1961, polyvinylidene fluoride for wire and cable, N. Capron (Pennsalt); *Industrie des Plastiques Modernes,* December 1960 and January 1961, PTFE, W. Wrobel; *Plastics Technology,* August 1960, extrusion of FEP, J. A. Blair *et al.* (du Pont); *Plastics,* January 1958, PTFE extrusion, E. M. Elliott (ICI); *SPE ANTEC* 1958, Teflon 100 (du Pont).

Ionomers: see **Polyethylene, ionomers.**

Nylon: see **Polyamides.**

Polyamides

SPE ANTEC 1967, C. D. Weiske (BASF), properties, screw-design, and conditions for nylon-6 and 66; *SPE Journal,* October 1963, R. Bonner (du Pont), screw design for nylon-66 and 610; *Plastics,* April and May 1963, properties, conditions, and equipment for nylon 11; *Journal of Colloid Science,* December 1962, discusses moisture content; *Kunststoffe,* May 1962, long article with many references; *Plastics Technology,* June 1957, R. L. Hughes (Spencer), covers nylon-6.

Polybutylene

Plastics Technology, February 1966, T. Reed (Petrotex).

Polycarbonates

SPE Journal, April 1966, H. J. Singer (General Electric).

Polyethylene, Cross-linked

Modern Plastics, November 1965, Raychem process; *SPE Journal,* September 1965 and *SPE ANTEC* 1965, C. H. Benning (Grace), peroxide linking agents.

Polyethylene, Low-density

Kunststoff & Gummi, February 1965, effect of reprocessing.

Polyethylene, High-density

Transactions of the Plastics Institute, April 1963, E. Overgage; *Plastics Technology,* July 1960, W. M. Mundy (NRM), screw-design.

Polyethylene, Ionomers

Noticiero del Plástico, November 1966, long article with chemical and permeability data; *SPE Journal,* September 1966; *Plastics Technology,* March 1965.

Polyethylene, Ultra-high Molecular Weight

Modern Plastics, July 1965. See also comments in Chapter 3 under "Special Techniques."

Polyethylene terephthalate (thermoplastic polyester)

British Plastics, November 1966, AKU material.

Polymethyl Methacrylate: see **Acrylics.**

Polypropylene

SPE Journal, January 1962.

PPO (Polyphenylene Oxide)

SPE Journal, May 1966, and *SPE ANTEC* 1966, E. S. Jaskot (General Electric).

Polystyrene, Super-impact

Plastics, July 1962, A. L. Griff, review of several commercial materials.

Polystyrene Foam

SPE Journal, July 1960, F. H. Collins (Dow), *Kunststoff-Berater*, May 1961. See also references at the end of Chapter 4.

Polysulfones

SPE Journal, January 1967 (screw-design), *SPE ANTEC* 1966 and *Plastics Technology*, October 1965: all these are by Union Carbide personnel.

Polyurethanes

British Plastics, May 1967, D. Mitchell (British Geon), processing conditions and properties; *SPE RETEC*, Boston, March 1, 1967, and *SPE Journal*, August 1967, W. E. Foster (Mobay), guide to conditions, problems and materials; *SPE ANTEC* 1967, F. L. Pittenger (Goodrich), equipment, recommendations, conditions and properties; *Poliplasti*, February 1966; *SPE Journal*, 1963 (Mobay information); *Modern Plastics*, October 1961 and *SPE ANTEC* 1961, C. A. Waugaman (Goodrich).

PVC

SPE ANTEC 1967, *SPE Journal*, July 1967, and *Chemical Engineering Progress*, December 1966, all articles by R. A. Paradis (Diamond Alkali), describing rheology experiments and their practical application; *SPE ANTEC* 1967, C. Guimon and F. Berger (Pechiney-St-Gobain), thorough study of resins, additives, conditions and equipment for powder blends; *SPE ANTEC* 1967, W. C. Johnson (Raychem), study of PVC behavior in extrusion dies; *British Plastics*, July and August 1966, H. Domininghaus (Hoechst); *Kunststoff Rundschau*, April and May 1966, also H. Domininghaus; *SPE Journal*, January 1966, F. R. Nissel (Prodex), vented extruders for dry-blends; *SPE ANTEC* 1965, P. V. McKinney (Armstrong Cork), Brabender study of variation of formulation on extrudability of plasticized compounds. *Materie Plastiche*, November and December 1964, long (36 pages) discussion of extrusion of plasticized PVC; *International Plastics Engineering*, October 1964, H. Von Rakowski (Reifenhäuser), general discussion; *SPE ANTEC* 1964, J. A. Rolls (Diamond Alkali), discusses dry-blends especially for pipe; *British Plastics*, April and May 1963, G. H. Burke and G. C. Portingell (British Geon), results of extrusion experiments with varying formulation and conditions. *Materie Plastiche*, December 1961 and *Modern Plastics*, October 1960, C. H. Brower and B. H. Maddock (Union Carbide), extrusion of plastisols; *International Plastics Engineering*, June 1961; *SPE ANTEC* 1961, N. T. Flathers *et al.* (Escambia), describes vacuum hopper and screens in dry-blend extrusion; *Kunststoff-Berater*, May 1961; and *Plastics Technology*, March and April 1959, information on foamed PVC; *SPE Journal,* August 1960, F. C. Schutz (Welding Engineers), use of Brabender tester in predicting PVC extrusion behavior; *SPE Journal,* March 1958, E. T. Hankey and R. D. Sackett (Monsanto), study of dry-blend extrusion variables.

Polyvinyl dichloride (PVDC)

SPE Journal, August 1966, Goodrich information.

BOOKS AND OTHER MAJOR REFERENCES

Books

"Grundzüge der Extruder-technik," by W. Mink (Battenfeld); Rudolf Zechner Verlag, Speyer am Rhein, West Germany, 1966, 210 pages. Practical, well-illustrated presentation of current technology in West Germany.

"Estrusione delle materie plastiche" (Extrusion of plastic materials), by A. L. Griff (Edison Technical Services, Inc.); ET/AS Editrice, Milan, 1965, 285 pages. Translation of the first edition of this book, which appeared in 1962.

"Extrusion of plastics," by E. G. Fisher (consultant); Iliffe Press, London, 1964, 271 pages. Second, much-expanded edition, covering construction and application of extruders and auxiliary equipment.

"Kunststoff-Extrudertechnik" (Plastics extrusion technology), by G. Schenkel (Troester, now at Stuttgart Technical College); Carl Hanser Verlag, Munich, 1963, 540 pages. Revised and expanded version of "Schneckenpressen für Kunststoffe," published by Hanser in 1959. An English translation of the revised (1963) edition has been made by L. A. H. Eastman, edited by B. S. Glyde (Metal Box), and entitled "Plastics extrusion technology and theory." It was published in 1966 by Iliffe Press in London and American Elsevier in New York, with 477 pages. In all versions, the book has over 400 illustrations and countless references, including patents. It covers the history and development of extrusion machinery, unusual variations, principles of construction of modern extruders, extrusion theory, and some information on application technology and die design.

"Grundlagen der Extrudertechnik," (Basic rules for extrusion), by H. R. Jacobi; Carl Hanser Verlag, Munich, 1960, 137 pages. English translation now available from Iliffe Press, London, 126 pages. Short book on theory of extrusion and plastic flow, with much mathematics.

"Processing of thermoplastic materials," edited by E. C. Bernhardt (du Pont); Reinhold Publishing Corp. (for the Society of Plastics Engineers), New York, 1959, 690 pages. Collected articles by specialists on various aspects of thermoplastics processing. The extrusion article (151 pages and 63 illustrations), written by a du Pont team, is good but quite theoretical, and requires some technical skill to understand and apply. Available also in Spanish.

Russian Books Concerned with Extrusion

"Chervyachnie Mashini dlya Pererabotki Plasticheskikh Mass i Resinovikh Smesei," (Screw Machines for Processing Plastic Materials and Resinous Mixtures) by D. Ryabinin. Moscow, 1965, Mashinostroyenie, 362 pages.

"Kabeli Svyazi v Plastmassye," (Cables for Communications in Plastics) by B. M. Volkov, Moscow, Svyaz, 1965, 190 pages.

"Vliyaniye Tekhnologicheskikh Parametrov Protzessa Extruzii na Kachestvo Izdelii iz Poliolefinov," (Influence of Technological Parameters in the Extrusion Process on the Quality of Polyolefin Products) by E. A. Starkov. Leningrad, 1965.

"Modernizatziya Oborudovaniya dlya Izgotovleniya Izdelii iz Plastmass," (Modernization of Equipment for Producing Articles from Plastics), Moscow, 1963, Moskva-Mashin, 202 pages.

"Oborudovaniye dlya Proizvodstva i Pererabotki Plasticheskikh Mass," (Equipment for Manufacture and Processing of Plastics) by N. A. Kozulina (editor). Leningrad, 1963, Gos. Nauchno-Tekhn. Izd-vo Khim. Lit-ri; 782 pages.

Other Major References

"Progress in extrusion technology in Europe," H. Domininghaus (Hoechst), *Der Plastverarbeiter,* October, November and December 1966. Three-part review of extrusion practices, application-oriented, with 32 pages and 106 illustrations. December portion covers experiments on single- and twin-screw machines. Similar review article by same author is in *Plastiques Modernes et Elastomères,* Vol. 18, Nos. 3, 4, and 5.

"Die Verarbeitung thermoplastischer Kunststoffe auf Einschneckenpressen" (The processing of thermoplastics in single-screw extruders), L. Kreft and Z. Doboczky, *Kunststoff & Gummi,* October 1965 through February 1967. Serialized discussion of principles and practice of extrusion, covering such topics as: Theory of Melting, Screw Design, Calculation and Prediction of Output, Temperature and Pressure Control, and Influence of Material. The series may still be continuing.

"Processing factors in extrusion," G. Schenkel (Troester, now at Stuttgart Technical College), *Kunststoffe,* August 1965. Review of new developments: improvements in each application area, especially in achieving better product quality. Includes 84 references and 33 illustrations. Long review of the status of extrusion technology, by the same author, is in *Kunststoffe,* November and December 1961, with 205 references and 27 illustrations.

"Vented single-screw extruders," G. Schenkel (Troester, now at Stuttgart Technical College), *International Plastics Engineering,* September, October and November 1962, and *Der Plastverarbeiter,* November and December 1962 and January 1963. Thorough discussion of theory and experience with vented extruders of various designs; includes 49 references.

"A guide to the literature and patents concerning PVC technology," edited by L. Whittington (National Latex) for the *SPE*. First edition 1962, revised and expanded edition 1964. Thousands of references, without further comments, divided by application area and processing method.

"Extrusion bibliography," L. Street (Welding Engineers, now with F. W. Egan), *SPE Journal*, June 1961. Over 300 listings, back to 1868, with thorough coverage of foreign works.

"Modern Plastics Encyclopedia," issued annually by *Modern Plastics* magazine, contains articles on each processing method, updated every one or two years. The extrusion article is a good introduction to the technology of the industry.

ADDRESSES OF LEADING PLASTICS JOURNALS OF THE WORLD

Australian Plastics & Rubber Journal: 243 Elizabeth St., Sydney, Australia.

British Plastics: Dorset House, Stamford Street, London S. E. 1, England.

Canadian Plastics: 1450 Don Mills Road, Don Mills, Ontario, Canada.

Connaissance des Plastiques: C.I.B., S.A., 7 rue Darboy, Paris 11, France.

Der Plastverarbeiter, see **Plastverarbeiter.**

Gummi/Asbest/Kunststoff: A. W. Gentner Verlag KG, Froststrasse 131, Postfach 688, Stuttgart, West Germany.

Indian Plastics Review: Plastics and Machinery Distributors, 4 Upper Chitpur Road, Calcutta 7, India.

Industrie des Plastiques Modernes (now **Plastiques Modernes et Elastomères):** 40, rue du Colisée, Paris 8, France.

International Plastics Engineering (now combined with **British Plastics):** Dorset House, Stamford Street, London S. E. 1, England.

Japan Plastics Age (formerly **Japan Plastics Age News):** 8,2-chrome, Kandakaji-cho, Chiyoda-ku, Tokyo, Japan.

Jornal de Plasticos: Rua Machado de Assis 334, Aclimação, São Paulo, Brazil.

Journal of Cellular Plastics: 202 North Street, Stamford, Conn., U.S.A.

Kunststoff & Gummi: Lessingstrasse 12–14, 65 Mainz, West Germany.

Kunststoff-Berater: Kettenhofweg 111, Frankfurt/Main, West Germany.

Kunststoffe: Carl Hanser Verlag GmbH, Leonard-Eck-Strasse 7, Munich 27, West Germany. Editors: Ernst Römer, Heinrich-Delp-Strasse 281, Darmstadt; and Erich Escales, 7595 Sasbachwalden, West Germany.

Kunststoffe-Plastics: Brummer Annoncen AG, Volt-Schild AG, Dornacherstrasse, Solothurn, Switzerland.

Kunststoff-Rundschau: Verlag Brunke & Garrels, Schloss-Strasse 2–6, Hamburg-Wandsbek, West Germany.

Materiale Plastice: c/o CARTIMEX, or Editura Tehnica, str. Stirbei Voda 37; both Bucharest, Rumania.

Materie Plastiche: 8 Via Farneti, Milan, Italy.

Modern Plastics: 1301 Avenue of the Americas, New York, N. Y. 10019, U.S.A.

Muanyag: c/o KULTURA, Budapest, Hungary.

Muoviviesti: Liisankatu 27, A-7, Helsinki, Finland.

Noticiero del Plástico: Leandro Alem 651, Buenos Aires, Argentina.

Officiel des Matières Plastiques: 12, Rue d'Isly, Paris 8, France.

Pensez Plastiques: Berne & Cie., 11 Blvd. des Batignolles, Paris 8, France.

Plaste und Kautschuk: VEB Deutscher Verlag für Grundstoffindustrie, Karl-Heine Strasse 27, 7031 Leipzig, East Germany.

Plastic: Teknisk Forlag A/S, 31 V. Farimagsgade, Copenhagen V, Denmark.

Plastica: Mathenesserlaan 310, Rotterdam, Netherlands; also c/o Kunststoffen-instituut TNO, Schoemakerstraat 97, Delft, Netherlands.

Plasticheskie Massi: Mezhdunarodnie Kniga, Smolenskaya Ploshchad 32-34, Moscow, USSR. (See also "Soviet Plastics")

Plasticke Hmoty a Kaučuk: c/o ARTIA, Ve Smeckach 30, Prague 1, Czechoslovakia.

Plásticos: Sarmiento 2494, Buenos Aires, Argentina.

Plasticos em Revista: Rua Marques de Itú 95, 3°-E, São Paulo, Brazil.

Plásticos Modernos Latinoamericanos: Hamburgo 238, Mexico D.F. 14.

Plásticos y Résinas: Bajio 335, despacho 603, México 7, Distrito Federal.

Plastics: 33–39 Bowling Green Lane, London E. C. 1, England. (Now merged with *British Plastics.*)

Plastics in Australia: Bell Publications Ltd., Sydney, Australia.

Plastics Design & Processing: Post Office Box 270, Libertyville, Illinois 60048, U.S.A.

Plastics Institute of Australia—News (PIA News): 30 Latrobe Street, Melbourne, Australia.

Plastics Institute Transactions and Journal: 11 Hobart Place, London S. W. 1, England.

Plastics Technology: 630 Third Avenue, New York, N. Y. 10017, U.S.A.

Plastics World: 3375 S. Bannock Street, Englewood, Colorado 80110, U.S.A.

Plastiques Modernes et Elastomères (formerly **Industrie des Plastiques Modernes):** 40, rue du Colisée, Paris 8, France.

Plastnytt: Universitetsgaten 14, Oslo, Norway.

Plastverarbeiter: Rudolf Zechner Verlag, Postfach 68, 672 Speyer am Rhein, West Germany.

Plastvärlden: 32 Norra Strandgatan, P.O. Box 183, Halsingborg, Sweden.

Popular Plastics: 126A Dhuruwadi off Dr. Nariman Rd., Bombay 28, India.

Progressive Plastics: 481 University Avenue, Toronto, Ontario, Canada.

Poliplasti: ET/AS Editrice, Via Mantegna 6, Milan, Italy.

Revue Belge des Matières Plastiques: 49 Square Marie-Louise, Brussels 4, Belgium.

Revista de Plásticos: Serrano 144, Madrid 6, Spain.

Rubber and Plastics Age: Gaywood House, Great Peter Street, London S. W. 1, England.

SPE ANTEC & RETEC: SPE conferences; papers available from SPE (Society of Plastics Engineers), 65 Prospect Street, Stamford, Connecticut 06902, U.S.A.

SPE Journal: 65 Prospect Street, Stamford, Connecticut 06902, U.S.A.

Soviet Plastics (translation of **Plasticheskie Massi**): Gaywood House, Great Peter Street, London S. W. 1, England.

Transactions of the Plastics Institute, see **Plastics Institute Transactions and Journal.**

Western Plastics: 274 Brannan Street, San Francisco, California 94107, U.S.A.

Index